The Inclusion-Based Boundary Element Method (iBEM)

The Inclusion-Based Boundary Element Method (iBEM)

Huiming Yin
Columbia University
New York, NY, United States

Gan Song
Columbia University
New York, NY, United States

Liangliang Zhang
China Agricultural University
Beijing, PR China

Chunlin Wu
Columbia University
New York, NY, United States

ELSEVIER
ACADEMIC PRESS
An imprint of Elsevier

Academic Press is an imprint of Elsevier
125 London Wall, London EC2Y 5AS, United Kingdom
525 B Street, Suite 1650, San Diego, CA 92101, United States
50 Hampshire Street, 5th Floor, Cambridge, MA 02139, United States
The Boulevard, Langford Lane, Kidlington, Oxford OX5 1GB, United Kingdom

ISBN: 978-0-12-819384-6

For information on all Academic Press publications
visit our website at https://www.elsevier.com/books-and-journals

Publisher: Matthew Deans
Acquisitions Editor: Brian Guerin
Editorial Project Manager: Isabella C. Silva
Production Project Manager: Nirmala Arumugam
Designer: Vicky Pearson

Typeset by VTeX

To my wife Yali Liu. - Huiming Yin

To my wife Ruiying Zhang and our parents Jinqi Song, Dongmei Sun, Xinhai Zhang, Jie Li. - Gan Song

To my fiancée Yifan Zhang and our family. - Liangliang Zhang

To my parents and girlfriend. - Chunlin Wu

Contents

List of figures

Biography

Huiming Yin

Huiming Yin, Associate Professor of Civil Engineering and Engineering Mechanics, specializes in the multiscale/physics modeling and characterization of civil engineering materials and structures with experimental, analytical, and numerical methods. He is the Site Director of NSF IUCRC Center for Energy Harvesting Materials and Systems at the Columbia University. He received a BS in engineering mechanics from Hohai University, China, in 1995, an MS in solid mechanics from Peking University, China, 1998, and a PhD in civil engineering from the University of Iowa in 2004.

Gan Song

Gan Song, PhD from Columbia University, specializes in analytical and numerical methods for characterizing mechanical properties and microstructures of composite materials. Currently he works as Quantitative Research VP at JPMorgan Chase on mathematical and numerical modelings. He won the best paper award from ASCE in EMI conference for the development work of iBEM. He also received the Mindlin scholar award from Department of Civil Engineering and Engineering Mechanics at Columbia University.

Liangliang Zhang

Liangliang Zhang, Assistant Professor of Department of Applied Mechanics, China Agricultural University, specializes in the multiscale/physics characterization of advanced quasicrystal composites and structures with analytical and numerical methods. He was the Associate Research Scientist in Columbia University and the Administrator of NSF IUCRC Center for Energy Harvesting Materials and Systems at the Columbia Site, 2017–2019. He received a BS in engineering mechanics in 2009, an MS in solid mechanics in 2012, and a PhD in mechanics engineering in 2015 from China Agricultural University.

Chunlin Wu

Chunlin Wu, PhD from Civil Engineering and Engineering Mechanics, specializes in the iBEM software developing. He received a BS in civil engineering from Tongji University, China, in 2017, MS in engineering mechanics from Columbia University, 2018, and PhD in engineering mechanics from Columbia University in October 2021. He received the Mindlin scholar award in Civil engineering and Engineering mechanics for his PhD studies.

Preface

This book stems from the course on Micromechanics of Composite Materials that I started in 2012 at Columbia University while I found the significant needs from students and researchers on composite material designs and modeling. At that time the classic micromechanics-based methods used the concept of representative volume element (RVE) to link the effective material behavior with the local solution. Therefore Eshelby's theory on a single particle in an infinite domain can play a role to predict the effective material behavior of a composite, which links mechanical behavior between the microscale of particles and the marcoscale of composites. However, with the advancement of experimental approach, we can control and manipulate the microstructure very well and fabricate very small specimens for advanced material characterization and optimization. Actually, the scale differentiation is not clear in experiments with small testing samples. Linking the fundamental material properties of each material phase with the sample testing results and then the effective composite material behavior becomes a challenging problem, which creates an urgent need for cross-scale modeling and virtual experiments that are crucial for new material development and discovery.

Then we came up with an idea to generalize Eshelby's method to consider the boundary effect on material behavior of composites in some experiments by changing the Green's function and indeed obtained some interesting results. However, it is not practical to find Green's function for every experiment. Eventually, I turned to the boundary integral method for general virtual experiments. At that time the second author Dr. Gan Song was still my PhD student. He implemented the algorithm with the boundary element method code, and we named the software package iBEM in 2015 as an abbreviation of the inclusion-based boundary element method. He applied iBEM to solve some challenging problems and completed an excellent PhD dissertation. He won the highest award in the Department, the Mindlin Award.

After Dr. Song graduated from Columbia, Dr. Liangliang Zhang took the main responsibility to maintain and develop the iBEM program, and Dr. Chunlin Wu joined the team in research and development of the algorithm. Because the early version of iBEM was developed for ellipsoidal particles only, the application is relatively narrow. Dr. Wu extended the program to two-dimensional (2D) case and generalized it to arbitrary particle shape for both 2D and 3D problems. Therefore iBEM can be a truly practical software for material engineers on material design and analysis, which brought significant interests from industry as well. The favorable response from both researchers and engineers encouraged us to work together and document the method into a book as a systemic record and study resource. Dr. Wu also won the Mindlin Award when he graduated in 2021.

Surely, this book does not intend to cover everything in this dynamic research area, but serves a start for a long journey of further development and applications of the iBEM. Although readers with micromechanics and boundary element method will feel easier to read the book, no particular background is required of readers because some necessary mathematics, mechanics, and physics are explained in the text and Appendices. Although we tried to be fair in citing the literature, due to the very rich research in this area, we have to apologize if some papers do not receive proper credit or are missing in citation. The chapters are arranged in a sequence of the learning process, and we recommend to read them one by one for beginners. However, the sections or subsections with an asterisk can be skipped as they could be in a branch area, which is relatively independent from other part, or an advanced topic for more focused or advanced readers.

During the preparation of this book, we have received help and support from our colleagues, students, and collaborators. First, we thank Yingjie Liu, Xin He, Fangliang Chen, Tengxiang Wang, Zifeng Yuan, and Zhenyu Shou for their contribution in this research topic at Columbia University. Junhe Cui read through the book and provided her feedback, which was very helpful for improving this book. I am grateful for my advisors and mentors, Lizhi Sun, Minzhong Wang, Glaucio Paulino, Bill Buttlar, George Weng, Frank DiMaggio, Rene Testa, and Jiun-Shyan (JS) Chen, who have provided guidance and encouragement in the course of my research and career progress toward the achievement of this book. I would like to thank my colleagues and collaborators, including Yuhong Wang, Yang Gao, Yingtao Zhao, Gautam Dasgupta, Raimondo Betti, Haim Waisman, Qiang Du, and Marco Giometto, for the insightful discussion. Some parts of this book are based on the publications and collaborative work with them.

In addition, this book has mainly been written in our spare time at night or weekend. We had weekly zoom meeting on Saturday morning in the past two years during the COVID pandemic period. We highly appreciate the support and consideration from our families.

Finally, we acknowledge the financial support from National Science Foundation (NSF) – Divisions of Civil, Mechanical and Manufacturing Innovation (CMMI) and Industrial Innovation and Partnerships (IIP), National Natural Science Foundation of China, Department of Defense (DOD) – Air Force Office of Scientific Research (AFOSR) and Office of Naval Research (ONR), and US Department of Agriculture – National Institute of Food and Agriculture. Without the research support, the work is impossible to be done.

Huiming Yin
August 23, 2021, at Alpine, New Jersey

Chapter 1

Introduction

Virtual experiments with iBEM

The inclusion-based boundary element method (iBEM) is a powerful tool, which is developed to efficiently solve the boundary value problems (BVPs) for a finite domain with multiple particles for modeling of composite materials. It is particularly suitable for virtual experiments with computer to simulate physical experiments and predict the test results. The equivalent inclusion method (EIM) and boundary element method (BEM) provide the foundation of the iBEM. Unlike traditional BEM, which requires to mesh all the surfaces of particles, each particle is treated as a source with an eigenfield, such as eigenstrain for elasticity, eigenstrain rate for Stokes' flow, or others for different BVPs, and the response can be obtained by analytical volume integrals for rapid computation and exactness of the integral. Different from discrete element method (DEM), iBEM is based on continuum mechanism and provides detailed local fields for analysis and design of advanced materials. Since iBEM adopts analytical integrals over the inclusion, it does not need to mesh the matrix domain and is able to converge to the exact solution more rapidly than the finite element method (FEM). Moreover, because iBEM solves the field variables on the boundary, which can directly describe the effective material response, iBEM can be an ideal tool for virtual experiments of composite materials and structures.

1.1 Virtual experiments

With the development of additive manufacturing and nanotechnologies, we can control the material fabrication process precisely to tailor and manipulate the microstructure of advanced materials for specific purposes. With the miniaturization of material characterization and testing, we can see crystals or particles and apply loads at the micron or even nanometer level. Cross-scale modeling is highly demanded for new material design and development. For the material preparation and testing at these levels, it often requires many trial and error iterations and versatile skills or experience to obtain meaningful results, which is generally time-consuming and expensive. In addition, another challenging task is interpreting, analyzing, and understanding the test results due to the randomness nature and test errors, which is commonly introduced by human factors and can be avoided by virtual experiments with the aid of computers.

The Inclusion-Based Boundary Element Method (iBEM). https://doi.org/10.1016/B978-0-12-819384-6.00009-7

Virtual experiments are proposed to reproduce the physical experiments on a computer with controlled physical parameters for material discovery, modeling, and prediction. It allows students and researchers to conduct experiments with microprocessor, computers, and high-performance clusters for different scales of problems. It is built on high-fidelity algorithms on microstructural realizations of materials and multiphysical modeling of material behavior. Particularly, the scale of material sampling and testing is very sensitive. If the number of particles is too large, then it is easy to lose the focus. On the contrary, it may not represent the material well. A specimen containing hundreds or thousands of particles of different shape, size, and distribution can be a common choice to meet the needs in virtual experiments. However, traditional multiscale modeling schemes may not work for this case, because the scale differentiation is not clear in those types of experiments with small testing samples. Linking the fundamental material properties of each material phase with the sample testing results and the effective composite material behavior becomes a challenging problem, which creates an urgent need for cross-scale modeling and virtual experiments that are crucial for new material development and discovery. Notice that the cross-scale problem covers a large size range from nano- and micro- to macroscales. For example, a small concrete specimen may contain sands, fine particles, and micropores crossing scales from micron to centimeter.

In this book, we introduce a computer-aided design paradigm of particulate composites by using the inclusion-based boundary element method (iBEM), which has been recently proposed to conduct high-fidelity virtual experiments of materials. It can be applied to potential flow, elastic, and fluid mechanics problems. Therefore the multiphysical and mechanical properties of a composite material can be predicted. The replacement of a number of laboratory tests by validated virtual experiments for material fabrication and characterization will significantly expedite the design and development progress of novel composites and save the costs in material manufacture process. The software being developed is applicable to general particulate composites and compatible to both personal computers and high-performance clusters.

The algorithm can integrate the recent work in digital image recognition and correlation of the composite microstructure into the preprocessor for model validation and high-fidelity simulation based on the state-of-the-art imaging technologies. Although research in multiscale modeling has been very active and fruitful in the past decade, it is still a challenging task to simultaneously characterize and simulate the composites with crossing scale features. For example, a nanocomposite contains micropores in micron and nanoparticles in nanometers dispersed in a matrix. It is challenging to quantify the direct interaction between micropores and nanoparticles. The iBEM-based virtual experiments enable a direct simulation of the composites crossing scales. This is a revolution compared with the traditional continuum methods, which may lead to an ill-conditioned mesh or extremely large DOFs.

Based on the algorithm of inclusion-based boundary element method, the software package iBEM is developed for virtual experiments and design of general particulate composites for two-dimensional (2D) or three-dimensional (3D) crossing scales. The recent development allows simulation of arbitrary shaped particles, which can be represented by ellipsoids or polyhedrons. It can be integrated with isogeometric analysis to reproduce particles with complex smooth surfaces, but it is not covered by this book. Although static problems are the main focus, it is possible to extend the program to transient problems or dynamic problems.

1.2 Inclusion and inhomogeneity

Although inclusion and inhomogeneity are often equally used to describe particle, in classic micromechanics, they have distinct meanings, which will be used in this book and clarified below.

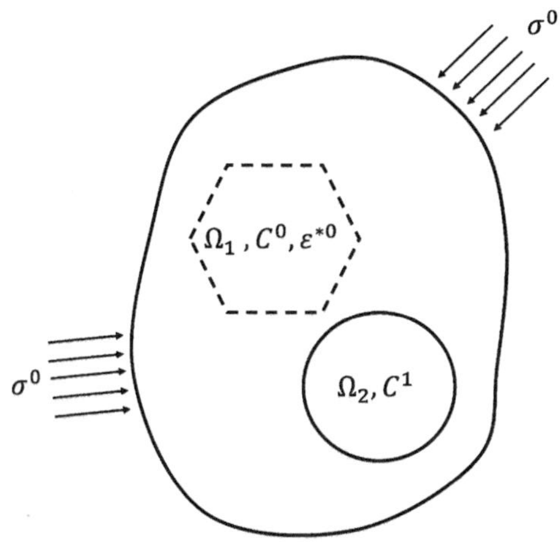

FIGURE 1.1 Schematic illustration of an infinite domain $\mathcal{D}$ containing two subdomains, Ω_1 of an inclusion and Ω_2 of an inhomogeneity under a far-field stress σ_{ij}^0.

Fig. 1.1 shows a homogeneous domain $\mathcal{D}$ containing two subdomains, Ω_1 and Ω_2, subjected to a far-field stress σ^0. Here Ω_1 shares the same stiffness of $\mathbf{C}^0$ as the matrix $\mathcal{D} - \Omega_1 - \Omega_2$ but subjected to a prescribed strain, which is also called eigenstrain, whereas Ω_2 exhibits a different stiffness of $\mathbf{C}^1$. When the subdomain Ω_1 has the same material properties as the rest region but includes a source field, it is defined as an inclusion. For example, when a circular sample is taken from a large steel plate, a circular steel bar with a slightly larger diameter can be used to fill the hole by the thermal fitting method, in which the steel bar is cooled to a certain temperature to fit into the hole. When temperature increases to the room temperature, the bar will experience a thermal expansion, so that the plate can be perfectly filled with the steel. Because both the plate and the circular bar have the same mechanical properties, the bar can be considered as an inclusion. The strain caused by temperature is not a mechanical strain caused by a stress; it is also called an eigenstrain. Therefore a subdomain with eigenstrain but the same stiffness as the surround material is an inclusion.

The source field does not limit to eigenstrain; it can be a distributed body force. This definition can be generalized to other physical problems such as heat conduction, Stokes flow, etc. Following Mura's definition, the source field of eigenstrain can be generalized to the prescribed eigenfield, such as eigentemperature gradient in heat conduction or eigenstrain rate in Stokes' flow [1]. The key point is that the inclusion shares the same material properties as the surround materials but is subjected to a source field.

On the other hand, when the subdomain Ω_2 has different mechanical properties from the rest region in the domain $\mathcal{D}$, it is called an inhomogeneity. A particle embedded in a matrix is a common example of an inhomogeneity. In the last example, if an aluminum bar replaces the steel bar to fit into the steel plate, then it becomes an inhomogeneity.

Inclusion and inhomogeneity may share the same geometry of a subdomain embedded in a large domain, but they are different from each other. An inclusion is defined by a prescribed field in the subdomain, whereas an inhomogeneity means a material difference in the subdomain.

When a test load is applied to the boundary $\partial\mathcal{D}$ of the domain, the static elastic boundary value problem can be set up as follows. The equilibrium equation keeps the same for both inclusion and inhomogeneity (or matrix):

$$\sigma_{ij,i} + b_j = 0 \tag{1.1}$$

where σ_{ij} is the Cauchy stress tensor, and the body force b_j is often considered separately to provide a homogeneous partial differential equation. For elastostatic problems, they vary with location $\mathbf{x}$ only.

The constitutive relation between stress and strain at any $\mathbf{x}$ is written as

$$\begin{aligned}\sigma_{ij} &= C^I_{ijkl}\left(\varepsilon_{kl} - \varepsilon^*_{kl}\right)\\ &= \lambda^I\delta_{ij}\left(\varepsilon_{kk} - \varepsilon^*_{kk}\right) + 2\mu^I\left(\varepsilon_{ij} - \varepsilon^*_{ij}\right)\end{aligned} \tag{1.2}$$

where C^I_{ijkl} is the stiffness tensor with the superscript I indicating the material type of the material point $\mathbf{x}$. For example, $I = 0$ signifies a matrix, and $I = 1$ an inhomogeneity. In this book, we focus on the isotropic elastic materials with $C_{ijkl} = \lambda\delta_{ij}\delta_{kl} + \mu\left(\delta_{ik}\delta_{jl} + \delta_{il}\delta_{jk}\right)$ with two Lame constants λ and μ. The strain ε_{kl} is calculated by the displacement field in Eq. (1.3). The superscript * in ε^*_{kl} indicates an eigenstrain caused by other reasons than elastic stress, such as thermal strain. There exist three cases depending on the location of $\mathbf{x}$ in Fig. 1.1:

1. If $\mathbf{x}$ is at the inclusion, then there exist ε^*_{kl} and the stiffness at C^0_{ijkl}.

2. If $\mathbf{x}$ is at the inhomogeneity, then ε_{kl}^* can be zero, but the stiffness is C_{ijkl}^1.
3. If $\mathbf{x}$ is at the matrix, then there exist $\varepsilon_{kl}^* = 0$ and the stiffness at C_{ijkl}^0.

The strain is related to the displacement

$$\varepsilon_{ij} = \frac{1}{2}\left(u_{i,j} + u_{j,i}\right) \tag{1.3}$$

The boundary condition can be given in terms of either displacement or traction or their relationship on the boundary points:

$$\begin{aligned} t_j^0 &= \sigma_{ij} n_i \quad \text{for} \quad x \in \partial\mathcal{D}^t \\ u_i^0 &= u_i \quad \text{for} \quad x \in \partial\mathcal{D}^u \end{aligned} \tag{1.4}$$

where the superscripts t or u indicate the boundary condition zone for traction and displacement, respectively.

Assuming no debonding on the interfaces of the inclusion or inhomogeneity with matrix, the interfacial continuity requires

$$\sigma_{ij}^+(\mathbf{x}) n_j^+ = \sigma_{ij}^-(\mathbf{x}) n_j^-, \quad u_i^+(\mathbf{x}) = u_i^-(\mathbf{x}) \quad \text{on } \partial\Omega_I \tag{1.5}$$

Overall, the inclusion or inhomogeneity problems keep the same except the constitutive relation (1.2) is different. Eshelby's original idea [2] is to introduce a material mismatch eigenstrain to transfer an inhomogeneity into an equivalent inclusion, so that homogeneous material system can be studied equivalently and Green's function technique can be systemically used. For simplicity, he investigated the unbounded matrix domain by introducing a uniform far-field stress summarized in the next section.

1.3 Equivalent inclusion method (EIM)

The mechanical response of a composite is definitely much more complicated than that of a homogeneous material, because the local field is significantly disturbed by the inhomogeneities. Particularly, when a composite material is subjected to a uniform far-field stress σ_{ij}^0, the distribution of stress and strain fields inside the material is not uniform anymore. To simulate this local material mismatch, the equivalent inclusion method (EIM) was first proposed by Eshelby [2,3], who investigated the elastic field inside and outside of one ellipsoidal inhomogeneity in the infinite domain. Mathematically, we can replace the inhomogeneity by an inclusion with the same material as the matrix, but on which the eigenstrains ε_{ij}^* are introduced as shown in Fig. 1.2. The eigenstrain is a kind of non-mechanical strains, which can be thought as plastic strains, misfitting strain, or strains caused by thermal expansion. By applying the stress equivalent conditions (1.6) inside the inhomogeneity, the unknown eigenstrain can be calculated, and thus the effect of the inhomogeneity can be simulated by an inclusion with an eigenstrain. With the aid of integrals of harmonic and

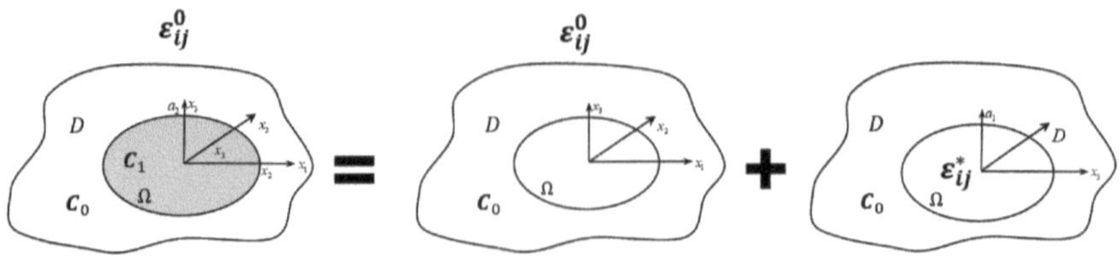

FIGURE 1.2 Illustration of equivalent condition in EIM.

biharmonic potentials, the eigenstrain field inside of the inhomogeneity can be analytically obtained with the stress equivalent conditions. Therefore the elastic field in the whole material domain can be derived by applying the Green's function technique.

$$C_{ijkl}^{0}\left(\varepsilon_{kl}^{0}+\varepsilon_{kl}^{\prime}-\varepsilon_{kl}^{*}\right)=C_{ijkl}^{1}\left(\varepsilon_{kl}^{0}+\varepsilon_{kl}^{\prime}\right) \tag{1.6}$$

where $\varepsilon_{kl}^{\prime}$ is the disturbed strain caused by the eigenstrain in the inclusion and can be evaluated by the Eshelby tensor [2] as follows:

$$\varepsilon_{ij}^{\prime}(\mathbf{x})=\int_{D} G_{ijkl}(\mathbf{x},\mathbf{x}^{\prime})\varepsilon_{kl}^{*}(\mathbf{x}^{\prime})d\mathbf{x}^{\prime} \tag{1.7}$$

where $G_{ijkl}(\mathbf{x},\mathbf{x}^{\prime})$ is the modified Green's function to show the response at $\mathbf{x}$ caused by a source at $\mathbf{x}^{\prime}$. It has been proved that the result from EIM for one ellipsoidal particle with uniform eigenstrain in the infinite domain subjected to far-field uniform stress is the exact solution of that problem. If more particles are considered and closed to each other, the uniform eigenstrain distribution is not sufficient to describe the particle interactions. Mura [1] has made a significant contribution to extending EIM to solve the case of multiple particles. Since the particle interactions are not negligible, higher-order polynomial approximations of the eigenstrain are indeed necessary to obtain acceptable results [4]. Because the shape of the inhomogeneities can be any ellipsoidal, we can easily adjust the geometrical parameters to analyze a spherical particle, fiber-like reinforcement or even penny-shaped crack embedded materials. Many researchers have extensively applied EIM to solve micromechanics problems [5–8], and a large number of papers in the literature have shown the potential applications of EIM in other physical problems, including steady-state heat conduction [9], Stokes flow [10], magnetoelasticity [11], and elastodynamics [12].

As an analytical method, EIM relies on the elliptical integrals of the fundamental solutions to certain boundary value problems (BVPs). For elastic problems, the fundamental solution of a point force in an infinite or semi-infinite domain has been used in EIM [13]. However, a general composite sample always exhibits a bounded boundary condition, which can be addressed by the boundary element method.

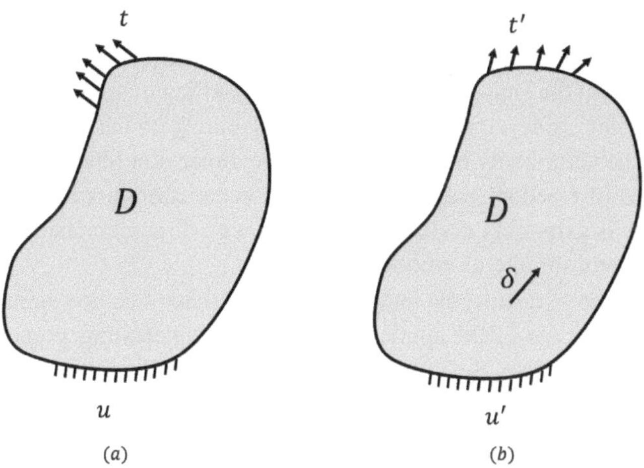

FIGURE 1.3 (a) True state with boundary conditions given on the boundary and (b) the state induced by a unit point force in an infinite domain with a displacement and traction on the corresponding boundary points provided by the fundamental solution.

1.4 Boundary element method (BEM)

Boundary element method (BEM) or boundary integral equation (BIE) is a well-developed numerical technique to solve physics problems in the past decades. It can be traced back to the 1970s [14,15]. When phenomena occurring in physics are governed by partial differential equations with prescribed boundary conditions, the boundary element method is an efficient numerical tool for solving BVPs, which has attracted a great deal of attention from researchers due to the mathematical elegance and numerical accuracy. It has been widely applied to a broad area of science and engineering including fluid mechanics [16], acoustics [17], electromagnetics [18], and fracture mechanics [19].

The reciprocal work theorem is a key to deriving the boundary element method. It can correlate one state (usually the true solution) of a problem to another state (usually referring to the fundamental solution for an infinite domain) by a reciprocal energy as shown in Fig. 1.3. The basic formulation in BEM of elastic problem is the following equation:

$$u_i\left(\mathbf{x}'\right) = \int_{\partial D} G_{ji}\left(\mathbf{x}', \mathbf{x}\right) t_j\left(\mathbf{x}'\right) \mathrm{d}\mathbf{x} - \int_{\partial D} T_{ji}\left(\mathbf{x}', \mathbf{x}\right) u_j\left(\mathbf{x}\right) \mathrm{d}\mathbf{x} \tag{1.8}$$

where G_{ji} and T_{ji} are the fundamental solutions of induced displacement and traction fields for a unit point force acting in the infinite domain. In this way, a linear equation system can be established to solve the responses on nodes of the surface elements after applying boundary conditions. Then the inner field can be evaluated by the surface integral again.

Compared with other existing numerical methods, such as the finite element method, the finite difference method or finite volume method, which requires discretization of the entire domain, BEM is more efficient due to the fact that the equation system is developed with the integral merely on the surface. Therefore it reduces the complexity of a problem by one dimension [20]. Fewer elements and degrees of freedom are required for the same configuration of problems. This feature is extremely useful when dealing with three-dimensional problems with large, semi-infinite or infinite domains.

To get accurate results, the integrals of kernel shape function should be carefully carried out. In BEM application, when the interesting point is located exactly at the boundary, the integrals are singular, including weakly singular integrals, such as functions of order $\ln r$ for 2D problems and $1/r$ for 3D problems, and strongly singular integrals, such as functions of order $1/r$ for 2D problems and $1/r^2$ for 3D problems [21]. A bad numerical integral scheme will definitely ruin the accuracy of boundary element method, which may make BEM to fail in practice [22]. Significant effort has been made to improve the performance of the singular integral by further subdividing the integral area and applying rigid body solutions [21].

For a heterogeneous material system, different fundamental solutions have to be used for the corresponding boundary value problems, and extra effort has to be conducted to keep the continuity of the local fields on the boundary. In the previous work, the traditional BEM adopts the multiregion [9,21,23], which meshes the boundary of subdomains and collocates the boundary nodes with continuity conditions of tractions and displacements. Proposed by Beer [24], the major disadvantage of the multiregion method is the introduction of additional equations and unknowns to the linear system of equations. Wu et al. [25] applied the fast multipole algorithm for two-dimensional nonlinear interface debonding problem. To avoid the multiregion issue, the revised boundary domain [26] and eigenfield-based boundary integral equations [27–29] are proposed. The dual reciprocity BEM (DRBEM), introduced by Nardini and Brebbia [30], avoids the volume/area integral of nonhomogeneous terms (i.e., body force) through approximation with basis functions [31,32]. Although DRBEM could solve the problem with BEM in original form, its numerical accuracy is highly determined by selection of basis functions and locations of the domain points [24,33]. Ingber et al. [34] conducted a comparison of domain integral with the DRBEM and particular solution method, and proved a better accuracy of the method with domain integral.

Combining EIM and BEM provides a viable way to avoid the problem by treating a particle as an inclusion with the same material properties as the matrix but using an eigenfield to simulate the material response caused by the material mismatch.

1.5 Inclusion-based boundary element method (iBEM)

EIM can successfully capture the disturbance of local fields around inhomogeneities in an infinite domain but neglecting the boundary effect, and BEM is an efficient numerical method to evaluate inner field only by integrals of boundary values, which has to assume that the material is homogeneous. Although different schemes have been used in BEM to address the material difference of inhomogeneities, they commonly require a discretization of the interface between particles and the matrix and solve the unknowns with interfacial continuity. The accuracy significantly depends on the discretization and geometry of particles. With the increase of the number of particles, it is challenging to balance the computational speed and accuracy. Especially, because the particle sizes can change over a large range, it will be challenging to find an appropriate element size for the boundary discretization and integration considering the numerical stability and geometric variation.

The iBEM aims to use the EIM in BEM [35] to fill the gap by introducing the concept of inclusions into the boundary integral formulas. Multiple inclusions can be simulated in the same fashion with the equivalence conditions. Then the original inhomogeneity problem can be successfully replaced by a homogeneous problem with resolved inclusions. With the aid of analytical forms of integrals of Green's functions over an elliptical domain, this method is very attractive due to its high computational efficiency for a large particle system and high accuracy for characterizing particle–particle and boundary–particle interactions comprehensively.

Based on some previous work on EIM [10,13,36,37], we developed the first version of the iBEM program and used to simulate lightweight concrete containing spherical particles [35,38] among other composite materials. The code has been extended to dynamic problems [39]. The iBEM is designed to efficiently solve multiinhomogeneity problems by using the boundary element method. Different from the traditional boundary element method, which should mesh all the surface including the interfaces of inhomogeneity, this solver only deals with the outer boundary element, and each inhomogeneity is simulated by an inclusion with eigenstrain. Wu and Yin [40] extended the inclusion-based boundary element method (iBEM) from the classic ellipsoidal inhomogeneities of a 3D problem to both 2D and 3D problems, from static analysis to transient and dynamic analysis [41] Therefore iBEM becomes a powerful tool for virtual composite with thousands of particles for mechanical and multiphysical behavior.

Although BEM demonstrates the versatile applications, it is not yet a mainstream software like FEM. The main drawbacks of BEM are the inconveniences to handle with multiregion domains or inelastic behavior as the BEM relies on the Green's function of a homogeneous material in an infinite uniform material. The recent breakthrough of iBEM has shown a practical approach to release these two constraints to BEM and to change BEM from Cinderella to a princess in computational mechanics. Two recent papers [42,43] enable the simulation of arbitrarily shaped inclusion domain with inelastic strains. Because the material mismatch can be simulated by an eigenstrain, and plastic strain is indeed an eigenstrain, we can use iBEM with boundary and volume/area integrals of a single Green's function. The polygonal and polyhedral inclusions can be used in domain discretization to approach the exact solution, which is similar to the FEM. The exact integral of polynomial eigenstrain or body force over a polygonal inclusion provides a straightforward and robust approach to the fascinating problem of stress singularity at the vertex.

When we look back of the development of iBEM, although it has been based on our research needs on virtual experiments of composites and the iBEM program was independently developed starting at Mura's work [1,4] with the aid of BEM [21,44], an early form of iBEM can be traced back to this Japanese group [45–47]. However, they assumed that the eigenstrain on the ellipsoidal particles are uniform despite the particle interaction and boundary effect, so that the volume integral can be directly obtained by Eshelby's tensor [2]. This assumption saves a lot of effort for volume integral, but the uniform eigenstrain is obviously not applicable to actual composites with high concentrations and arbitrary shapes of particles. As a consequence, the work was not continued and applied to actual problems. The iBEM started later than this previous work at a polynomial eigenstrain distribution over ellipsoidal inhomogeneities and reached much higher accuracy when particles are close to each other or to boundary [13,35–37], and the recent work extended the iBEM to arbitrary particles with domain discretization, and therefore the iBEM can be used for mechanical and multiphysical modeling of general composite materials.

The detail implementation of iBEM is illustrated as follows. In the preprocessing procedure, the geometric information of the composite materials and elliptical shape particles is required. Only the outer surface of the composite domain is meshed into small elements. The material properties of matrix and particles need to be provided for generating a complete input file. In the next step, boundary element formulas are developed, which include not only the surface integrals but also the effect from inclusions on all particles. Then the equivalence conditions on flux at each center of the particle are assembled into the global equations. The surface integrals from boundary values will appear in equivalent inclusion formulas to take the boundary effect into account. In the solving procedure, boundary conditions are applied, and Gauss elimination method with pivot is deployed to accurately solve the linear equation system. After that, the field responses inside the domain can be evaluated by the equivalent inclusion problem with both known boundary values and eigenfield in the postprocessing procedure.

Compared with other existing numerical methods such as FEM, BEM, etc., iBEM possess the following remarkable properties:

1. The iBEM does not need mesh inside the composite material or the interface for the virtual experiments although domain discretization can be used for angular particles for higher accuracy;

2. The iBEM predicts the stress discontinuity or singularity by Eshelby's tensor with high stability and convergence for angular particles, so that a small number of elements can provide high accuracy;
3. The iBEM can provide rapid prediction of local fields for composites containing arbitrarily shaped particles with analytical integrals of triangular and elliptical inclusions, so that a large number of particles can be simulated in virtual experiments;
4. Because Eshelby's tensor is size independent, the iBEM can simulate a large-size range of particles for cross-scale modeling and simulations;
5. Eshelby's tensor handles the singularity and discontinuity analytically with the Green's function technique, which exhibits a small influence zone. The calculation is very stable and can be scaled up or down flexibly with the speed or accuracy needs;
6. No mesh is needed in the matrix, so that iBEM can simulate the particle motion or microstructural evolution without the regeneration of mesh;
7. When the local field is not required, the iBEM can accurately predict the effective material behaviors with a single or very few elements on each particle, so that virtual experiments of actual samples can be conducted.

Since the meshes for ellipsoidal or elliptical particles are not necessary, this algorithm has efficiently simulated thousands of particles [40]. The boundary–particle and particle–particle interactions are fully considered. The effect of each inclusion appears in BEM formulas, and the boundary integrals are incorporated into the EIM formulas as well. This is a great extension of EIM to the bounded domain case. The iBEM will perform better than many other models which assume that particles only interact pairwise in the infinite domain with neglecting any boundary effects. iBEM can simulate virtual experiments of composite for various physics problems by simply changing the underlying fundamental solution. This feature enables iBEM to be potentially applied to a wide range of engineering problems.

Up to today, the EIM has been mostly applied to the ellipsoidal/elliptical inhomogeneities. For polygonal inclusions, a number of pioneering works on the classic Eshelby's tensor can be found in the literature [48–55], which focus on inclusion problems with a constant eigenstrain. Because the Eshelby's tensor is not constant, it is hard to extend it to inhomogeneity problems. Recently, Wu and Yin [42] extended the Eshelby's tensor for a polynomial form of eigenstrain and used it with EIM to solve for the elastic field of an infinite domain containing polygonal inhomogeneities. However, there are two challenging issues to use the method for actual experiments of composites with arbitrarily shaped inhomogeneities:

a) Because singularity effects exist on the vertexes of a polygon, a single polynomial form of eigenstrain over the whole polygon cannot accurately approach the local solution;
b) Because actual composite samples exhibit a certain size, the interactions between inhomogeneities and the boundary cannot be quantified.

For the first challenging issue of singularity effect, to improve the accuracy of the EIM on polygonal inhomogeneities, two general approaches were used in the literature:

a) Increasing the order of eigenstrain field [1,4,42] or the strain field [54];
b) Discretization over the inhomogeneities with small/elementary elements [8, 56,57].

As for the first approach, based on the continuity of the eigenstrain field, it can be expanded in Taylor series [4]. However, Wu and Yin [42] have shown that even with the quadratic order eigenstrain, the prediction of stress fields of one triangular inhomogeneity under uniform far-field load may not be accurate enough due to the singularity. Regarding to the second approach, Nakasone et al. [8] and Hou et al. [57] discretized the inhomogeneity domain with triangular elements and applied the numerical EIM. Zhou et al. [56] discretized the subdomain with small enough rectangle elements, where the eigenstrain field is enforced to be uniform. A certain level of agreement was obtained in their work. However, the continuity of eigenstrain is lost under the assumption of uniform eigenstrain in the basic elements. In addition, even had the assumption be acceptable with small elements, to approach the boundary of the polygonal subdomains may lead to a dramatic increase of the number of elements, which causes inefficiency of computation. We use the analytical Eshelby's tensor for piecewise continuous eigenstrain at the C^0 continuity for a high accuracy and rapid convergence to the exact solution by refining the mesh.

For the second challenging issue, it is important to consider the boundary effect on the local field because different loads can be applied to the boundary in physical experiments and an actual specimen always exhibits a finite size. The infinite domain is a theoretical assumption. To reproduce physical experiments in a computer, the inclusion-based boundary element method combines the EIM and BEM [35,40,58] to consider the effects from both the inhomogeneities and the boundary. The advantages of iBEM over BEM are as follows:

1) For ellipsoidal or elliptical inhomogeneities, an internal mesh is required neither for the matrix nor for the subdomains, and a polynomial eigenstrain can provide tailorable accuracy to the exact solution;
2) For arbitrary polygonal inhomogeneities, a piecewise eigenstrain on a discretized inhomogeneity ensures a high accuracy with the C^0 continuity of the eigenstrain field. Refining particle discretization can approach the exact solution for the inhomogeneity problems;
3) Because no mesh is required in the matrix, and Eshelby tensors on inhomogeneities are size independent, the ill-conditioned mesh is avoided, and the size gradation of inhomogeneities can be simulated given the dimensions and locations.

Notice that although the ellipsoidal or elliptical inhomogeneities have been simulated with traditional Eshelby's tensors [40], it can also be numerically reproduced by a number of polyhedrons or polygons [42] or analytically represented by the integral of infinitesimal polygonal elements.

With the above advantages, the iBEM is particularly suitable for modeling the effective material behavior containing a large variety of inhomogeneities in size, shape, and material properties. It can be used for particle interaction crossing size-scales and virtual experiments of a physical composite specimen. It can predict both the local field and nonlocal behavior of a composite specimen. Some notes of iBEM related to some existing methods in the literature are provided as follows:

1) The EIM has been applied to simulate the elastic fields of ellipsoidal or elliptical inhomogeneities, which served as the foundation for several micromechanics models, such as the Mori–Tanaka method, the self-consistent scheme, and the dilute model [58–60] using various homogenization assumptions without directly considering the boundary effect and the effect of particle–particle interactions. The iBEM easily addressed the boundary effect by the boundary integral; moreover, it extended the particle shape to arbitrary shapes of inhomogeneities. Hence iBEM is not only applicable to micromechanics, but also serves as a versatile computational method for materials or structures with heterogeneous microstructure.

2) Due to the limitations of micromechanics models, several numerical methods were developed to conduct the virtual experiments of composite under various boundary conditions (BCs). In the previous work, the traditional BEM adopts the multiregion method [9,23,61], which meshes the boundary of subdomains and collocates the boundary nodes with continuity conditions of tractions and displacements. Based on Beer [24], the major disadvantage of the multiregion method is the introduction of additional equations and unknowns to the linear system of equations. Therefore it is expensive and uncertain to obtain the stable and convergent solution for angular shape of particles, particularly for a large number of particles. The present iBEM uses the same fundamental solution for the whole domain with stress equivalence and avoids the interface continuity equations.

3) Helsing [62] proposed an adaptive algorithm based on the Fredholm integral equation for multiple cracks and investigated the effective elastic properties under 10,000 cracks in 2D problem. Fu et al. [63], Fu and Rodin [64], and Lai and Rodin [65] developed the fast Galerkin BEM for domain containing many particles. Liu [66] and Liu and Shen [67] applied the fast multipole algorithm. Wu et al. [25] extended it for a two-dimensional nonlinear interface debonding problem. Although the fast multipole method (FMM) can significantly save the computational cost, it still relies on the interface continuity equations for the multiregion domain. On the contrary, the iBEM can use very few or even merely one element for each particle with analytical integral of particle domain to achieve high accuracy. The concept of FMM has also been used in iBEM to define an influence zone or particle interaction cutoff given a certain error limit, so that iBEM can theoretically achieve the same level of computational efficiency.

4) To avoid the multiregion issue, the revised boundary-domain [26] and eigenfield-based boundary integral equations [27–29] are proposed. The dual reciprocity BEM (DRBEM), introduced by Nardini and Brebbia [30], avoids the volume/area integral of nonhomogeneous terms (i.e., body force) through approximation with basis functions [31,32]. Although DRBEM can solve the problem with BEM in original form, its numerical accuracy is highly determined by selections of basis functions and locations of the domain points [24,33]. Ingber et al. [34] conducted a comparison of domain integral with the DRBEM and particular solution method, and proved a better accuracy for the method with domain integral. However, all those methods rely on the numerical integrals over the particle, which bring difficulties for different shapes of particles. For example, a spherical particle needs a lot of linear elements to match the geometry. The iBEM integrates the analytical domain integral for ellipsoidal/elliptical and polyhedral/polygonal particles into the computation and represents the highest accuracy for the domain integral.
5) Another extension of the FEM / BEM is a combination of isogeometric analysis (IGA) [68], namely IGAFEM [69–71] and IGABEM [24,72–75], whose advantage is the seamless integration with CAD. Beer et al. [24,76] proposed an iterative scheme to simulate the inclusion problem, which converts the mismatch initial stress to body force and traction increments. Such body force increments in a general inclusion can only be treated by a numerical domain integral of the Green's function. Compared with FEM or FVM, iBEM exhibits advantages in a much lower degree of freedom (DOF); different from IGABEM [76], the iBEM is based on the closed-form area (2D) or volume (3D) integrals of the Green's function and avoids the numerical stability and convergence in iterations.
6) Moreover, besides the BEM, most numerical methods require the discretization of the entire domain, such as the finite element method (FEM) and the finite volume method (FVM). Sukumar et al. [77] and Du et al. [78] proposed the extended finite element method (xFEM) to solve the inhomogeneity problems, in which the FEM approximation is enriched by additional discontinuous functions. Benowitz and Waisman [79] later used the numerical spline-based enrichment function with inclusion internal boundary discretization. The iBEM contains the singularity in the Eshelby's tensor as well, but only the integral over the particle is required. Therefore much fewer degrees of freedom are needed to solve the problem.

The iBEM can provide smooth elastic fields with Green's function. It is very convenient to extend it to simulate the multiphysical behavior and analyze the effective material properties of composites for different microstructures and volume fractions of particles. As the recent research on polygonal and polyhedral inclusions enables the modeling of composites with arbitrary shape of inhomogeneities, it becomes a powerful tool for the simulation and modeling of arbitrary composites. In Wu and Yin's recent work [40] the effective material behavior does change with shapes (i.e., aspect ratio, orientations), size gradation (i.e., dimension), and distribution of the inhomogeneities. The iBEM can capture the difference caused by these microstructural features. The iBEM using analytical integrals on inhomogeneities can avoid the numerical integral issues and provide higher accuracy in comparison with the above BEM methods.

1.6 Case study

To show the concept of inclusion and inhomogeneity and illustrate the iBEM, we present a case study. Consider a square structural steel plate with edge 0.2 m as an example in Fig. 1.4.

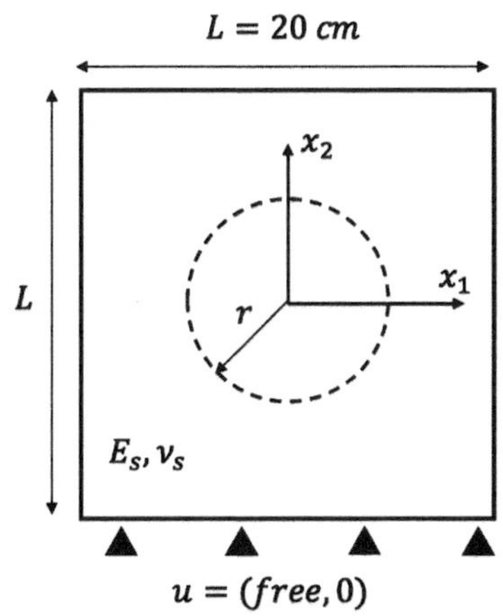

FIGURE 1.4 Schematic plot of a structural steel square plate with edge $L = 20\,\text{cm}$ containing a subdomain centered at the origin with radius equaling $1, 4.5, 9\,\text{cm}$ under traction-free boundary conditions.

It exhibits Young's modulus $E_s = 200\,\text{GPa}$, Poisson's ratio $\nu_s = 0.30$, and the thermal expansion coefficient $\alpha_s = 1.2 \times 10^{-5}\,\frac{\text{m}}{\text{m}\cdot{}^{\circ}\text{C}}$. When a circular sample is taken out at the center of the plate, to repair the hole, we can cool down a slightly larger circle with same material to the perfect matched size of the hole to fill it. When the temperature reaches equilibrium, the expansion of the circular steel piece will make the inclusion to tightly fill the hole without welding. This is an inclusion problem.

Consider the temperature change $-20\,^{\circ}\text{C}$ to perform the thermal fitting. It is corresponding to an eigenstrain of $\varepsilon^*_{ij} = 2.4 \times 10^{-4}\delta_{ij}$. Fig. 1.5 shows the stress distribution along the x_2 axis. When the radius is as small as 1 cm, the stress is nearly uniform on the inclusion because the boundary effect is negligible and Eshelby's tensor describes the stress. However, when the radius increases, the stress on the inclusion is not uniform anymore. The larger the inclusion, the smaller the stress on the inclusion, because the stress is relaxed by the free boundary.

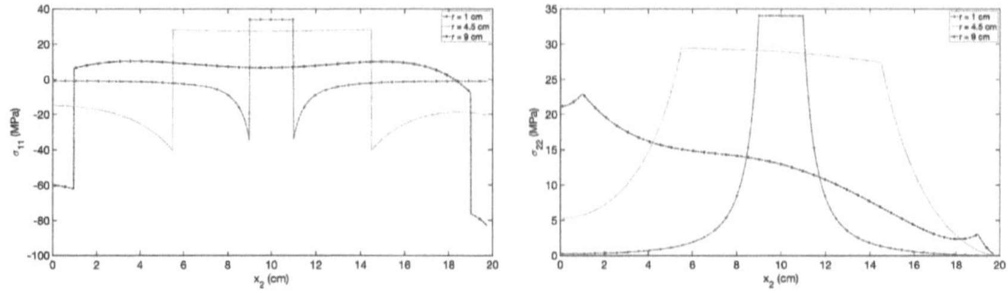

(a) (b)

FIGURE 1.5 Variation of stresses along the x_2 axis (a) σ_{11} and (b) σ_{22} with prescribed thermal strain $2.4 \times 10^{-4}\delta_{ij}$ in a circular inclusion with radii $r = 1, 4.5$, and 9 cm.

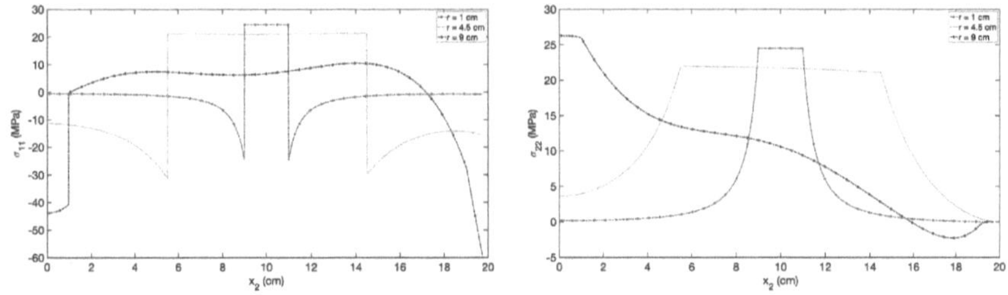

(a) (b)

FIGURE 1.6 Variation of stresses along the x_2 axis (a) σ_{11} and (b) σ_{22} with prescribed thermal strain $2.4 \times 10^{-4}\delta_{ij}$ in a circular aluminum inhomogeneity with radii $r = 1, 4.5$, and 9 cm.

If the hole is filled by an aluminum piece with the same misfitted strain at $\varepsilon^*_{ij} = 2.4 \times 10^{-4}\delta_{ij}$, then the aluminum exhibits different elastic constants at Young's modulus $E_a = 70\,\text{GPa}$ and Poisson's ratio $\nu_a = 0.33$. The filler becomes an inhomogeneity. Eshelby's EIM can be used to solve for the stress field. Similarly, Fig. 1.6 shows the stress distribution along the x_2 axis. When the radius is as small as 1 cm, the stress is still nearly uniform on the inclusion because the boundary effect is negligible and Eshelby's tensor is a constant. However, when the radius increases, the stress on the inclusion is not uniform anymore. Compared with Fig. 1.5, the stress variation in the inhomogeneity is slightly higher because the eigenstrain due to the material mismatch of the inhomogeneity is not constant. We use a quadratic eigenstrain distribution to consider the boundary effect. Because aluminum is more compliant than steel, the stress in aluminum inhomogeneity is smaller than that in the steel inclusion.

When a uniform external load (downward pressure) is applied to the boundary (top line), say $\sigma^0_{22} = -1\,\text{MPa}$, the internal stress caused by the load will be different for the inclusion and inhomogeneity problems. For the inclusion problem in Fig. 1.5, because there exists only one material phase, a uniform internal stress will be induced. However, for the inhomogeneity problem in Fig. 1.6, the local stress will be disturbed by the material mismatch at the neighborhood of the inhomogeneity. In the third case, Fig. 1.7 indicates the stress caused by ex-

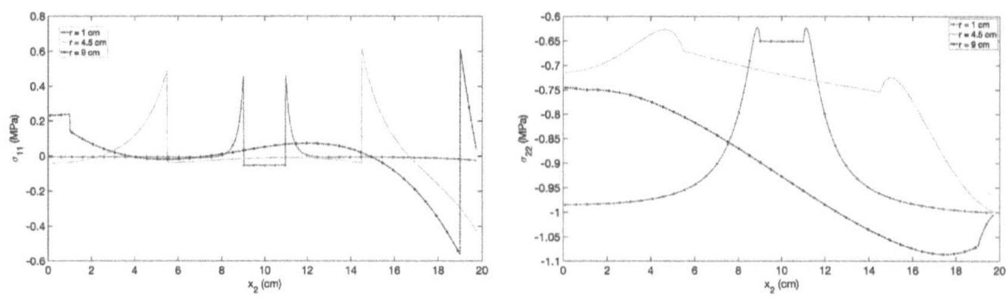

(a) (b)

FIGURE 1.7 Variation of stresses along the x_2 axis (a) σ_{11} and (b) σ_{22} with a circular inhomogeneity with radius $r = 1, 4.5$, and 9 cm under the uniform downward pressure $\sigma_{22}^0 = -1$ MPa applied on the top line

ternal load for the three cases of inhomogeneity radii. Due to the displacement constraint at the bottom, the larger the inhomogeneity, the less the stress discontinuity amplitude observed in the neighborhood of the bottom. In addition, even the top and bottom boundary conditions are not the same, symmetry of stress field exists for the case $r = 1$ cm, whereas the other two cases exhibit opposite trend, which could be interpreted as the boundary interactions or effect. Later in Chapter 5, such effects are introduced by the boundary integral equations.

1.7 Scope of this book

This book starts with the fundamental solutions for boundary value problems (BVP). Using an elastic problem as an example, the fundamental solution is the Green's function for a point force applied to the infinite domain. The Green's function has been used for both volume integral for inclusion problems and boundary integral for BVPs with bounded domain. Using the equivalent inclusion method, a material inhomogeneity is simulated with an inclusion for local elastic field without the boundary effect. When the Green's function for a semi-infinite is considered, the boundary effect to the local fields can be evaluated. For virtual experiments, the inclusion-based boundary element method is developed to solve the BVPs with general boundary conditions. Polynomial eigenstrains are used to reach tailorable accuracy for arbitrary shape of particles. However, a polynomial eigenstrain distribution is not sufficient for a complex shape of a particle, particularly, for polyhedral particles with singularity effect. Piecewise interpolation of the eigenstrain through the shape functions is used to approach the exact solution by the particle domain discretization.

Chapter 2 reviews different BVPs with potential flow, elastic, and Stokes flow equations and provides the fundamental solutions. Chapter 3 derives the volume integral of the fundamental solution over an inclusion with distributed eigenstrain field. Both ellipsoidal and polyhedral inclusions are considered with polynomial eigenstrains. The method is reduced to 2D problems as well. The analytical integrals provide the exact solution for inclusion problems. Chapter 4 introduces the Eshelby's equivalent inclusion method (EIM) for an ellipsoidal inhomogeneity in an infinite domain and generalizes it for polyhedral inhomogeneity. It provides an exact solution for ellipsoidal particles but approximate one for polyhedral particles. Particle discretization is introduced to obtain higher accuracy. To show the boundary effect, the Green's function for a semi-infinite domain is used in the EIM to study the local field for particles at different distance to the surface.

Virtual experiments use material specimens with finite volume. To simulate the elastic response of a composite with an arbitrary boundary, Chapter 5 introduces the inclusion-based boundary element method to calculate the elastic field with both boundary and volume integrals. First, using a polynomial form of eigenstrains in inclusions, the elastic field can be solved for inclusions embedded in a finite domain with polyhedral and ellipsoidal shapes. Then the EIM is used to consider the material mismatch. Because a polynomial eigenstrain may not provide sufficient accuracy for irregular shape of particles, Chapter 6 extends the iBEM to linear piecewise interpolation of the eigenstrain by a particle discretization, which may approach the exact solution by increasing the resolution of the mesh.

Chapters 7 and 8 extend the elastic analysis in Chapters 3–6 to the potential and Stokes' flows, respectively. Both three-dimensional (3D) and two-dimensional (2D) formulations are provided. Chapter 9 extends the static analysis of the above three types of BVP to transient or dynamic problems considering the time factor. Time-dependent material behavior is also introduced and modeled with the elastic–viscoelastic correspondence principle. Chapter 10 presents a case study to use the iBEM for multiphysical virtual experiments, which integrate different boundary value problems on material modeling and analysis. Finally, some recent development of iBEM for virtual experiments of composites and multiscale/physical modeling of structures and materials is presented.

Appendix 1.A Index notation of vectors and tensors

In this book, we use the index notation for derivation and calculation. In this section, we provide an overview of the index notation for the algebra operations and calculus of vectors and tensors. In general, we use a lower case letters to represent vectors and upper case letters for tensors. The bold font shows the symbolic form, and the italic font with index indicates the component form.

1.A.1 Index notation of vectors and tensors

A vector can be graphically depicted by a directed line in two- or three-dimensional Euclidean space. In a three-dimensional Cartesian coordinate $(0, x_1, x_2, x_3)$, the unit base vector along each coordinate axis is written as $\mathbf{e}_i$ $(i = 1, 2, 3)$. Then an arbitrary vector $\mathbf{v}$ could be expressed in terms of the projections in three orthogonal directions:

$$\mathbf{v} = \sum_{i=1}^{3} v_i \mathbf{e}_i = v_i \mathbf{e}_i \tag{1.9}$$

where v_i is the component of $\mathbf{v}$ relative to the spatial basis $\mathbf{e}$. The summation in the three directions can be concisely written in the index notation with the following rules: 1. The range of the index is from 1 to n. Here n is of the dimension of the coordinate, say 2 or 3 for 2D or 3D problems, respectively. 2. The repeated index means summation of the term over the range of the index. For example, $v_i \mathbf{e}_i\, v_1\mathbf{e}_1 + v_2\mathbf{e}_2 + v_3\mathbf{e}_3$. The repeated index is also called a dummy index because it can be replaced by another symbol without changing its outcome. 3. An index should *not* be repeated more than once. For example, $v_i b_i \mathbf{e}_i$ is not defined and should be avoided. For brevity, a vector $\mathbf{v}$ can also be shortened by the index form v_i by dropping the base vector.

Now let $\left(0, x_1', x_2', x_3'\right)$ be another coordinate with base vectors $\mathbf{e}_i'$ $(i = 1, 2, 3)$. The relation between $\mathbf{e}_i'$ and $\mathbf{e}_j$ can be written as

$$\mathbf{e}_i = Q_{ij}\mathbf{e}_j' \tag{1.10}$$

where $Q_{ij} = \cos\left(\mathbf{e}_i, \mathbf{e}_j'\right)$, which means the projection of vector $\mathbf{e}_i$ on the new base vector $\mathbf{e}_j'$. The matrix Q_{ij} is also called a transformation matrix. Inversely, we can also write

$$\mathbf{e}_i' = Q_{ij}'\mathbf{e}_j \tag{1.11}$$

with $Q_{ij}' = \cos\left(\mathbf{e}_i', \mathbf{e}_j\right)$. Obviously, when Q_{ij} and Q_{ij}' are written as matrices, $\mathbf{Q}' = \mathbf{Q}^T$. Here the superscript T denotes the transpose of the matrix or the swap of the indices. Using the relation of base vectors in Eq. (1.10), we can write the vector in Eq. (1.9) as

$$\mathbf{v} = v_i Q_{ij}\mathbf{e}_j' \tag{1.12}$$

So the component of $\mathbf{v}$ relative to the new basis $\mathbf{e}_i'$ is written as

$$v_i' = Q_{ji}v_j = Q_{ij}^T v_j \tag{1.13}$$

The algebra definition of a vector can be given as an array with the components (a_1, a_2, a_3) that satisfies Eq. (1.13) during the coordinate transformation. In continuum mechanics, some variables may require specification of two or more vectors for their description. For example, a stress component may be defined in a plane with surface normal vector along a certain force orientation, which is related to the surface normal vector and the force orientation vector. It will be a second-rank tensor because two vectors are involved. In addition, when a vector is transformed to another vector through a linear transformation, a second-rank tensor is typically used. In the Cartesian coordinate the components of a second-rank tensor will be associated with a dyadic product of two base vectors, say $\mathbf{e}_i \otimes \mathbf{e}_j$, which is also called a unit dyad. Mathematically, a second-rank tensor $\mathbf{A}$ can be written in terms of a linear combination of unit dyads, that is,

$$\mathbf{A} = A_{ij}\mathbf{e}_i \otimes \mathbf{e}_j \tag{1.14}$$

Similarly to vectors, during the coordinate transformation, the component of a tensor needs to satisfy

$$A'_{ij} = Q_{mi}Q_{nj}A_{mn} = Q^T_{im}A_{mn}Q_{nj} \tag{1.15}$$

The algebra definition of a second-rank tensor can be given as a matrix A_{ij} that satisfies Eq. (1.15) during the coordinate transformation.

In general, the component of a tensor may be associated with n vectors in terms of polyadic products of base vectors, say $\mathbf{e}_{i_1} \otimes \mathbf{e}_{i_2} \ldots \otimes \mathbf{e}_{i_n}$. Then Eq. (1.15) becomes

$$A'_{i_1 i_2 \ldots i_n} = Q_{j_1 i_1}Q_{j_2 i_2}\ldots Q_{j_n i_n}A_{j_1 j_2 \ldots j_n} \tag{1.16}$$

When $n = 0$, the tensor becomes a scalar, which does not change with coordinate transformation at all; when $n = 1$, the tensor becomes a vector; when $n = 3$, the unit polyad is called triad; and when $n = 4$, the unit polyad is called tetrad. The algebra definition of an n-rank tensor can be given by Eq. (1.16).

We provide two special tensors. Kronecker's delta δ_{ij} is defined as

$$\delta_{ij} = \begin{cases} 1 & i = j \\ 0 & i \neq j \end{cases} \tag{1.17}$$

which is also called the unit second-rank tensor, written as $\mathbf{I}$. A unique feature of δ_{ij} is its substitution property: if one index of Kronecker's delta is repeated with another tensor, then the index of that tensor can be replaced by the other index of Kronecker's delta. For example, $A_{ij}\delta_{jk} = A_{ik}$ and $A_{ij}\delta_{ji} = A_{ii} = A_{jj}$.

The Levi-Civita symbol, also called the permutation symbol, e_{ijk} is defined

$$e_{ijk} = \begin{cases} 1 & \text{even permutation of } i, j, k \\ 0 & \text{two or more repeated indices} \\ -1 & \text{odd permutation of } i, j, k \end{cases} \tag{1.18}$$

For example, $e_{111} = 0$, $e_{123} = 1$, $e_{213} = -1$. A very useful identity of e_{ijk} is

$$e_{ijk}e_{imn} = \delta_{jm}\delta_{kn} - \delta_{jn}\delta_{km} \tag{1.19}$$

which will be proved later.

1.A.2 Algebra operations of vectors and tensors

We introduce four common algebra operations.

1. Dot product, also called the scalar product or inner product, is defined as

$$\mathbf{a}\cdot\mathbf{b} = |\mathbf{a}|\,.\,|\mathbf{b}|\cos(\mathbf{a},\mathbf{b}) \tag{1.20}$$

Obviously, for a set of base vectors, their dot product can be written as

$$\mathbf{e}_i\cdot\mathbf{e}_j = \delta_{ij} \tag{1.21}$$

Eq. (1.20) can be rewritten as

$$\mathbf{a}\cdot\mathbf{b} = (a_i\mathbf{e}_i)\cdot\left(b_j\mathbf{e}_j\right) = a_ib_j\delta_{ij} = a_ib_i \tag{1.22}$$

2. Cross product, also called the vector product, is defined as

$$\mathbf{a}\times\mathbf{b} = |\mathbf{a}|\,.\,|\mathbf{b}|\sin(\mathbf{a},\mathbf{b})\mathbf{e}_c \tag{1.23}$$

where $\mathbf{e}_c$ is a unit vector perpendicular to the plane of **a** and **b**, positive in the direction of a right-handed screw turning **a** toward **b**. Therefore $\mathbf{a}\times\mathbf{b} = -\mathbf{b}\times\mathbf{a}$. For a set of base vectors, their cross product can be written as

$$\mathbf{e}_i\times\mathbf{e}_j = e_{ijk}\mathbf{e}_k \tag{1.24}$$

Eq. (1.23) can be rewritten as

$$\mathbf{a}\times\mathbf{b} = (a_i\mathbf{e}_i)\times\left(b_j\mathbf{e}_j\right) = a_ib_je_{ijk}\mathbf{e}_k \tag{1.25}$$

3. Scalar triple product, also called the triple product, is defined as

$$(\mathbf{a}\times\mathbf{b})\cdot\mathbf{c} = e_{ijk}a_ib_jc_k \tag{1.26}$$

which equals the volume of the parallelepiped box formed by vectors **a**, **b**, and **c**, positive in the case that the directional vector of a right-handed screw turning **a** toward **b** has the angle with **c** less than $\pi/2$. It can be written as the determinant of a matrix as follows:

$$(\mathbf{a}\times\mathbf{b})\cdot\mathbf{c} = \begin{vmatrix} a_1 & a_2 & a_3 \\ b_1 & b_2 & b_3 \\ c_1 & c_2 & c_3 \end{vmatrix} \tag{1.27}$$

The multiplication of two scalar triple products can be obtained by the multiplication of two matrices as follows:

$$
\begin{aligned}
[(\mathbf{a}\times\mathbf{b})\cdot\mathbf{c}]\,[(\mathbf{u}\times\mathbf{v})\cdot\mathbf{w}] &= \begin{vmatrix} a_1 & a_2 & a_3 \\ b_1 & b_2 & b_3 \\ c_1 & c_2 & c_3 \end{vmatrix} \cdot \begin{vmatrix} u_1 & v_1 & w_1 \\ u_2 & v_2 & w_2 \\ u_3 & v_3 & w_3 \end{vmatrix} \\
&= \left| \begin{pmatrix} a_1 & a_2 & a_3 \\ b_1 & b_2 & b_3 \\ c_1 & c_2 & c_3 \end{pmatrix} \cdot \begin{pmatrix} u_1 & v_1 & w_1 \\ u_2 & v_2 & w_2 \\ u_3 & v_3 & w_3 \end{pmatrix} \right| \\
&= \begin{vmatrix} \mathbf{a}\cdot\mathbf{u} & \mathbf{a}\cdot\mathbf{v} & \mathbf{a}\cdot\mathbf{w} \\ \mathbf{b}\cdot\mathbf{u} & \mathbf{b}\cdot\mathbf{v} & \mathbf{b}\cdot\mathbf{w} \\ \mathbf{c}\cdot\mathbf{u} & \mathbf{c}\cdot\mathbf{v} & \mathbf{c}\cdot\mathbf{w} \end{vmatrix}
\end{aligned} \tag{1.28}
$$

For a set of base vectors, their scalar triple product can be written as

$$
\mathbf{e}_i \times \mathbf{e}_j \cdot \mathbf{e}_k = e_{ijm}\mathbf{e}_m \cdot \mathbf{e}_k = e_{ijm}\delta_{mk} = e_{ijk} \tag{1.29}
$$

Therefore substitution of Eq. (1.29) into Eq. (1.28) yields

$$
\begin{aligned}
e_{ijk}e_{pqr} = (\mathbf{e}_i \times \mathbf{e}_j \cdot \mathbf{e}_k)(\mathbf{e}_p \times \mathbf{e}_q \cdot \mathbf{e}_r) &= \begin{vmatrix} \mathbf{e}_i\cdot\mathbf{e}_p & \mathbf{e}_i\cdot\mathbf{e}_q & \mathbf{e}_i\cdot\mathbf{e}_r \\ \mathbf{e}_j\cdot\mathbf{e}_p & \mathbf{e}_j\cdot\mathbf{e}_q & \mathbf{e}_j\cdot\mathbf{e}_r \\ \mathbf{e}_k\cdot\mathbf{e}_p & \mathbf{e}_k\cdot\mathbf{e}_q & \mathbf{e}_k\cdot\mathbf{e}_r \end{vmatrix} \\
&= \begin{vmatrix} \delta_{ip} & \delta_{iq} & \delta_{ir} \\ \delta_{jp} & \delta_{jq} & \delta_{jr} \\ \delta_{kp} & \delta_{kq} & \delta_{kr} \end{vmatrix}
\end{aligned} \tag{1.30}
$$

Eq. (1.30) implies

$$
\begin{aligned}
e_{ijk}e_{imn} &= \begin{vmatrix} \delta_{ii} & \delta_{im} & \delta_{in} \\ \delta_{ji} & \delta_{jm} & \delta_{jn} \\ \delta_{ki} & \delta_{km} & \delta_{kn} \end{vmatrix} \\
&= 3\begin{vmatrix} \delta_{jm} & \delta_{jn} \\ \delta_{km} & \delta_{kn} \end{vmatrix} - \delta_{im}\begin{vmatrix} \delta_{ji} & \delta_{jn} \\ \delta_{ki} & \delta_{kn} \end{vmatrix} + \delta_{in}\begin{vmatrix} \delta_{ji} & \delta_{jm} \\ \delta_{ki} & \delta_{km} \end{vmatrix} \\
&= \delta_{jm}\delta_{kn} - \delta_{jn}\delta_{km}
\end{aligned} \tag{1.31}
$$

4. Dyadic product, also called the outer product, is defined as

$$
\mathbf{a}\otimes\mathbf{b} = (a_i\mathbf{e}_i)\otimes(b_j\mathbf{e}_j) = a_i b_j \mathbf{e}_i \otimes \mathbf{e}_j \tag{1.32}
$$

which makes a second-rank tensor.

The rules of the multiplication of vectors can be easily extended to tensors as vectors are in fact first-rank tensors. We introduce some operations of tensors, which will be used later in the following chapter.

The outer product of two tensors is written as

$$\mathbf{A}\otimes\mathbf{B}=\left(A_{ij}\mathbf{e}_i\otimes\mathbf{e}_j\right)\otimes\left(B_{mn}\mathbf{e}_m\otimes\mathbf{e}_n\right)=A_{ij}B_{mn}\mathbf{e}_i\otimes\mathbf{e}_j\otimes\mathbf{e}_m\otimes\mathbf{e}_n \tag{1.33}$$

which makes a fourth-rank tensor. If **B** is a vector, this equation is reduced to

$$\mathbf{A}\otimes\mathbf{b}=\left(A_{ij}\mathbf{e}_i\otimes\mathbf{e}_j\right)\otimes\left(b_m\mathbf{e}_m\right)=A_{ij}b_m\mathbf{e}_i\otimes\mathbf{e}_j\otimes\mathbf{e}_m \tag{1.34}$$

If the order of two tensors is swapped, the result is different.

The inner product of two tensors is written as

$$\begin{aligned}\mathbf{A}\cdot\mathbf{B} &= \left(A_{ij}\mathbf{e}_i\otimes\mathbf{e}_j\right)\cdot\left(B_{mn}\mathbf{e}_m\otimes\mathbf{e}_n\right) = A_{ij}B_{mn}\mathbf{e}_i\otimes\left(\mathbf{e}_j\cdot\mathbf{e}_m\right)\mathbf{e}_n\\ &= A_{ij}B_{mn}\delta_{jm}\mathbf{e}_i\otimes\mathbf{e}_n = A_{ij}B_{jn}\mathbf{e}_i\otimes\mathbf{e}_n\end{aligned} \tag{1.35}$$

If **B** is a vector, then this equation is reduced to

$$\mathbf{A}\cdot\mathbf{b}=\left(A_{ij}\mathbf{e}_i\otimes\mathbf{e}_j\right)\cdot\left(b_m\mathbf{e}_m\right)=A_{ij}b_j\mathbf{e}_i \tag{1.36}$$

The double dot product of two tensors is

$$\mathbf{A}:\mathbf{B}=\left(A_{ij}\mathbf{e}_i\otimes\mathbf{e}_j\right):\left(B_{mn}\mathbf{e}_m\otimes\mathbf{e}_n\right)=A_{ij}B_{mn}\left(\mathbf{e}_i\cdot\mathbf{e}_m\right)\left(\mathbf{e}_j\cdot\mathbf{e}_n\right)=A_{ij}B_{ij} \tag{1.37}$$

which makes a scalar. If **A** is a fourth-rank tensor, then the last two base vectors are used in the dot product as follows:

$$\begin{aligned}\left(A_{ijkl}\mathbf{e}_i\otimes\mathbf{e}_j\otimes\mathbf{e}_k\otimes\mathbf{e}_l\right):\left(B_{mn}\mathbf{e}_m\otimes\mathbf{e}_n\right)&=A_{ijkl}B_{mn}\delta_{km}\delta_{ln}\mathbf{e}_i\otimes\mathbf{e}_j\\ &=A_{ijkl}B_{kl}\mathbf{e}_i\otimes\mathbf{e}_j\end{aligned} \tag{1.38}$$

The double dot product of a second-rank tensor **A** with unit tensor **I** is also called the trace of **A**, that is,

$$Tr\left(\mathbf{A}\right)=\mathbf{A}:\mathbf{I}=\left(A_{ij}\mathbf{e}_i\otimes\mathbf{e}_j\right):\left(\delta_{mn}\mathbf{e}_m\otimes\mathbf{e}_n\right)=A_{ii} \tag{1.39}$$

1.A.3 Calculus of vector and tensor fields

In a domain Ω a vector or tensor field indicates a vector or tensor function at any point $\mathbf{x}$ of Ω, which can be represented by $\mathbf{a}(\mathbf{x})$ for a vector field or $\mathbf{A}(\mathbf{x})$ for a tensor field. For dynamic problems, a vector or tensor field may change with time. For example, a speed field in a fluid can be a vector field; the stiffness of a graded material can be a tensor field. In this section, vector or tensor fields are considered to be continuous and differentiable.

1.A.3.1 Del operator and operations

The del operator ∇, also called the nabla or Hamilton operator, is written as $\partial_i\mathbf{e}_i$ or $\frac{\partial}{\partial x_i}\mathbf{e}_i$, which is a vector operator. It follows the rules of vector multiplication.

1. Gradient. For a differentiable scalar field $\phi(\mathbf{x})$, the gradient of ϕ is defined as

$$grad(\phi) = \nabla\phi = \frac{\partial \phi}{\partial x_i}\mathbf{e}_i = \phi_{,i}\mathbf{e}_i \tag{1.40}$$

which is a vector indicating the greatest change rate and direction of the scalar field.

The gradient of a vector field $\mathbf{a}(\mathbf{x})$ is defined as

$$grad(\mathbf{a}) = \nabla\mathbf{a} = \partial_i \mathbf{e}_i \otimes a_j(\mathbf{x})\mathbf{e}_j = \partial_i a_j \mathbf{e}_i \otimes \mathbf{e}_j = a_{j,i}\mathbf{e}_i \otimes \mathbf{e}_j \tag{1.41}$$

Similarly, the gradient of a tensor field $\mathbf{A}(\mathbf{x})$ is defined as

$$grad(\mathbf{A}) = \nabla \otimes \mathbf{A} = \partial_i A_{jk}\mathbf{e}_i \otimes \mathbf{e}_j \otimes \mathbf{e}_k = A_{jk,i}\mathbf{e}_i \otimes \mathbf{e}_j \otimes \mathbf{e}_k \tag{1.42}$$

In fact, the gradient of a vector or tensor field is the outer product of the del operator and a vector or tensor field. The rank will increase by one. Following this definition, we can write

$$\mathbf{a}\nabla = \partial_j a_i \mathbf{e}_i \otimes \mathbf{e}_j = a_{i,j}\mathbf{e}_i \otimes \mathbf{e}_j \tag{1.43}$$

which is different from definition (1.41).

2. Divergence. For a vector field $\mathbf{a}(\mathbf{x})$, the divergence of $\mathbf{a}$ is defined as

$$div(\mathbf{a}) = \nabla \cdot \mathbf{a} = \partial_i \mathbf{e}_i \cdot a_j \mathbf{e}_j = \partial_i a_i = a_{i,i} \tag{1.44}$$

which indicates the volume density of the outward flux of the vector field from an infinitesimal volume around the point. It is the inner product (dot product) of the del operator and a vector field. For tensor fields, the same rule is applied, and the rank of a tensor will reduce by one.

3. Curl. For a vector field $\mathbf{a}(\mathbf{x})$, the curl of $\mathbf{a}$ is defined as

$$curl(\mathbf{a}) = \nabla \times \mathbf{a} = \partial_i \mathbf{e}_i \times a_j \mathbf{e}_j = \partial_i a_j e_{ijk}\mathbf{e}_k = a_{j,i}e_{ijk}\mathbf{e}_k \tag{1.45}$$

which describes the infinitesimal rotation of the vector field. It is the cross product (vector product) of the del operator and a vector field. For tensor fields, the same rule is applied.

From the above definition, given a vector field, we can find that $\nabla\mathbf{a} = (\mathbf{a}\nabla)^T$, $\nabla \cdot \mathbf{a} = \mathbf{a} \cdot \nabla$, and $\nabla \times \mathbf{a} = -\mathbf{a} \times \nabla$. In addition, we can see the following identities:

$$\nabla \times (\nabla\phi) = \mathbf{0} \tag{1.46}$$

$$\nabla \cdot (\nabla \times \mathbf{a}) = 0 \tag{1.47}$$

$$\nabla \times (\nabla \times \mathbf{a}) = \nabla(\nabla \cdot \mathbf{a}) - \nabla^2\mathbf{a} \tag{1.48}$$

and

$$\nabla \cdot (\nabla \phi) = \nabla^2 \phi = \phi_{,ii} \tag{1.49}$$

where $\nabla^2 = \nabla \cdot \nabla = \partial_i \partial_i$ is the Laplace operator.

1.A.3.2 Gauss' theorem

The Gauss theorem, also called the divergence theorem, transforms a volume integral of the divergence of a vector field $\mathbf{a}(\mathbf{x})$ with continuous derivatives in a domain Ω to the surface flux across the boundary $\partial\Omega$:

$$\int_{\Omega} \nabla \cdot \mathbf{a}\, dV = \int_{\partial\Omega} \mathbf{n} \cdot \mathbf{a}\, dS \tag{1.50}$$

where dV and dS represent the volume and surface infinitesimal elements for volume and surface integrals. The Gauss theorem can be generalized to other operations of the del operator and other types of fields, namely scalar fields or tensor fields. Therefore we can write the generalized Gauss theorem [80] as follows:

$$\int_{\Omega} \nabla * \mathbf{T}\, dv = \int_{\partial\Omega} \mathbf{n} * \mathbf{T}\, dS \tag{1.51}$$

where $*$ indicates an operation, such as inner, outer, or cross product, and $\mathbf{T}$ can be a tensor field of certain rank, that is, zeroth rank for a scalar, first rank for a vector, and second or higher ranks for general tensors. If the order of ∇ and $\mathbf{T}$ is switched, then the order of $\mathbf{n}$ and $\mathbf{T}$ will also be switched.

1.A.3.3 Green's and Stokes' theorems

In a simply connected two-dimensional domain S in the x_1–x_2 coordinate system with boundary ∂S, let U and V be continuous functions with continuous derivatives. The Green's theorem states that

$$\int_{\partial S} U dx_1 + V dx_2 = \int_S \left(\frac{\partial V}{\partial x_1} - \frac{\partial U}{\partial x_2} \right) dx_1 dx_2 \tag{1.52}$$

where ∂S is positive along the counterclockwise direction.

Green's theorem transforms an integral over a two-dimensional domain into a line integral over its boundary. Extending it to the three-dimensional domain, a similar equation can be obtained, which is Stokes' theorem. For a simply connected surface S in the x_1, x_2, x_3 coordinate system with boundary ∂S, let $\mathbf{a}$ be a vector field with continuous derivatives. Stokes' theorem states that

$$\int_{\partial S} \mathbf{a} \cdot d\mathbf{l} = \int_S (\nabla \times \mathbf{a}) \cdot \mathbf{n}\, dS \tag{1.53}$$

where the positive direction of $d\mathbf{l}$ is along the right-handed screw direction of the surface normal vector $\mathbf{n}$.

Appendix 1.B Two generalized functions

The generalized functions are objects extending the notion of functions, and one of its main usages is enabling analytical operations of singular functions, such as Green's functions, which exhibit extensive applications in physics and mathematics [81]. In the following, we introduce two widely used functions, the Dirac delta and Heaviside functions.

1.B.1 Definitions of two generalized functions

The Dirac delta function [82] in 1D problem is defined as

$$\delta(x) = \begin{cases} 0, & x \neq 0 \\ \infty, & x = 0 \end{cases} \tag{1.54}$$

which is also known as the unit impulse symbol, a generalized function with zero value everywhere except at zero. Its integral over the entire real line or over a neighborhood of 0 is equal to one:

$$\int_{-\infty}^{\infty} \delta(x)dx \equiv 1 \tag{1.55}$$

As shown in Eq. (1.54), the Dirac delta function is not a function in classic sense as it does not exhibit a certain value at $x = 0$.

The Heaviside function [83] is the piecewise function defined as

$$H(x) = \begin{cases} 0, & x < 0 \\ 1, & x \geq 0 \end{cases} \tag{1.56}$$

which is also known as the unit step function, and its derivative is the Dirac delta function, that is,

$$H'(x) = \delta(x) \tag{1.57}$$

Therefore it can also written as the integral of $\delta(x)$:

$$H(x) = \int_{-\infty}^{x} \delta(x)dx \tag{1.58}$$

The integral of the Heaviside function can be written as a ramp function as

$$R(x) = \int_{-\infty}^{x} H(x)dx = xH(x) \tag{1.59}$$

which can be extended to higher-order terms by iterating integrals.

The Dirac and Heaviside functions show the unit impulse and step at the origin, which can be shifted to other location x' with $\delta(x - x')$ and $H(x - x')$.

Therefore the combination of multiple generalized functions can illustrate the multiple impulses and steps in a 1D problem. For example, $H(x) - H(x-1)$ shows a distribution with zero value except the range of $x \in (0, 1)$ with a value of 1.

They can be generalized to higher dimensions by the multiplication operation. For 3D problems, a unit impulse at $\mathbf{x}'$ can be written as

$$\delta(\mathbf{x} - \mathbf{x}') = \delta(x_1 - x_1')\delta(x_2 - x_2')\delta(x_3 - x_3') \tag{1.60}$$

which exhibits similar properties:

$$\delta(\mathbf{x} - \mathbf{x}') = \begin{cases} 0, & \mathbf{x} \neq \mathbf{x}' \\ \infty, & \mathbf{x} = \mathbf{x}' \end{cases} \tag{1.61}$$

and

$$\int_{\mathcal{D}} \delta(\mathbf{x} - \mathbf{x}')d\mathbf{x} = \begin{cases} 0, & \mathbf{x}' \notin \mathcal{D} \\ 1, & \mathbf{x}' \in \mathcal{D} \end{cases} \tag{1.62}$$

For 2D problems, a similar operation can be used as well.

1.B.2 Properties of two generalized functions

To elaborate the method of Green's function in Chapter 2, we list some properties.

1. Dirac measure. One way to rigorously capture the notion of the Dirac delta function is to define a measure as follows:

$$\int_{-\infty}^{\infty} \delta(x - x_0) f(x)dx = f(x_0) \tag{1.63}$$

because $f(x)\delta(x - x_0) = f(x_0)\delta(x - x_0)$ as $\delta(x - x_0)$ is zero for $x \neq x_0$. These properties can extended to higher dimensions with the same form. It satisfies the scaling rule

$$\int_{-\infty}^{\infty} f(x)\delta(ux - v)dx = \frac{1}{|u|} f(\frac{v}{u}) \tag{1.64}$$

which can be proved by replacement of variable $t = ux - v$.

2. Distributional derivatives. If there exist nth-order derivatives $f^n(0)$ of a function $f(x)$ at $x = 0$, then

$$\int_{-\infty}^{\infty} f(x)\delta^n(x)dx = (-1)^n f^n(0) \tag{1.65}$$

which can be shown step by step by the following identities:

$$\int_{-\infty}^{\infty} f(x)\delta'(x)dx = \int_{-\infty}^{\infty} f(x)d\delta(x) = 0 - \int_{-\infty}^{\infty} \delta(x)df(x) = -f'(0) \tag{1.66}$$

considering that the values of $\delta(x)$ and its derivatives at $x = \pm\infty$ are zero.

3. Fourier transform. The Fourier transform of the Dirac delta function can be well-defined in the Fourier space as

$$\tilde{\delta}(\xi) = \mathcal{F}(\delta(x - x')) = \frac{1}{2\pi}\int_{-\infty}^{\infty} \delta(x - x')e^{-i\xi x}dx = \frac{1}{2\pi}e^{-i\xi x'} \tag{1.67}$$

where for $x' = 0$, $\mathcal{F}(\delta(x)) = \frac{1}{2\pi}$. The inverse Fourier transform of a uniform function in the Fourier space defines the delta function:

$$\int_{-\infty}^{\infty} \left(\frac{1}{2\pi}\right) e^{i\xi x} d\xi = \delta(x) \tag{1.68}$$

Similarly, we can also write

$$\int_{-\infty}^{\infty} \left(\frac{1}{2\pi}e^{-i\xi x'}\right) e^{i\xi x}dk = \left(\frac{1}{2\pi}\right)\int_{-\infty}^{\infty} e^{i\xi(x-x')}dk = \delta(x - x') \tag{1.69}$$

For the 3D case, the above equation can be rewritten as

$$\delta(\mathbf{x}-\mathbf{x}') = \delta(x_1 - x_1')\delta(x_2 - x_2')\delta(x_3 - x_3') = \left(\frac{1}{2\pi}\right)^3 \int_{-\infty}^{\infty} e^{i\boldsymbol{\xi}\cdot(\mathbf{x}-\mathbf{x}')}d\boldsymbol{\xi} \tag{1.70}$$

where $d\boldsymbol{\xi} = d\xi_1 d\xi_2 d\xi_3$.

4. Generalization in multiple dimensions. In Chapter 9, to obtain the fundamental solutions to a 3D transient problem caused by a impulse point heat source at $\mathbf{x} = \mathbf{x}_0$ and $t = t_0$, the method of separation of variables will be applicable, and the Dirac delta function can be written as follows:

$$\delta(\mathbf{x} - \mathbf{x}_0, t - t_0) = \delta(x - x_0)\delta(y - y_0)\delta(z - z_0)\delta(t - t_0) \tag{1.71}$$

where x_0, y_0, and z_0 are the components of the distance vector $\mathbf{x} - \mathbf{x}_0$ in Cartesian coordinate. Notice that its extension to other coordinates, say the spherical coordinate, should be written as

$$\delta(\mathbf{x} - \mathbf{x}_0, t - t_0) = \frac{1}{\rho^2 \sin(\theta)}\delta(\rho - \rho_0)\delta(\theta - \theta_0)\delta(\beta - \beta_0)\delta(t - t_0) \tag{1.72}$$

where ρ, θ, and β are the radial distance, inclination angle, and azimuth angle, respectively.

Chapter 2

Fundamental solutions

Potential flow, elastic, and Stokes flow problems

In this chapter, we first introduce the three types of boundary value problems. When an infinite domain is subjected to a point source, the response can be represented by a Green's function, which is the fundamental solution. The fundamental solutions for potential flow, elastic, and Stokes flow equations are provided for three-dimensional (3D) and two-dimensional (2D) cases. When the infinite domain contains two materials with a plane interface, the fundamental solution is obtained in the Appendix. The static solution is extended to time-dependent loading and time-dependent material behavior in Chapter 9.

2.1 Introduction to boundary value problems

Boundary value problems are formulated with a set of differential equations together with a set of additional constraints. The differential equations govern the behaviors of variables in the field. The additional constraints called boundary conditions ensure the problem has a unique solution and is well posed. This type of problems widely exists in physics. The scalar or vector field variable u, governed by a set of differential equations, can be a temperature field in the potential flow, a velocity field in the Stokes equations, a displacement field in the Navier–Cauchy equations, etc. By satisfying all the boundary conditions on D the differential equations form a complete system $f(u, E) = 0$ with the constitutive law, which characterizes the field response with different material properties E.

Let us give an example of an elastic problem. Fig. 2.1 shows multiple inhomogeneities Ω^I $(I = 1, 2, \ldots)$ embedded in a bounded domain $\mathcal{D}$ subjected to various boundary conditions in the Cartesian coordinate of the x_1–x_2 plane. The matrix domain is denoted by Ω^0, and therefore $\mathcal{D} = \Omega^0 \bigcup_{I=1,2,\ldots} \Omega^I$. It can be either a plane stress or plane strain problem. This section uses a plane strain for the formulation in which the inhomogeneities can be considered as long prisms or fibers with the corresponding cross-sections. However, the formulation can be used for plane stress problems by replacing the elastic moduli [58]. Here t_i and u_i are prescribed tractions and displacements, respectively. An inelastic strain may exist on Ω^I, written as ε_{ij}^{*I}, which can, for example, be a thermal strain or misfit strain. It is also called an eigenstrain [1]. For an isotropic elastic medium,

The Inclusion-Based Boundary Element Method (iBEM). https://doi.org/10.1016/B978-0-12-819384-6.00010-3

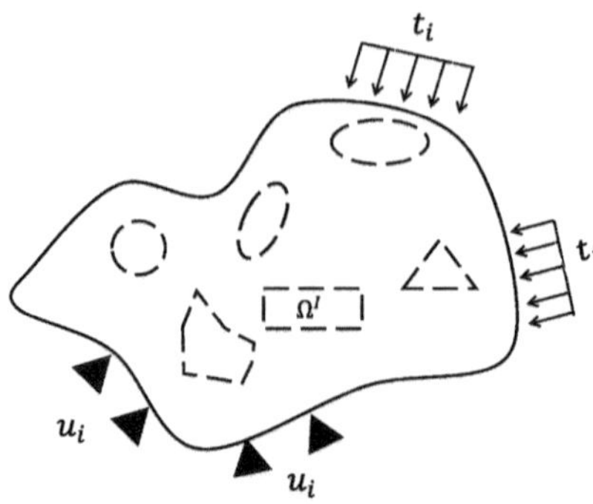

FIGURE 2.1 Schematic illustration of multiple inhomogeneities embedded in a domain $\mathcal{D}$ with prescribed boundary conditions.

let $C^I_{ijkl} = \lambda^I \delta_{ij}\delta_{kl} + \mu^I(\delta_{ik}\delta_{jl} + \delta_{il}\delta_{jk})$, where λ^I and μ^I denote the Lame constants of the Ith phase of the composite, where $I = 0$ and $I = 1, 2, \ldots$ represent the matrix and the Ith inhomogeneity phase, respectively. The governing equation can be written as

$$C^I_{ijkl}(u_{k,li} - \varepsilon^{*I}_{kl,i}) + b_j = o \quad \text{for } \mathbf{x} \in \Omega^I \tag{2.1}$$

which is the Navier equation in each material domain of $I = 0, 1, 2, \ldots$. Here the displacement field u_k is continuous and differentiable, whereas the eigenstrain ε^{*I}_{kl} is piecewise continuous. For simplicity, the body force b_j is taken as zero unless it is particularly specified.

Assuming no debonding on the interfaces, the boundary value problem of inhomogeneities can be enacted with the continuity equations of displacements and tractions in the equation

$$\sigma^+_{ij}(\mathbf{x})n^+_j = \sigma^-_{ij}(\mathbf{x})n^-_j, \quad u^+_i(\mathbf{x}) = u^-_i(\mathbf{x}) \quad \text{on } \partial\Omega^I \tag{2.2}$$

where $\partial\Omega^I$ represents the interface between the Ith subdomain and the matrix, superscripts + and − denote the outer and inner surfaces of the interface, respectively, and n_j is the unit normal vector of the surface.

Considering the traction and displacement boundary conditions on $\partial\mathcal{D}$, this BVP can be solved by the multiregion BEM with two challenges:

1) the prescribed eigenstrain on the inhomogeneities needs to be carefully addressed;
2) the singularity on the vertices of a polygon requires very fine elements to obtain accurate results, which leads to high computational costs.

The iBEM avoids the multiregion BEM by using an eigenstrain to simulate the material mismatch. Both the boundary integral of the source field on the boundary and volume integral of the source field on the inclusion rely on the fundamental solution, which is also the Green's function caused by a unit point source.

We further introduce the fundamental solutions for elastic, potential flow, and Stoke's flow problems, which include 3D and 2D cases and a bimaterial infinite domain.

2.2 Fundamental solution for elastic problems

Consider the Navier governing equation (2.1) for a point force in an infinite domain. The governing equation can be rewritten as

$$(\lambda^0 + \mu^0)u_{i,ij} + \mu^0 u_{j,ii} + b_j \delta(\mathbf{x}') = 0 \tag{2.3}$$

where $\delta(\mathbf{x}')$ denotes a point force at $\mathbf{x}'$. The fundamental solution was given by Kelvin [84] as

$$u_i(\mathbf{x}) = G_{ij}(\mathbf{x}, \mathbf{x}')b_j \tag{2.4}$$

where $G_{ij}(\mathbf{x}, \mathbf{x}')$ is the Green's function to describe the field at $\mathbf{x}$ caused by a point source field at $\mathbf{x}'$, which is written as

$$G_{ij}(\mathbf{x}, \mathbf{x}') = \frac{1}{4\pi\mu^0}\delta_{ij}\phi - \frac{1}{16\pi\mu^0(1-\nu^0)}\psi_{,ij} \tag{2.5}$$

where ϕ and ψ are the harmonic and biharmonic potentials. For 3D problems,

$$\phi = \frac{1}{|\mathbf{x} - \mathbf{x}'|}, \quad \psi = |\mathbf{x} - \mathbf{x}'| \tag{2.6}$$

and for 2D problems,

$$\phi = -\ln|\mathbf{x} - \mathbf{x}'|^2, \quad \psi = -|\mathbf{x} - \mathbf{x}'|^2 \frac{\ln|\mathbf{x} - \mathbf{x}'|^2 - 1}{2} \tag{2.7}$$

which can be obtained by the integral of the 3D Green's function along x_3. Notice that the Poisson ratio is related to the Lame constants as $\nu^0 = \frac{\lambda^0}{2(\lambda^0+\mu^0)}$.

When dynamic effects are considered, the governing equation will include the inertia force with an acceleration term, so that the potentials need to be reconstructed with the time variable, which will be elaborated in Chapter 9.

2.3 Fundamental solution for potential flows

Potential flow problems commonly exist in multiphysical analysis of material behavior, for example, heat flow, magnetic flux, electric flow, or water pressure head. Here the steady-state heat conduction is used as an example to illustrate our I-BEM algorithm. Compared with the elastic problem, the displacement u_i is replaced with the temperature field u inside the domain, which is a scalar field. The material property is the conductivity K. The governing equation in absence of heat source or sink is the Laplace equation

$$K\nabla^2 u + q\delta(\mathbf{x}') = 0 \tag{2.8}$$

where $q\delta(\mathbf{x}')$ is the heat source at $\mathbf{x}'$. The Green's function G is written as

$$G(\mathbf{x}, \mathbf{x}') = \frac{1}{4\pi K}\phi \tag{2.9}$$

where the harmonic potential ϕ for 3D and 2D problems can be found from Eqs. (2.6) and (2.7).

For transient heat conduction problems, the governing equation includes the effect of internal heat energy changing with time, so that the potentials need to be reconstructed with the time variable, which will be elaborated in Chapter 9.

2.4 Fundamental solution for the Stokes flows

For the Stokes flow problem, the field quantity u_i is the velocity of the fluid, which is a vector similar to the displacement field in the elastic problem. The material property is the viscosity μ. Mathematically, replacing the displacement in the elastic problem with the velocity and the elastic constants with viscosity and $\nu = 0.5$, the elastic problem is equivalent to the Stokes flow with a pressure field and continuity equation as follows:

$$\mu\nabla^2 u_i - p_{,i} + b_i\delta(\mathbf{x}') = 0, \quad u_{i,i} = 0 \tag{2.10}$$

where p is the pressure field, and b_i is the point force in the Stokes flow. The velocity field can be written in terms of the Green's function G written as

$$G_{ij}(\mathbf{x}, \mathbf{x}') = \frac{1}{4\pi\mu}\left(\delta_{ij}\phi - \frac{1}{2}\psi_{,ij}\right) \tag{2.11}$$

and the pressure field can be written as $p(\mathbf{x}) = P_i(\mathbf{x}, \mathbf{x}')b_i$ with

$$P_i(\mathbf{x}, \mathbf{x}') = -\frac{1}{4\pi}\frac{\partial\phi}{\partial x_i} \tag{2.12}$$

where the harmonic potential ϕ and biharmonic potential ψ for 3D and 2D problems can be found in Eqs. (2.6) and (2.7).

For different boundary value problems, different fundamental solutions or Green's functions can be obtained. Here we consider the fundamental solution for a point source in the infinite domain, and the boundary integral method can address arbitrary boundary conditions with any geometric shapes. In this chapter, we provide three types of fundamental solutions for both homogeneous 2D and 3D domains. In the following Appendix, we consider an infinite domain containing two materials with a plane interface and can reduce the fundamental solution to homogeneous materials by taking the two materials identical. When one material is taken as a vacuum with material coefficients at zero or perfectly rigid/conductive material with material coefficient at the infinite problem, the latter can be reduced to a semi-infinite problem as the local field in the first material domain is trivial.

In addition, in Chapter 9, we provide fundamental solutions for transient or dynamic problems and for viscoelastic material behavior.

Appendix 2.A Extension to bimaterial infinite domain

Consider an infinite two-phase elastic solid in Fig. 2.2, which consists of two different semi-infinite, homogeneous, and isotropic halves with a fully bonded interface. The Cartesian coordinate has been set up with the origin on the interface of the x_1–x_2 plane at $x_3 = 0$. When a point source is applied in this infinite domain, the solution can also be obtained through the fundamental solutions for elastic, potential flow, and Stokes' flow problems.

2.A.1 Elastic fundamental solution for bimaterials

A concentrated force $\mathbf{F}$ is applied in the interior of the upper half at $\mathbf{x}' = \left(x_1', x_2', x_3'\right)$. The stiffness tensors for the two phases of materials are denoted by $\mathbf{C}^0$ and $\bar{\mathbf{C}}^0$ for $x_3 \geq 0$ and $x_3 < 0$, respectively. Specifically, the shear modulus and Poisson's ratio are written as μ^0 and υ^0 for $x_3 \geq 0$ and $\bar{\mu}^0$ and $\bar{\upsilon}^0$ for $x_3 < 0$, respectively.

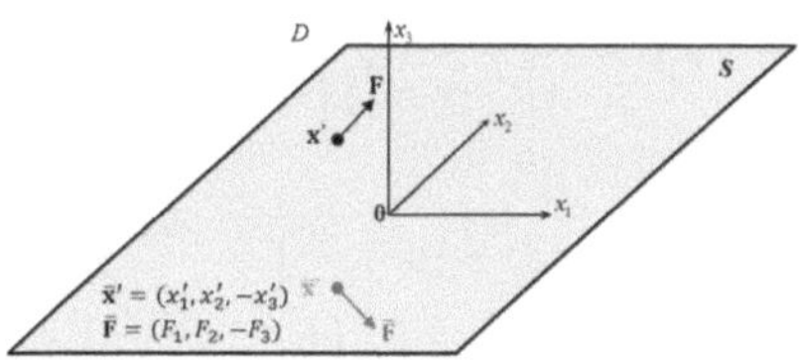

FIGURE 2.2 Two joined semi-infinite domains ($x_3 \geq 0$) containing a concentrated force $\mathbf{F}$ at $\mathbf{x}'$ with the image force $\bar{\mathbf{F}}$ at $\bar{\mathbf{x}}'$.

The image point of $\mathbf{x}'$ is written as $\bar{\mathbf{x}}'$ with coordinate $\left(x_1', x_2', -x_3'\right)$. The corresponding image force is $\bar{\mathbf{F}} = (F_1, F_2, -F_3)$. Notice that the image of a vector, say $\mathbf{F}$ and $\mathbf{x}'$, can be written by a mirror projection as

$$(\bar{\bullet}) = \mathbf{Q} \cdot (\bullet) \text{ with } \mathbf{Q} = \begin{bmatrix} 1 & 0 & 0 \\ 0 & 1 & 0 \\ 0 & 0 & -1 \end{bmatrix} \tag{2.13}$$

where $(\bullet)$ and $(\bar{\bullet})$ mean any vectors in the original and mirror projected coordinates, such as $\bar{F}_i = Q_{ji} F_j$ and $F_i = Q_{ij} \bar{F}_j$.

In [85], the displacement at $\mathbf{x}$ in the infinite two-phase solid can be written in terms of Green's function as,

$$u_i(\mathbf{x}) = G_{ij}\left(\mathbf{x}, \mathbf{x}'\right) F_j\left(\mathbf{x}'\right) \tag{2.14}$$

where the form of $G_{ij}\left(\mathbf{x}, \mathbf{x}'\right)$ depends on the location of $\mathbf{x}$. However, the fundamental solution in [85] only considers the case of $x_3' > 0$. For $x_3 \geq 0$ [36],

$$\begin{aligned}4\pi\mu^0 G_{ij}\left(\mathbf{x}, \mathbf{x}'\right) = &\left[\phi\delta_{ij} - \frac{1}{4\left(1-v^0\right)}\psi_{,ij}\right] + A\bar{\phi}\delta_{ij} + B\left(\delta_{i3}\delta_{jk} - \delta_{ik}\delta_{j3}\right)\overline{\alpha}_{,k}\\ &- Cx_3\Big[Q_{jm}\bar{\psi}_{,im3} + 4\left(1-v^0\right)\delta_{j3}\bar{\phi}_{,i}\\ &+ 2\left(1-2v^0\right)\delta_{i3}Q_{jm}\bar{\phi}_{,m} - Q_{jm}x_3\bar{\phi}_{,im}\Big]\\ &- DQ_{im}Q_{jn}\bar{\psi}_{,mn} - (G+B)Q_{jm}\overline{\beta}_{,im}\end{aligned} \tag{2.15}$$

and for $x_3 < 0$,

$$\begin{aligned}4\pi\mu^0 G_{ij}\left(\mathbf{x}, \mathbf{x}'\right) = &\left[\phi\delta_{ij} - \frac{1}{4\left(1-v^0\right)}\psi_{,ij}\right] + A\phi\delta_{ij} + B\left(\delta_{i3}\delta_{jk} - \delta_{ik}\delta_{j3}\right)\alpha_{,k}\\ &- D\psi_{,ij} - x_3F\alpha_{,ij} - (G+B)Q_{im}\beta_{,jm}\end{aligned} \tag{2.16}$$

where

$$\begin{aligned}A &= \frac{\mu^0 - \bar{\mu}^0}{\mu^0 + \bar{\mu}^0}, \quad B = \frac{2\mu^0\left(1-2v^0\right)\left(\mu^0 - \bar{\mu}^0\right)}{\left(\mu^0 + \bar{\mu}^0\right)\left[\mu^0 + \bar{\mu}^0\left(3-4v^0\right)\right]}\\ C &= \frac{\mu^0 - \bar{\mu}^0}{2\left(1-v^0\right)\left(\mu^0 + \left(3-4v^0\right)\bar{\mu}^0\right)}, \quad D = \frac{\left(3-4v^0\right)}{2}C\\ F &= \frac{2\mu^0\left[\mu^0\left(1-2\bar{v}^0\right) - \bar{\mu}^0\left(1-2v^0\right)\right]}{\left[\mu^0 + \bar{\mu}^0\left(3-4v^0\right)\right]\left[\bar{\mu}^0 + \mu^0\left(3-4\bar{v}^0\right)\right]}\\ G &= \frac{\mu^0\left[\bar{\mu}^0\left(1-2\bar{v}^0\right)\left(3-4v^0\right) - \mu^0\left(1-2v^0\right)\left(3-4\bar{v}^0\right)\right]}{\left[\mu^0 + \bar{\mu}^0\left(3-4v^0\right)\right]\left[\bar{\mu}^0 + \mu^0\left(3-4\bar{v}^0\right)\right]}\end{aligned}$$

and

$$\phi = 1/\left|\mathbf{x} - \mathbf{x}'\right|, \quad \overline{\phi} = 1/\left|\mathbf{x} - \bar{\mathbf{x}}'\right| \tag{2.17}$$

$$\psi = \left|\mathbf{x} - \mathbf{x}'\right|, \quad \overline{\psi} = \left|\mathbf{x} - \bar{\mathbf{x}}'\right| \tag{2.18}$$

$$\alpha = \ln\left(-x_3 + x_3' + \psi\right), \quad \overline{\alpha} = \ln\left(x_3 + x_3' + \overline{\psi}\right) \tag{2.19}$$

$$\begin{aligned}\beta &= \left(-x_3 + x_3'\right)\ln\left(-x_3 + x_3' + \psi\right) - \psi\\ \overline{\beta} &= \left(x_3 + x_3'\right)\ln\left(x_3 + x_3' + \overline{\psi}\right) - \overline{\psi}\end{aligned} \tag{2.20}$$

When the source $x_3' < 0$, the Green's function can be derived in similar fashion as Walpole or using the symmetric properties that $G_{ij}(\mathbf{x}, \mathbf{x}') = G_{ij}(-\mathbf{x}, -\mathbf{x}')$. Notice that "-" interchange the phases of materials, therefore, the coefficients shown in Eq. (2.16) should be modified accordingly, i.e., $A = \frac{\mu^0 - \overline{\mu}^0}{\mu^0 - \overline{\mu}^0}$ becomes $A^l = \frac{\overline{\mu}^0 - \mu^0}{\mu^0 - \overline{\mu}^0}$. For $x_3 < 0$,

$$\begin{aligned}
4\pi \overline{\mu^0} G_{ij}\left(\mathbf{x}, \mathbf{x}'\right) = & \left[\phi \delta_{ij} - \frac{1}{4\left(1 - \overline{v}^0\right)} \psi_{,ij}\right] + A^l \bar{\phi}^l \delta_{ij} - B\left(\delta_{i3}\delta_{jk} - \delta_{ik}\delta_{j3}\right) \overline{\alpha}^l_{,k} \\
& - C x_3 \Big[Q_{jm} \bar{\psi}_{,im3} + 4\left(1 - \overline{v}^0\right) \delta_{j3} \bar{\phi}_{,i} \\
& + 2\left(1 - 2\overline{v}^0\right) \delta_{i3} Q_{jm} \bar{\phi}^l_{,m} - Q_{jm} x_3 \bar{\phi}^l_{,im} \Big] \\
& - D Q_{im} Q_{jn} \bar{\psi}_{,mn} - (G + B) Q_{jm} \overline{\beta}^l_{,im}
\end{aligned} \tag{2.21}$$

and for $x_3 > 0$,

$$\begin{aligned}
4\pi \overline{\mu}^0 G_{ij}\left(\mathbf{x}, \mathbf{x}'\right) = & \left[\phi \delta_{ij} - \frac{1}{4\left(1 - \overline{v}^0\right)} \psi_{,ij}\right] + A \phi \delta_{ij} - B\left(\delta_{i3}\delta_{jk} - \delta_{ik}\delta_{j3}\right) \alpha^l_{,k} \\
& - D \psi_{,ij} + x_3 F \alpha^l_{,ij} - (G + B) Q_{im} \beta^l_{,jm}
\end{aligned} \tag{2.22}$$

where α^l and β^l change as,

$$\begin{aligned}
\alpha^l = \ln(x_3' - x_3 + \psi), \quad \overline{\alpha}^l &= \ln(-x_3' - x_3 + \overline{\psi}) \\
\beta^l &= (x_3' - x_3) \ln(x_3' - x_3 + \psi) - \psi \\
\overline{\beta}^l &= -(x_3' + x_3) \ln(-x_3' - x_3 + \psi) - \overline{\psi}
\end{aligned}$$

$\overline{\alpha}$ and $\overline{\beta}$ can be written as the following line integrals from $(x_1, x_2, +\infty)$ to (x_1, x_2, x_3), which means that the potential at the far-field is zero and thus leads to an integral constant term, but it disappears when taking a derivative, so it is ignored in the equations:

$$\overline{\alpha} = \int_{+\infty}^{x_3} \frac{1}{\sqrt{\left(x_1 - x_1'\right)^2 + \left(x_2 - x_2'\right)^2 + \left(t + x_3'\right)^2}} dt = \int_{+\infty}^{x_3} \overline{\phi}(\mathbf{x}^t, \mathbf{x}') dt \tag{2.23}$$

and

$$\begin{aligned}
\overline{\beta} &= \int_{+\infty}^{x_3} \int_{+\infty}^{t} \frac{1}{\sqrt{\left(x_1 - x_1'\right)^2 + \left(x_2 - x_2'\right)^2 + \left(s + x_3'\right)^2}} ds dt \\
&= \int_{+\infty}^{x_3} \int_{+\infty}^{t} \overline{\phi}(\mathbf{x}^s, \mathbf{x}') ds dt
\end{aligned} \tag{2.24}$$

where x^t and x^s mean (x_1, x_2, t) and (x_1, x_2, s), and thus $\overline{\phi}(\mathbf{x}^t, \bar{\mathbf{x}}')$ is defined as follows:

$$\overline{\phi}(\mathbf{x}^t, \bar{\mathbf{x}}') = \frac{1}{\sqrt{\left(x_1 - x_1'\right)^2 + \left(x_2 - x_2'\right)^2 + \left(t + x_3'\right)^2}} \tag{2.25}$$

Similarly, α and β can be written as

$$\alpha = -\int_{-\infty}^{x_3} \phi(\mathbf{x}^t, \mathbf{x}')dt \tag{2.26}$$

and

$$\beta = \int_{-\infty}^{x_3}\int_{-\infty}^{t} \phi(\mathbf{x}^s, \mathbf{x}')dsdt \tag{2.27}$$

When $\bar{\mu}^0 = 0$, the above Green's function is reduced to the Mindlin solution [86] for a semi-infinite domain with a free surface:

$$\begin{aligned} 4\pi\mu^0 G_{ij}\left(\mathbf{x}, \mathbf{x}'\right) = {} & \phi\delta_{ij} - \frac{1}{4\left(1 - \upsilon^0\right)}\psi_{,ij} + \bar{\phi}\delta_{ij} - \frac{\left(3 - 4\upsilon^0\right)}{4\left(1 - \upsilon^0\right)} Q_{im}Q_{jn}\bar{\psi}_{,mn} \\ & - \frac{x_3}{2\left(1 - \upsilon^0\right)}\left[Q_{jm}\bar{\psi}_{,im3} + 4\left(1 - \upsilon^0\right)\delta_{j3}\bar{\phi}_{,i}\right. \\ & \left. + 2\left(1 - 2\upsilon^0\right)\delta_{i3}Q_{jm}\bar{\phi}_{,m} - Q_{jm}x_3\bar{\phi}_{,im}\right] \\ & + 2\left(1 - 2\upsilon^0\right)\left(\delta_{i3}\delta_{jk} - \delta_{ik}\delta_{j3}\right)\overline{\alpha}_{,k} - \left(1 - 2\upsilon^0\right)Q_{jm}\overline{\beta}_{,im} \end{aligned} \tag{2.28}$$

Notice that this mathematical form of the Mindlin solution is more concise than the original form of the Mindlin solution although they are consistent with each other [86–89]. As $\bar{\mu}^0 \to \infty$, the above Green's function is reduced to the Rongved's solution [13,87] for a semi-infinite domain with a fixed surface:

$$\begin{aligned} 4\pi\mu^0 G_{ij}\left(\mathbf{x}, \mathbf{x}'\right) = {} & (\phi - \bar{\phi})\delta_{ij} - \frac{1}{4(1 - \nu^0)}(\psi_{,ij} - Q_{im}Q_{jn}\bar{\psi}_{,mn}) \\ & + \frac{x_3}{2(1 - \nu^0)(3 - 4\nu^0)}\left[\begin{array}{c} (Q_{jm}\bar{\psi}_{,im3} + 4(1 - \nu^0)\delta_{j3}\bar{\phi}_{,i} \\ +2(1 - 2\nu^0)\delta_{i3}Q_{jm}\bar{\phi}_{,m} - Q_{jm}x_3\bar{\phi}_{,im} \end{array}\right] \end{aligned} \tag{2.29}$$

To obtain the two-dimensional Green's function for a concentrated force in an infinite bimaterial, we use the three-dimensional Green's function for a concentrated force in an infinite bimaterial and extend the point force from $-\infty$ to $+\infty$ along the x_2 axis [90]. In this way, the Green's function for two-dimensional case is derived by a direct integral along one axis. Therefore the two-dimensional Green's function is given by

$$4\pi \mu^0 G_{ij}^{2D}(\mathbf{x}, \mathbf{x}') = 4\pi \mu^0 \int_{-\infty}^{\infty} G_{ij}(\mathbf{x}, \mathbf{x}') dx_2' \tag{2.30}$$

Switching x_3 to x_2, we can write the plane strain fundamental solution in 2D for the bimaterial infinite domain as follows: For $x_2 \geq 0$,

$$\begin{aligned} 4\pi \mu^0 G_{ij}(\mathbf{x}, \mathbf{x}') = & \left[\phi \delta_{ij} - \frac{1}{4(1-v^0)} \psi_{,ij} \right] + A\bar{\phi}\delta_{ij} + B(\delta_{i2}\delta_{jk} - \delta_{ik}\delta_{j2})\overline{\alpha}_{,k} \\ & - Cx_2 \Big[Q_{jm}\bar{\psi}_{,im2} + 4\left(1-v^0\right)\delta_{j2}\bar{\phi}_{,i} \\ & + 2\left(1-2v^0\right)\delta_{i2}Q_{jm}\bar{\phi}_{,m} - Q_{jm}x_2\bar{\phi}_{,im} \Big] \\ & - DQ_{im}Q_{jn}\bar{\psi}_{,mn} - (G+B)Q_{jm}\overline{\beta}_{,im} \end{aligned} \tag{2.31}$$

and for $x_2 < 0$,

$$\begin{aligned} 4\pi \mu^0 G_{ij}(\mathbf{x}, \mathbf{x}') = & A\phi\delta_{ij} + B(\delta_{i2}\delta_{jk} - \delta_{ik}\delta_{j2})\alpha_{,k} - D\psi_{,ij} - x_2 F\alpha, ij \\ & - (G+B)Q_{im}\beta_{,jm} \end{aligned} \tag{2.32}$$

where the harmonic potentials ϕ and $\overline{\phi}$ and the biharmonic potentials ψ and $\overline{\psi}$ for 2D problems can still be obtained by Eq. (2.7), and α, β, $\overline{\alpha}$, and $\overline{\beta}$ can also be written in terms of the integral of the harmonic potential in the 2D case with the integral constants disregarded as Eq. (2.7):

$$\begin{aligned} \overline{\alpha} &= \int_{+\infty}^{x_2} \overline{\phi}(\mathbf{x}^t, \mathbf{x}')dt = \int_{+\infty}^{x_2} -2\ln\sqrt{(x_1 - x_1')^2 + (t + x_2')^2}dt \\ &= (x_2 + x_2')(\overline{\phi} + 2) + 2(x_1 - x_1')\arctan\frac{x_1 - x_1'}{x_2 + x_2'} \end{aligned} \tag{2.33}$$

$$\begin{aligned} \overline{\beta} &= \int_{+\infty}^{x_2}\int_{+\infty}^{t} \overline{\phi}(\mathbf{x}^s, \mathbf{x}')dsdt \\ &= \int_{+\infty}^{x_2}\int_{+\infty}^{t} -2\ln\sqrt{(x_1 - x_1')^2 + (t + x_2')^2}dsdt \\ &= \frac{1}{2}\left[-(x_1 - x_1')^2 + (x_2 + x_2')^2\right]\overline{\phi} + \frac{3}{2}(x_2 + x_2')^2 \\ &+ 2(x_1 - x_1')(x_2 + x_2')\arctan\frac{x_1 - x_1'}{x_2 + x_2'} \end{aligned} \tag{2.34}$$

$$\begin{aligned} \alpha &= -\int_{-\infty}^{x_2} \phi(\mathbf{x}^t, \mathbf{x}')dt \\ &= -(-x_2 + x_2')(\phi + 2) - (x_1 - x_1')\arctan\frac{x_1 - x_1'}{-x_2 + x_2'} \end{aligned} \tag{2.35}$$

and

$$\beta = \int_{-\infty}^{x_2}\int_{-\infty}^{t} \phi(\mathbf{x}^s, \mathbf{x}')dsdt$$
$$= \frac{1}{2}\left[-(x_1 - x_1')^2 + (x_2 - x_2')^2\right]\phi + \frac{3}{2}(x_2 - x_2')^2$$
$$+ 2(x_1 - x_1')(x_2 - x_2') \arctan \frac{x_1 - x_1'}{x_2 - x_2'} \tag{2.36}$$

When $\bar{\mu}^0 = 0$, the above Green's function is reduced to the Mindlin solution for the 2D space as

$$4\pi\mu^0 G_{ij}(\mathbf{x}, \mathbf{x}') = \phi\delta_{ij} - \frac{1}{4(1-\nu^0)}\psi_{,ij} + \bar{\phi}\delta_{ij} - \frac{(2-4\nu^0)}{4(1-\nu^0)}Q_{im}Q_{jn}\bar{\psi}_{,mn}$$
$$- \frac{x_2}{2(1-\nu^0)}\Big[Q_{jm}\bar{\psi}_{,im2} + 4\left(1-\nu^0\right)\delta_{j2}\bar{\phi}_{,i}$$
$$+2\left(1-2\nu^0\right)\delta_{i2}Q_{jm}\bar{\phi}_{,m} - Q_{jm}x_2\bar{\phi}_{,im}\Big]$$
$$+2\left(1-2\nu^0\right)(\delta_{i2}\delta_{jk} - \delta_{ik}\delta_{j2})\overline{\alpha}_{,k} - \left(1-2\nu^0\right)Q_{jm}\overline{\beta}_{,im} \tag{2.37}$$

As $\bar{\mu}^0 \to \infty$, the above Green's function is reduced to the Rongved solution [13,87] as

$$4\pi\mu^0 G_{ij}(\mathbf{x}, \mathbf{x}') = (\phi - \bar{\phi})\delta_{ij} - \frac{1}{4(1-\nu^0}(\psi_{,ij} - Q_{im}Q_{jn}\bar{\psi}_{,mn})$$
$$+ \frac{x_2}{2(1-\nu^0)(3-4\nu^0)}\begin{bmatrix}(Q_{jm}\bar{\psi}_{,im2} + 4(1-\nu^0)\delta_{j2}\bar{\phi}_{,i} \\ +2(1-2\nu^0)\delta_{i2}Q_{jm}\bar{\phi}_{,m} - Q_{jm}x_2\bar{\phi}_{,im}\end{bmatrix} \tag{2.38}$$

2.A.2 Fundamental solution for a potential flow in bimaterials

Consider the infinite bimaterial constituted of upper and lower parts, divided by the interface **S** of $x_3 = 0$, where the upper semi-infinite matrix domain **D** exhibits the homogeneous thermal conductivity K^0, and the lower semi-infinite domain **L** the homogeneous thermal conductivity $\bar{K}^0$. The source point is denoted as $\mathbf{x}' = (x_1', x_2', x_3')$, whereas the field point of interest is denoted as $\mathbf{x} = (x_1, x_2, x_3)$.

Following the imaging displacement setting [36] and modifying Walpole's formulation [85], the temperature field at in the infinite domain can be written in terms of Green's function as

$$u(\mathbf{x}) = G(\mathbf{x}, \mathbf{x}')\mathbf{q}(\mathbf{x}') \tag{2.39}$$

where $\mathbf{q}(\mathbf{x}')$ is the point heat source at $\mathbf{x}'$, $G(\mathbf{x}, \mathbf{x}')$ is the Green's function needed to be developed in virtue of the image method. In general, the form of $G(\mathbf{x}, \mathbf{x}')$ depends on the location of $\mathbf{x}$ and $\mathbf{x}'$.

Therefore referring to a heat source $\mathbf{q}$ at the coordinate $\mathbf{x}' = (x_1', x_2', x_3')$ above the interface where $x_3 > 0$ and its image point at $\bar{\mathbf{x}}' = (x_1', x_2', -x_3')$, the Green's function can be written explicitly as follows: For $x_3 \geq 0$,

$$G(\mathbf{x}, \mathbf{x}') = \frac{1}{4\pi K'}\phi + \frac{K' - K''}{K' + K''}\frac{1}{4\pi K'}\overline{\phi} \tag{2.40}$$

and for $x_3 < 0$,

$$G(\mathbf{x}, \mathbf{x}') = 1/2\pi(K' + K'')\phi \tag{2.41}$$

where the harmonic potentials ϕ and $\overline{\phi}$ can be obtained from Eqs. (2.6) and (2.7) for the 3D and 2D cases, respectively. Similarly to elastic problems, we can set $\bar{K}^0 = 0$ or ∞ to consider a semi-infinite domain with an adiabatic and fixed temperature boundary condition on the surface, respectively.

2.A.3 Fundamental solution for a Stokes flow in bimaterials

Similarly, replacing the displacement in the elastic problem with the velocity and the elastic constants with viscosity and $\nu = 0.5$, the elastic problem is equivalent to the Stokes flow with a pressure field and continuity equation. For a bimaterial fluid, the upper semi-infinite matrix domain $\mathbf{D}$ exhibits the viscosity μ^0, and the lower semi-infinite domain $\mathbf{L}$ the viscosity $\bar{\mu}^0$. The fundamental solution can be similarly written as follows: For $x_3 \geq 0$,

$$\begin{aligned} 4\pi\mu^0 G_{ij}(\mathbf{x}, \mathbf{x}') = &\left[\phi\delta_{ij} - \frac{1}{2}\psi_{,ij}\right] + A\bar{\phi}\delta_{ij} - DQ_{im}Q_{jn}\bar{\psi}_{,mn} \\ &- Cx_3\left[Q_{jm}\bar{\psi}_{,im3} + 2\delta_{j3}\bar{\phi}_{,i} - Q_{jm}x_3\bar{\phi}_{,im}\right] \end{aligned} \tag{2.42}$$

and for $x_3 < 0$,

$$4\pi\mu^0 G_{ij}(\mathbf{x}, \mathbf{x}') = A\phi\delta_{ij} + B\left(\delta_{i3}\delta_{jk} - \delta_{ik}\delta_{j3}\right)\eta_{,k} - D\psi_{,ij} \tag{2.43}$$

where $A = \frac{\mu^0 - \bar{\mu}^0}{\mu^0 + \bar{\mu}^0}$, $C = \frac{\mu^0 - \bar{\mu}^0}{\mu^0 + \bar{\mu}^0}$, and $D = \frac{1}{2}C$. When $\bar{\mu}^0 = 0$, the above Green's function is reduced to the Mindlin solution as

$$\begin{aligned} 4\pi\mu^0 G_{ij}(\mathbf{x}, \mathbf{x}') = &\left[\phi\delta_{ij} - \frac{1}{2}\psi_{,ij}\right] + \bar{\phi}\delta_{ij} - \frac{1}{2}Q_{im}Q_{jn}\bar{\psi}_{,mn} \\ &- x_3\left[Q_{jm}\bar{\psi}_{,im3} + 2\delta_{j3}\bar{\phi}_{,i} - Q_{jm}x_3\bar{\phi}_{,im}\right] \end{aligned} \tag{2.44}$$

As $\bar{\mu}^0 \to \infty$, the above Green's function is reduced to the Rongved solution [13,87] as

$$4\pi\mu^0 G_{ij}\left(\mathbf{x},\mathbf{x}'\right) = (\phi - \bar{\phi})\delta_{ij} - \frac{1}{2}(\psi_{,ij} - Q_{im}Q_{jn}\bar{\psi}_{,mn}) \\ + x_2\left[(Q_{jm}\bar{\psi}_{,im2} + 2\delta_{j2}\bar{\phi}_{,i} - Q_{jm}x_2\bar{\phi}_{,im}\right] \tag{2.45}$$

Similarly, the above equation can be reduced to the 2D problems as well by replacing the harmonic and biharmonic potentials with Eq. (2.7). Notice that compared with the elastic problem, the terms of α and β do not exist.

Chapter 3

Integrals of Green's functions and their derivatives

Eshelby's tensor for elastic inclusion problems

This chapter derives the volume integral of the fundamental solution over an inclusion with a distributed eigenstrain field. Both ellipsoidal and polyhedral inclusions are considered with polynomial eigenstrains. The method is extended to 2D problems as well. The analytical integrals provide the exact solution to inclusion problems.

3.1 Introduction to inclusion problems

Following Eshelby's procedure [2], we can consider an inclusion problem where the particle exhibits the same stiffness as the matrix of C^0_{ijkl} but is subjected to a body force b_j and a polynomial eigenstrain. Here the eigenstrain is a generic term that refers to a nonelastic strain, such as a phase transformation, misfit, or thermal strain [1]. Now the governing equation can be rewritten as

$$(\lambda^0 + \mu^0)u_{i,ij} + \mu^0 u_{j,ii} - C^0_{ijkl}\varepsilon^p_{kl,i} + b_j = 0 \tag{3.1}$$

where the eigenstrain ε^p_{kl} is continuous on the inclusion Ω only and is zero on the matrix Ω. Eq. (3.1) is uniform over the whole domain $\mathcal{D}$ with the sources of eigenstrain and a body force on the inclusion Ω. Because the material is uniform, the far-field stress creates a uniform stress over the whole domain, which can be superposed to the elastic field caused by the source on the inclusion. The displacement caused by the body force and prescribed eigenstrain can be written as

$$u_i(\mathbf{x}) = \int_D G_{ij}(\mathbf{x}, \mathbf{x}')[b_j(\mathbf{x}') - C^0_{mjkl}\varepsilon^p_{kl,m'}(\mathbf{x}')]d\mathbf{x}' \tag{3.2}$$

where $G_{ij}(\mathbf{x}, \mathbf{x}')$ is the Green's function or fundamental solution to describe the field at $\mathbf{x}$ caused by a point source field at $\mathbf{x}$', which is elaborated in the last chapter. It can be rewritten as

$$G_{ij}(\mathbf{x}, \mathbf{x}') = \frac{1}{4\pi\mu^0}\delta_{ij}\phi - \frac{1}{16\pi\mu^0(1-\nu^0)}\psi_{,ij} \tag{3.3}$$

The Inclusion-Based Boundary Element Method (iBEM). https://doi.org/10.1016/B978-0-12-819384-6.00011-5

Note that the Poisson ratio is related to the Lame constants as $\nu^0 = \frac{\lambda^0}{2(\lambda^0+\mu^0)}$. Using the Gauss theorem on the last item of Eq. (3.2), we can write

$$\begin{aligned}
&\int_D G_{ij}(\mathbf{x},\mathbf{x}')C^0_{mjkl}\varepsilon^p_{kl,m'}(\mathbf{x}')d\mathbf{x}' \\
&= \int_{\partial D} G_{ij}(\mathbf{x},\mathbf{x}')C^0_{mjkl}\varepsilon^p_{kl}(\mathbf{x}')'n'_m d\mathbf{x}' - \int_D G_{ij,m'}(\mathbf{x},\mathbf{x}')C^0_{mjkl}\varepsilon^p_{kl}(\mathbf{x}')d\mathbf{x}' \\
&= \int_\Omega G_{ij,m}(\mathbf{x},\mathbf{x}')C^0_{mjkl}\varepsilon^p_{kl}(\mathbf{x}')d\mathbf{x}'
\end{aligned} \tag{3.4}$$

where $\varepsilon^p_{kl} = 0$ on ∂D, and the identity $G_{ij,m} = \frac{\partial G_{ij}}{\partial x_m} = -\frac{\partial G_{ij}}{\partial x'_m} = -G_{ij,m'}$ is applied. Therefore Eq. (3.2) can be rewritten as

$$u_i(\mathbf{x}) = \int_\Omega G_{ij}(\mathbf{x},\mathbf{x}')b_j(\mathbf{x}') - G_{ij,m}(\mathbf{x},\mathbf{x}')C^0_{mjkl}\varepsilon^p_{kl}(\mathbf{x}')d\mathbf{x}' \tag{3.5}$$

Note that the integral domain $\mathcal{D}$ is replaced by the subdomain Ω as the eigenstrain and body force are zero otherwise. As the fundamental solution is given in terms of the harmonic and biharmonic potentials, whereas the source is given in a polynomial form of field on Ω, the integral of the potentials over the source field immediately provides the local solution. It depends on the variation of eigenstrain, the shape of the inclusion, and the distance to the inclusion. The local solutions to 3D or 2D problems can be obtained in a similar fashion through the domain integral. In the following, we introduce the integral of polynomial eigenstrains over ellipsoidal, elliptical, polyhedral, and polygonal inclusions. Basically, they are represented by the derivatives of the potentials of distributed sources on a 2D or 3D domain and can be obtained by the potential theory.

3.2 Eshelby's tensors for polynomial eigenstrains of an ellipsoidal or elliptical inclusion

3.2.1 Ellipsoidal inclusion with uniform eigenstrain

We assume that an eigenstrain is uniform on an ellipsoidal inclusion Ω as shown in Fig. 3.1, where

$$\Omega: \frac{x_1^2}{a_1^2} + \frac{x_2^2}{a_2^2} + \frac{x_3^2}{a_3^2} \leq 1 \tag{3.6}$$

The displacement at any point $\mathbf{x}$ can be written as

$$\begin{aligned}
u_i(\mathbf{x}) &= -C_{jkmn}\varepsilon^*_{mn}\int_\Omega G_{ij,k}\left(\mathbf{x},\mathbf{x}'\right)d\mathbf{x}' \\
&= -\frac{\varepsilon^*_{mn}}{8\pi(1-\nu)}\int_\Omega g_{imn}(\mathbf{n})\frac{d\mathbf{x}'}{|\mathbf{x}-\mathbf{x}'|^2}
\end{aligned} \tag{3.7}$$

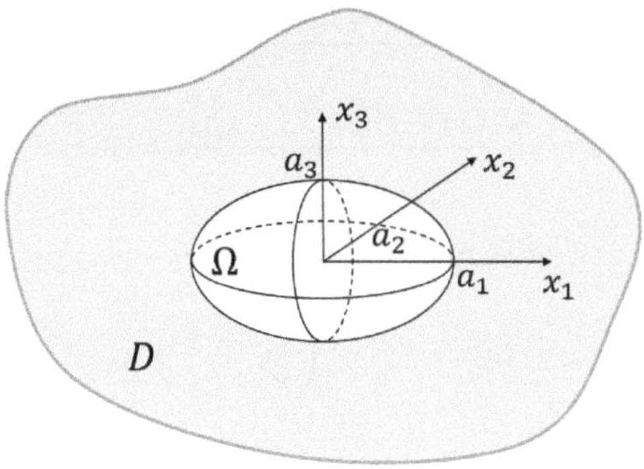

FIGURE 3.1 An ellipsoidal inclusion in the infinitely extended domain with radii a_1, a_2, and a_3.

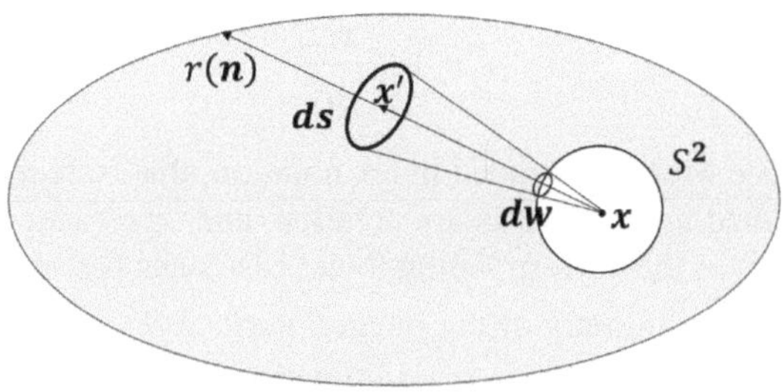

FIGURE 3.2 Integral of the Green's function over an ellipsoidal inclusion.

where

$$g_{imn} = (1 - 2\nu)(\delta_{im} n_n + \delta_{in} n_m - \delta_{mn} n_i) + 3 n_i n_m n_n \tag{3.8}$$

which can be verified as

$$\frac{g_{imn}}{8\pi(1-\nu)|\mathbf{x}-\mathbf{x}'|^2} = -C_{jkmn} G_{ij,k} \tag{3.9}$$

For $\mathbf{x} \in \Omega$,

$$d\mathbf{x} = dr \cdot r^2 \cdot d\omega = dr \cdot \left|\mathbf{x}' - \mathbf{x}\right|^2 \cdot d\omega \tag{3.10}$$

where ω is the unit spherical shell originated at $\mathbf{x}$. As shown in Fig. 3.2, Eq. (3.7) can be rewritten as

$$\begin{aligned} u_i &= -\frac{\varepsilon^*_{mn}}{8\pi(1-\nu)} \int_{S^2} \int_0^{r(\mathbf{n})} dr . g_{imn}(\mathbf{n})\, d\omega \\ &= -\frac{\varepsilon^*_{mn}}{8\pi(1-\nu)} \int_{S^2} r(\mathbf{n}) . g_{imn}(\mathbf{n})\, d\omega \end{aligned} \tag{3.11}$$

where $r(\mathbf{n})$ is the largest positive root of

$$\frac{(x_i+rn_i)(x_i+rn_i)}{a_I^2}=\frac{(x_1+rn_1)(x_1+rn_1)}{a_1^2}+\frac{(x_2+rn_2)(x_2+rn_2)}{a_2^2}+\frac{(x_3+rn_3)(x_3+rn_3)}{a_3^2}=1 \tag{3.12}$$

and can be further explicitly written as

$$r(\mathbf{n})=-\frac{f}{g}+\left(\frac{f^2}{g^2}+\frac{e}{g}\right)^{1/2} \tag{3.13}$$

where

$$\begin{aligned} g&=\frac{n_i n_i}{a_I^2} \\ e&=1-\frac{x_i x_i}{a_I^2} \end{aligned} \tag{3.14}$$

Notice that we use Mura's extended index notation, that is, repeated lower case indices are summed up as usual index notation, and upper case indices take on the same numbers as the corresponding lower case ones but are not summed.

Considering the symmetry of the integral surface S^2, because $\frac{f^2}{g^2}+\frac{e}{g}$ is an even function whereas $g_{imn}(\mathbf{n})$ is an odd function of $\mathbf{n}$, the displacement can be rewritten as

$$u_i(\mathbf{x})=\frac{\varepsilon^*_{mn}}{8\pi(1-\nu)}\int_{S^2}\frac{f}{g}\cdot g_{imn}(\mathbf{n})\,d\omega=\frac{x_p\varepsilon^*_{mn}}{8\pi(1-\nu)}\int_{\Sigma}\frac{\lambda_p g_{imn}}{g}d\omega \tag{3.15}$$

where $\lambda_i=\frac{n_i}{a_I^2}$. Thus the strain field can be written as

$$\varepsilon_{ij}=\frac{\varepsilon^*_{mn}}{16\pi(1-\nu)}\int_{\Sigma}\frac{\lambda_i g_{jmn}+\lambda_j g_{imn}}{g}d\omega \tag{3.16}$$

Using the definition of $g_{imn}(\mathbf{n})$, we can write the above integral in terms of $\frac{n_i^2}{a_I^2 g}$, $\frac{n_i^4}{a_I^4 g}$, and so on, which can be obtained through the following identities of the elliptic integrals. For interior points, the integrals are provided as

$$\begin{cases} I_I=\int_{\Sigma}\frac{n_i^2}{a_I^2 g}d\omega=2\pi a_1a_2a_3\int_0^{\infty}\frac{ds}{(a_I^2+s)\Delta(s)} \\ I_{II}=\int_{\Sigma}\frac{n_i^4}{a_I^4 g}d\omega=2\pi a_1a_2a_3\int_0^{\infty}\frac{ds}{(a_I^2+s)^2\Delta(s)} \\ I_{IJ}=\int_{\Sigma}\frac{n_i^2n_j^2}{a_I^2a_J^2 g}d\omega=2\pi a_1a_2a_3\int_0^{\infty}\frac{ds}{(a_I^2+s)(a_J^2+s)\Delta(s)} \end{cases} \quad \text{for } x\in\Omega \tag{3.17}$$

where

$$\Delta S=\sqrt{(a_1^2+s)(a_2^2+s)(a_3^2+s)} \tag{3.18}$$

For exterior points, the integrals are similarly provided:

$$\begin{cases} I = 2\pi a_1 a_2 a_3 \int_\lambda^\infty \frac{ds}{\Delta(s)} \\ I_I = 2\pi a_1 a_2 a_3 \int_\lambda^\infty \frac{ds}{\left(a_I^2+s\right)^2 \Delta(s)} \\ I_{IJ} = 2\pi a_1 a_2 a_3 \int_\lambda^\infty \frac{ds}{\left(a_I^2+s\right)\left(a_J^2+s\right)\Delta(s)} \end{cases} \quad \text{for } x \in D - \Omega \tag{3.19}$$

where λ is defined as the largest root of the equation

$$\frac{x_i x_i}{\left(a_I^2+\lambda\right)} = \frac{x_1 x_1}{a_1^2+\lambda} + \frac{x_2 x_2}{a_2^2+\lambda} + \frac{x_3 x_3}{a_3^2+\lambda} = 1 \tag{3.20}$$

(a) For a spherical domain, i.e., $a_1 = a_2 = a_3 = a$, we can write

$$\begin{cases} I(\lambda) = 4\pi a^3 / \left(a^2+\lambda\right)^{1/2} \\ I_i(\lambda) = 4\pi a^3 / 3\left(a^2+\lambda\right)^{3/2} \\ I_{\underbrace{ij\ldots k}_{n}}(\lambda) = \frac{4\pi a^3}{(2n+1)3\left(a^2+\lambda\right)^{n+1/2}} \end{cases} \tag{3.21}$$

(b) For different shape of ellipsoids ($a_1 \neq a_2 \neq a_3$),

1. Fiber: $a_1 = a_2$, $a_3 \to \infty$
2. Penny: $a_1 = a_2$, $a_3 \to 0$
3. Spheroid: $a_1 = a_2 \neq a_3$ $\begin{cases} \text{oblate spheroid,} & a_1 = a_2 > a_3 \\ \text{prolate spheroid,} & a_1 = a_2 < a_3 \end{cases}$
4. Cracks: $\begin{cases} \text{penny shape,} & a_1 = a_2,\ a_3 = 0 \\ \text{slit-like,} & a_1 \neq 2a,\ a_2 = \infty,\ a_3 = 0 \\ \text{flat ellipsoidal crack,} & a_1 = a_2 \neq 0,\ a_3 = 0 \end{cases}$

The details are described in pages 93–95 of Mura's book [1]. The relationship between the eigenstrain and induced strain in Eq. (3.16) can be written in terms of the famous Eshelby tensor [91] as follows:

$$S_{ijkl} = \begin{cases} \begin{bmatrix} S_{IK}^1(\lambda)\delta_{ij}\delta_{kl} + S_{IJ}^2(\lambda)\left(\delta_{ik}\delta_{jl} + \delta_{il}\delta_{jk}\right) + S_I^3(\lambda)\delta_{ij}\hat{n}_k\hat{n}_l \\ +S_K^4(\lambda)\delta_{kl}\hat{n}_i\hat{n}_j + S_I^5(\lambda)\left(\delta_{ik}\hat{n}_j\hat{n}_l + \delta_{il}\hat{n}_j\hat{n}_k\right) \\ +S_J^6(\lambda)\left(\delta_{jk}\hat{n}_i\hat{n}_l + \delta_{jl}\hat{n}_i\hat{n}_k\right) + S_{IJKL}^7(\lambda)\hat{n}_i\hat{n}_j\hat{n}_k\hat{n}_l \end{bmatrix} \\ \text{for } r > a \\ S_{IK}^1(0)\delta_{ij}\delta_{kl} + S_{IJ}^2(0)\left(\delta_{ik}\delta_{jl} + \delta_{il}\delta_{jk}\right) \quad \text{for } r \leq a \end{cases} \tag{3.22}$$

where $\hat{\mathbf{n}}$ denotes the outward unit normal vector on the imaginary ellipsoidal surface,

$$\hat{n}_i = \frac{x_i}{\left(A_I^2+\lambda\right)\Theta_i(\lambda)} \tag{3.23}$$

in which

$$\Theta(\lambda) = \sqrt{\Theta_i(\lambda)\,\Theta_i(\lambda)} \tag{3.24}$$

with

$$\Theta_i(\lambda) = \frac{x_i}{\left(A_I^2 + \lambda\right)} \tag{3.25}$$

The tensor coefficients are provided in Appendix 3.A. More specifically, for a spherical inclusion, Eshelby's tensor can be written as

$$S_{ijkl} = \begin{cases} \frac{5\nu-1}{15(1-\nu)}\delta_{kl}\delta_{mn} + \frac{4-5\nu}{15(1-\nu)}(\delta_{km}\delta_{ln} + \delta_{kn}\delta_{lm}) & \text{for } r > a \\ \frac{\rho^3}{30(1-\nu)}[(3\rho^2 + 10\nu - 5)\delta_{kl}\delta_{mn} & \\ + (3\rho^2 - 10\nu + 5)(\delta_{km}\delta_{ln} + \delta_{kn}\delta_{lm}) & \\ + 15(1-\rho^2)\delta_{kl}n_m n_n + 15(1 - 2\nu - \rho^2)\delta_{mn}n_k n_l & \\ + 15(\nu - \rho^2)(\delta_{km}n_n n_l + \delta_{lm}n_n n_k + \delta_{kn}n_m n_l + \delta_{ln}n_m n_k) & \\ + 15(7\rho^2 - 5)n_k n_l n_m n_n] & \text{for } r \le a \end{cases} \tag{3.26}$$

3.2.2 Ellipsoidal inclusion with a polynomial eigenstrain

A continuous eigenstrain field over an ellipsoidal inclusion Ωt can be approximated by the Taylor series in a polynomial form such as

$$\varepsilon^*_{ij}(\mathbf{x}) = \begin{cases} E_{ij} + E_{ijk}x_k + E_{ijkl}x_k x_l + \cdots, & \mathbf{x} \in \Omega \\ 0, & \mathbf{x} \in D - \Omega \end{cases} \tag{3.27}$$

Then the displacement and strain fields can be written as

$$u_i = -\int_\Omega G_{ik,l}\left(\mathbf{x}, \mathbf{x}'\right) C_{klmn}\varepsilon^*_{mn}(\mathbf{x}')d\mathbf{x}' \tag{3.28}$$

and

$$\varepsilon_{ij} = -\int_\Omega \Gamma_{ijkl}\left(\mathbf{x}, \mathbf{x}'\right) C_{klmn}\varepsilon^*_{mn}(\mathbf{x}')d\mathbf{x}' \tag{3.29}$$

For an isotropic material, the integrand in the above equation can be written as

$$C_{klmn}\varepsilon^*_{mn} = \lambda\delta_{kl}\varepsilon^*_{mm} + 2\mu\varepsilon^*_{kl} \tag{3.30}$$

Rewriting the fundamental solutions as $G_{ij}\left(\mathbf{x}, \mathbf{x}'\right) = \frac{1}{4\pi\mu}\delta_{ij}\phi - \frac{1}{16\pi\mu(1-\upsilon)}\psi_{,ij}$, we can use the following equations to straightforwardly obtain the integral of the Green's function over the ellipsoidal domain Ω:

$$\begin{aligned} u_i = \frac{1}{8\pi(1-\upsilon)}&\left[\Psi_{,ikl}E_{kl} - 2\upsilon\Phi_{,i}E_{mm} - 4(1-\upsilon)\Phi_{,m}E_{im}\right. \\ &\left. + \Psi_{m,ikl}E_{klm} - 2\upsilon\Phi_{k,i}E_{mmk} - 4(1-\upsilon)\Phi_{k,m}E_{imk} + \cdots\right] \end{aligned} \tag{3.31}$$

and

$$\varepsilon_{ij} = \frac{1}{8\pi(1-v)}\left[F_{ijkl}E_{kl} + F_{ijklm}E_{klm} + \cdots\right] \tag{3.32}$$

with

$$F_{ijkl} = \frac{1}{8\pi(1-v)}\begin{bmatrix} \Psi_{,ijkl} - 2v\Phi_{,ij}\delta_{kl} \\ -(1-v)\left(\Phi_{,kj}\delta_{il} + \Phi_{,ki}\delta_{jl} + \Phi_{,lj}\delta_{ik} + \Phi_{,li}\delta_{jk}\right) \end{bmatrix} \tag{3.33}$$

and

$$F_{ijklm} = \frac{1}{8\pi(1-v)} \times \begin{bmatrix} \Psi_{m,ijkl} - 2v\Phi_{m,ij}\delta_{kl} \\ -(1-v)\left(\Phi_{m,kj}\delta_{il} + \Phi_{m,ki}\delta_{jl} + \Phi_{m,lj}\delta_{ik} + \Phi_{m,li}\delta_{jk}\right) \end{bmatrix} \tag{3.34}$$

where Φ, Ψ, Φ_p, and Ψ_p denote the integrals of the functions ϕ, ψ, $\phi x'_p$, and $\psi x'_p$ over an ellipsoidal inclusion Ω, that is,

$$\begin{cases} \Phi = \int_\Omega \phi \, dx' \\ \Psi = \int_\Omega \psi \, dx' \\ \Phi_p = \int_\Omega \phi x'_p \, dx' \\ \Psi_p = \int_\Omega \psi x'_p \, dx' \end{cases} \tag{3.35}$$

In the following, the integral terms of Φ, Ψ, Φ_p, and Ψ_p and their derivative terms over an ellipsoidal domain are derived and expressed in Appendix 3.A. The method can be extended to the higher-order form of eigenstrains in the same fashion for higher accuracy but with more complex formulation. It has been extended to quadratic terms in the iBEM software package with the closed-form solution [40]. It is recommended using numerical differentiation for high-order (cubic or higher-order) form eigenstrains to avoid long derivation and expedite the computation process for the future work.

3.2.3 Ellipsoidal inclusion with body force

Considering the body force, the equilibrium equation can be written as

$$C_{ijkl}u_{k,li} = -b_j \tag{3.36}$$

A continuously distributed body force on the ellipsoidal inclusion Ω can be approximated in the polynomial form as

$$b_j(\mathbf{x}) = \begin{cases} B_j + B_{jk}x_k + \cdots, & \mathbf{x} \in \Omega \\ 0, & \mathbf{x} \in D - \Omega \end{cases} \tag{3.37}$$

Then the displacement can be rewritten as

$$u_i(\mathbf{x}) = \int_{\Omega} \left(\frac{1}{4\pi\mu} \delta_{ij}\phi - \frac{1}{16\pi\mu(1-\upsilon)} \psi_{,ij} \right) \left(B_j + B_{jk}x_k + \cdots \right) d\mathbf{x}' \quad (3.38)$$

and then we obtain

$$u_i = \frac{1}{16\pi\mu(1-\upsilon)} \left[-\Psi_{,ij}B_j - \Psi_{k,ij}B_{jk} + 4(1-\upsilon)\left(\Phi B_j + \Phi_k B_{jk}\right) + \cdots \right] \quad (3.39)$$

Then the strain and stress fields can be straightforwardly obtained from the integral of the potentials in Appendix 3.A.

3.2.4 Elliptical inclusion with a polynomial eigenstrain

In Section 3.2.2, the continuous eigenstrain field over an ellipsoidal inclusion Ω is expressed in the polynomial form of the Taylor series. In Section 2.2, the 2D fundamental solution is obtained through the Hadmard regularization, and thus the elliptical case can be considered as a particular form of the ellipsoidal case:

(1) For the plane strain problem (elliptical cylinder), a_1, a_2 are bounded, and $a_3 \to \infty$. Based on the Hadmard regularization, Eshelby's tensor for plane strain problem can be solved as a particular case of a cylindrical inclusion.
(2) For plane stress problem with a penny-shape inclusion, a_1, a_2 are bounded, and $a_3 \to 0$. By switching Young's modulus and Poisson's ratio for the equivalence between the plane stress and plane strain problems, the Eshelby tensor for the plane stress problem can be obtained from the plane strain problem. Therefore the explicit expression of Eshelby's tensor is exactly same as Eqs. (3.33) and (3.34).

The implementation of iBEM for 2D cases can be found in [40], which shows 3D and 2D cases in a parallel fashion. The key formulations are provided in Appendix 3.A.

For a circular cylindrical inclusion, taking the limits of the I integrals at $a_1 = a_2 = a$ and $a_3 = \infty$ yield

$$\begin{cases} I(\lambda) = \frac{4\pi a^2 a_3}{\sqrt{a_3^2 - a^2}} \cosh^{-1}\left(\sqrt{\frac{a_3^2+\lambda}{a^2+\lambda}}\right) \\ I_i(\lambda) = 4\pi a^2 / 2\left(a^2+\lambda\right)^{2/2} \\ I_{\underbrace{ij\ldots k}_{n}}(\lambda) = \frac{4\pi a^2}{(2n)\left(a^2+\lambda\right)^{2n/2}} \end{cases} \quad (3.40)$$

For a circular integral domain centered at the origin, Eshelby's tensor is written in two zones:

i. For $\mathbf{x}$ inside the circle of radius r_0, $|\mathbf{x}| \leq r_0$, it is the constant

$$S_{klmn} = \frac{4\nu - 1}{8(1-\nu)}\delta_{kl}\delta_{mn} + \frac{3-4\nu}{8(1-\nu)}(\delta_{km}\delta_{ln} + \delta_{kn}\delta_{lm}) \tag{3.41}$$

ii. For $|\mathbf{x}| > r_0$, it depends on the distance and orientation and is written as

$$S_{klmn}(\mathbf{x}') = \frac{\rho^2}{8(1-\nu)} \times \begin{bmatrix} (\rho^2 + 4\nu - 2)\delta_{kl}\delta_{mn} + (\rho^2 - 4\nu + 2)(\delta_{km}\delta_{ln} + \delta_{kn}\delta_{lm}) \\ +4(1-\rho^2)\delta_{kl}n_m n_n + 4(1 - 2\nu - \rho^2)\delta_{mn}n_k n_l \\ +4(\nu - \rho^2)(\delta_{km}n_n n_l + \delta_{lm}n_n n_k + \delta_{kn}n_m n_l + \delta_{ln}n_m n_k) \\ +8(3\rho^2 - 2)n_k n_l n_m n_n \end{bmatrix} \tag{3.42}$$

Notice that Eshelby's tensor is a fourth-rank tensor, which requires the partial differentiation of the potentials. For a cylindrical inclusion, since the third axis a_3 is infinite, the I integral is an infinite constant. Therefore it is not possible to evaluate the biharmonic potential Ψ directly, but its derivatives can be provided by the finite part.

3.3 Eshelby's tensors for polynomial eigenstrains at an polyhedral inclusion

Eshelby's pioneer work focused on the uniform eigenstrain on a single ellipsoidal particle embedded in the infinite domain. Because Eshelby's tensor is constant for ellipsoidal inclusions and the strain field on the inclusion is uniform, for inhomogeneity problems, a uniform eigenstrain can exactly simulate the material mismatch by the equivalent inclusion method (EIM) and yields the exact solution. However, the actual physical problems commonly exhibit nonuniform eigenstrains to satisfy the stress equivalent conditions and boundary conditions. Particularly, for polyhedral or polygonal inclusions for 3D and 2D problems, respectively, the eigenstrain caused by the material mismatch is not uniform even for a single particle in the infinite domain, so that polynomial eigenstrains should be considered. In the following, we consider the case of polyhedral inclusions.

3.3.1 Eshelby's tensor for a uniform eigenstrain on a polyhedral inclusion

Consider an arbitrarily shaped polyhedral inclusion Ω embedded in the infinite homogeneous isotropic elastic medium $\mathcal{D}$. The classic Eshelby tensor for displacement (g_{ikl}) and strain (S_{ijkl}) [1,2] in the three-dimensional space is derived

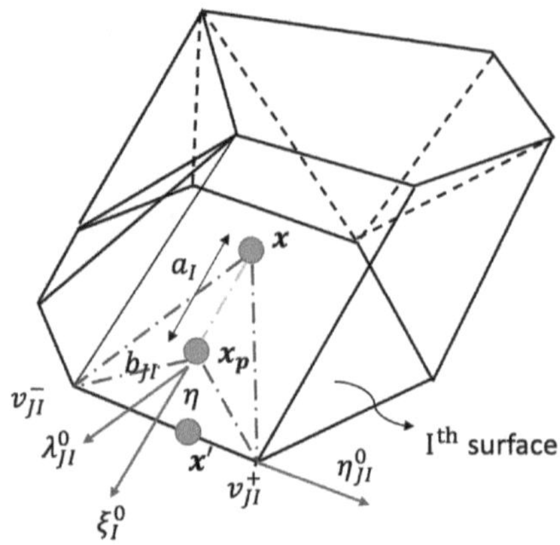

FIGURE 3.3 Schematic plot of the transformed geometry of an arbitrarily shaped polyhedral inclusion.

from the volume integral over the inclusion domain Ω in Eq. (3.43), and it can be split into biharmonic Ψ and harmonic Φ potentials:

$$\begin{aligned} g_{ikl} &= \frac{1}{8\pi(1-\nu)}[\Psi_{,ikl} - 2\nu\delta_{kl}\Phi_{,i} - 2(1-\nu)(\delta_{il}\Phi_{,k} + \delta_{ik}\Phi_{,l})] \\ S_{ijkl} &= \frac{1}{8\pi(1-\nu)}[\Psi_{,klij} - 2\nu\delta_{kl}\Phi_{,ij} - (1-\nu)(\delta_{il}\Phi_{,jk} + \delta_{,ik}\Phi_{,jl} \\ &\quad + \delta_{jl}\Phi_{,ik} + \delta_{jk}\Phi_{,il})] \end{aligned} \tag{3.43}$$

where the integrals of the potential functions are written as $\Psi = \int_\Omega |\mathbf{x}-\mathbf{x}'|d\mathbf{x}'$ and $\Phi = \int_\Omega \frac{1}{|\mathbf{x}-\mathbf{x}'|}d\mathbf{x}'$. An explicit form solution of the integrals over ellipsoidal inclusions can be found in Mura's work [1]. In the following, the integrals over polyhedral inclusions will be derived. Without the loss of generality, we consider a domain Ω containing N_I surfaces and the Ith surface containing N_{JI} edges. Here $\mathbf{x}$ and $\mathbf{x}'$ denote the observing field point and the source point, respectively. Nazaki and Taya [51], Rodin [49], and Trotta [92] have proposed the closed-form Eshelby tensor based on three different approaches. Among their methods, Rodin applied the Gauss' theorem in the partial differentiation process, which provided simple forms of expressions to be implemented in programming. Liu and Gao [36] further simplified the integral through the application of the Stokes theorem by defining some constructed potentials. Since it is not straightforward to derive such potentials for higher-order Eshelby's tensor, this section follows Rodin's approach.

Shown in Fig. 3.3, a transformed coordinate (TC) system for the polyhedral is constructed upon each edge with the right-hand basis. Here I and JI denote the Ith surface and the Jth edge belonging to the Ith surface, and thus we can name the base vectors of the TC accordingly; ξ^0_I, λ^0_{JI}, and η^0_{JI} are the unit outward vector of the Ith surface normal, outward normal vector of the Jth

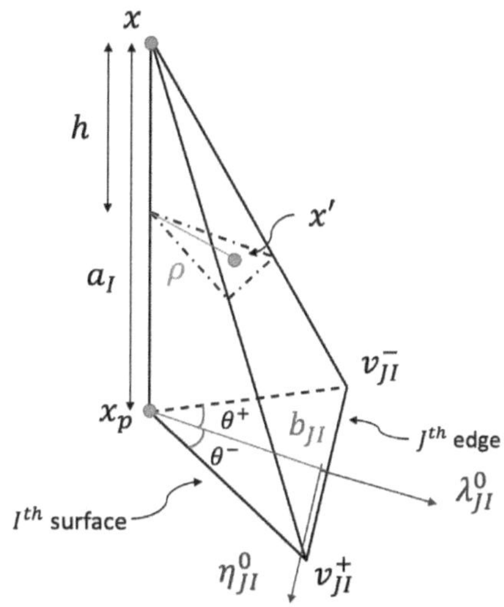

FIGURE 3.4 Schematic plot of the direct volume integral of a piecewise smooth function with integral limits.

edge, and directional vector of the Jth edge in the right-hand basis; and $\mathbf{x}_p$ is the projection of $\mathbf{x}$ on the Ith surface. The geometry parameters for the Ith surface and Jth edge between two vertices $v^{\pm}_{JI}$ are expressed as

$$a_I = (\xi_I^0)_i((v^+_{JI})_i - x_i),\ \ b_{JI} = (\lambda^0_{JI})_i((v^+_{JI})_i - x_i),\ \ l^{\pm}_{JI} = (\eta^0_{JI})_i((v^{\pm}_{JI})_i - x_i) \tag{3.44}$$

where a_I is the perpendicular distance between the observing point to the Ith surface, b_{JI} is the perpendicular distance between the projections to the Jth edge, and $l^{\pm}_{JI}$ are the distances between the projections to the two vertices $v^{\pm}_{JI}$. To derive Eshelby's tensor, partial derivatives of the potentials Ψ and Φ are required, so it can be reformulated by the Gauss theorem. However, in this section, we provide a scheme to complete the domain integral over a polyhedral. Shown in Fig. 3.4, the domain integral of any arbitrary piecewise smooth function $\mathcal{F}(|\mathbf{x}-\mathbf{x}'|)$ can be performed as

$$\int_\Omega \mathcal{F}(|\mathbf{x}-\mathbf{x}'|)\,d\mathbf{x}' = \int_{\tan^{-1}[l^-_{JI}/b_{JI}]}^{\tan^{-1}[l^+_{JI}/b_{JI}]} \int_0^{a_I} \int_0^{\frac{hb_{JI}}{a_I}\sqrt{1+\tan[\theta]^2}} \mathcal{F}(\sqrt{h^2+\rho^2})\rho\, d\rho dh d\theta \tag{3.45}$$

where the integral limits, illustrated in Fig. 3.4, are as follows:

(1) $\rho \in [0, \frac{hb_{JI}\sqrt{1+\tan[\theta]^2}]}{a_I}]$, where ρ is the triangular polar radius;
(2) $h \in [0, a_I]$, where h is the height parameter of the pyramid;
(3) $\theta \in [\tan^{-1}[\frac{l^-_{JI}}{b_{JI}}], \tan^{-1}[\frac{l^+_{JI}}{b_{JI}}]]$, where θ is the triangular polar angle.

With $\mathcal{F}_\Phi = \frac{1}{|\mathbf{x}-\mathbf{x}'|}$ and $\mathcal{F}_\Psi = |\mathbf{x} - \mathbf{x}'|$, the analytical solutions to the harmonic and biharmonic potentials are obtained in Eq. (3.121) and Eq. (3.122), respectively. Note that $\sqrt{a_I^2 + b_{JI}^2 + (l_{JI}^{\pm})^2}$ is the distance $|(v_{JI}^{\pm}) - x|$ between the observing point and vertices. At this stage, the two potentials are expressed in terms of the parameters in the transformed coordinate, where to obtain their partial derivatives, we can use

$$\mathcal{G}_{,i}(a_I, b_{JI}, l_{JI}^+, l_{JI}^-) = -\frac{\partial \mathcal{G}}{\partial a_I}(\xi_I^0)_i - \frac{\partial \mathcal{G}}{\partial b_{JI}}(\lambda_0^I)_i - \left(\frac{\partial \mathcal{G}}{\partial l_{JI}^+} + \frac{\partial \mathcal{G}}{\partial l_{JI}^-}\right)(\eta_{JI}^0)_i \tag{3.46}$$

where the vectors are generated from the derivatives of a_I, b_{JI}, and $l_{JI}^{\pm}$, respectively. Following the differentiation scheme, Eshelby's tensor for strain can be obtained (the fourth derivative of Ψ and second derivative of Φ). Alternatively, as mentioned at the beginning of the section, using Gauss' theorem in the differentiation process, we can transfer the volume integral to the surface. Taking the harmonic potential Φ as an example, we can write

$$\Phi_{,i} = -\sum_{I=1}^{N_I}\sum_{J=1}^{N_{JI}}(\xi_I^0)_i \int_{A_I} \frac{1}{|\mathbf{x}-\mathbf{x}'|} dA_I(\mathbf{x}') \tag{3.47}$$

where A_I is the Ith surface of the polyhedron. Because the internal surface integrals will be eliminated with each other with inverse surface normal, only the surface of the polyhedron is needed for the surface integral.

The integral on the surface can be similarly conducted in the transformed coordinate in 2D surface along ρ and θ, which reduces to the two limits along the bottom edge of two vertices [43] as follows:

$$\int_{A_I} \frac{1}{|\mathbf{x}-\mathbf{x}'|} dA_I(\mathbf{x}') = \mathcal{F}_1(a_I, b_{JI}, l_{JI}^+) - \mathcal{F}_1(a_I, b_{JI}, l_{JI}^-) \tag{3.48}$$

The original integral expressions of $\mathcal{F}_x$ functions are listed as Table 3.1, and their explicit forms are provided in Appendix 3.C.

3.3.2 Eshelby's tensor for higher-order eigenstrains

For higher-order Eshelby's tensor, the integral formula contains the source term x_i'. Although the domain integral can be solved with direct volume approach, using the Gauss theorem would save efforts in differentiation process. Based on the Gauss theorem, the source term now exists in the surfaces of the polyhedron. The key is expressing the source term x_i' in terms of the transformed coordinate. The distance vector $\mathbf{x} - \mathbf{x}' = -a_I \boldsymbol{\xi}_I^0 - \rho\cos\theta \boldsymbol{\lambda}_{JI}^0 - \rho \sin\theta \boldsymbol{\eta}_{JI}^0$, where ρ and θ are the triangular radius and angle in the polar coordinate, is shown in Fig. 3.4. Separating the source terms into the observing points and

TABLE 3.1 Table of the integrand functions listing from $\mathcal{F}_1$ to $\mathcal{F}_{12}$.

$\mathcal{F}_i$ functions	Integrand in TC
$\mathcal{F}_1$	$\frac{\rho}{\sqrt{a_I^2+\rho^2}}$
$\mathcal{F}_2$	$\frac{\rho^2\cos\theta}{\sqrt{a_I^2+\rho^2}}$
$\mathcal{F}_3$	$\frac{\rho^2\sin\theta}{\sqrt{a_I^2+\rho^2}}$
$\mathcal{F}_4$	$\rho\sqrt{a_I^2+\rho^2}$
$\mathcal{F}_5$	$(\rho^2\cos\theta)\sqrt{a_I^2+\rho^2}$
$\mathcal{F}_6$	$(\rho^2\sin\theta)\sqrt{a_I^2+\rho^2}$
$\mathcal{F}_7$	$\frac{\rho^3\cos^2\theta}{\sqrt{a_I^2+\rho^2}}$
$\mathcal{F}_8$	$\frac{\rho^3\cos\theta\sin\theta}{\sqrt{a_I^2+\rho^2}}$
$\mathcal{F}_9$	$\frac{\rho^3\sin^2\theta}{\sqrt{a_I^2+\rho^2}}$
$\mathcal{F}_{10}$	$(\rho^3\cos^2\theta)\sqrt{a_I^2+\rho^2}$
$\mathcal{F}_{11}$	$(\rho^3\cos\theta\sin\theta)\sqrt{a_I^2+\rho^2}$
$\mathcal{F}_{12}$	$(\rho^3\sin^2\theta)\sqrt{a_I^2+\rho^2}$

components of the distance vector, we can write

$$\Psi_p = \int_\Omega |\mathbf{x}-\mathbf{x}'|(x_p + x'_p - x_p)d\mathbf{x}' = x_p\Psi + \int_\Omega (x'_p - x_p)|\mathbf{x}-\mathbf{x}'|d\mathbf{x}' \tag{3.49a}$$

$$\Phi_p = \int_\Omega \frac{x_p + x'_p - x_p}{|\mathbf{x}-\mathbf{x}'|}d\mathbf{x}' = x_p\Phi + \int_\Omega \frac{x'_p - x_p}{|\mathbf{x}-\mathbf{x}'|}d\mathbf{x}' \tag{3.49b}$$

Using the Gauss' theorem, the first-order derivatives of Ψ_p and Φ_p are simplified to sums of the surface integrals:

$$\Psi_{p,i} = \delta_{ip}\Psi + x_p\Psi_{,i} - \sum_{I=1}^{N_I}\sum_{J=1}^{N_{JI}}(\xi_I^0)_i \int_{A_I} |\mathbf{x}-\mathbf{x}'|(x'_p - x_p)d\mathbf{x}' \tag{3.50a}$$

$$\Phi_{p,i} = \delta_{ip}\Phi + x_p\Phi_{,i} - \sum_{I=1}^{N_I}\sum_{J=1}^{N_{JI}}(\xi_I^0)_i \int_{A_I} \frac{x'_p - x_p}{|\mathbf{x}-\mathbf{x}'|}d\mathbf{x}' \tag{3.50b}$$

where, in the TC, the component of the distance vector is expressed as $x'_p - x_p = a_I(\xi_I^0)_p + \rho\cos\theta(\lambda_{JI}^0)_p + \rho\sin\theta(\eta_{JI}^0)_p$. Here ρ and θ are illustrated in Fig. 3.4, where the source point $\mathbf{x}'$ is in the Ith surface. The second terms in Eqs. (3.50a) and (3.50b) can be rewritten as

$$\Phi_{,i} = \sum_{I=1}^{N_I}\sum_{J=1}^{N_{JI}}(-\xi_I^0)_i\mathcal{H}_1(a_I, b_{JI}, l_{JI}^+, l_{JI}^-),$$
$$\Psi_{,i} = \sum_{I=1}^{N_I}\sum_{J=1}^{N_{JI}}(-\xi_I^0)_i\mathcal{H}_4(a_I, b_{JI}, l_{JI}^+, l_{JI}^-) \tag{3.51}$$

where

$$\mathcal{H}_n(a_I, b_{JI}, l_{JI}^+, l_{JI}^-) = \mathcal{F}_n(a_I, b_{JI}, l_{JI}^+) - \mathcal{F}_n(a_I, b_{JI}, l_{JI}^-), \quad n = 1, 2, \ldots, 12 \tag{3.52}$$

in which the explicit expressions of integrand function $\mathcal{F}_n$ ($n = 1, 2, \ldots, 12$) are provided in Appendix 3.C. The integrals of the third term in Eqs. (3.50a), (3.50b) can be written as

$$\int_{A_I}\frac{x'_p - x_p}{|\mathbf{x}-\mathbf{x}'|}d\mathbf{x}' = \int_{A_I}\frac{a_I(\xi_I^0)_p + \rho\cos\theta(\lambda_{JI}^0)_p + \rho\sin\theta(\eta_{JI}^0)_p}{\sqrt{a_I^2+\rho^2}}\rho\, d\rho d\theta$$
$$= a_I(\xi_I^0)_p\mathcal{H}_1(a_I, b_{JI}, l_{JI}^+, l_{JI}^-) + (\lambda_{JI}^0)_p\mathcal{H}_2(a_I, b_{JI}, l_{JI}^+, l_{JI}^-)$$
$$+ (\eta_{JI}^0)_p\mathcal{H}_3(a_I, b_{JI}, l_{JI}^+, l_{JI}^-) \tag{3.53a}$$

$$\int_{A_I}(x'_p - x_p)|\mathbf{x}-\mathbf{x}'|d\mathbf{x}'$$
$$= \int_{A_I}(a_I(\xi_I^0)_p + \rho\cos\theta(\lambda_{JI}^0)_p + \rho\sin\theta(\eta_{JI}^0)_p)(\sqrt{a_I^2+\rho^2})\, d\rho d\theta$$
$$= a_I(\xi_I^0)_p\mathcal{H}_4(a_I, b_{JI}, l_{JI}^+, l_{JI}^-) + (\lambda_{JI}^0)_p\mathcal{H}_5(a_I, b_{JI}, l_{JI}^+, l_{JI}^-)$$
$$+ (\eta_{JI}^0)_p\mathcal{H}_6(a_I, b_{JI}, l_{JI}^+, l_{JI}^-) \tag{3.53b}$$

With the Gauss theorem, the source terms move on the surfaces, and thus the integral limits are adjusted as

(1) $\rho \in [0, b_{JI}\sqrt{1+\tan^2[\theta]}]$;
(2) $\theta \in [\tan^{-1}[\frac{l_{JI}^-}{b_{JI}}], \tan^{-1}[\frac{l_{JI}^+}{b_{JI}}]]$.

The partial differentiation chain rule is proposed in [49], and therefore we below list only the fourth-derivatives of Ψ_p and second-derivatives of Φ_p, whereas we provide the components of other derivatives in Appendix 3.C.1. To simplify the notations, we write the reduction functions $\mathcal{H}_n(a_I, b_{JI}, l_{JI}^+, l_{JI}^-)$ in the concise form $\mathcal{H}_n$.

$$\begin{aligned}\Psi_{p,ijkl} &= \delta_{ip}\Psi_{,jkl} + \delta_{jp}\Psi_{,ikl} + \delta_{kp}\Psi_{,ijl} + \delta_{lp}\Psi_{,ijk} \\ &\quad - \sum_{I=1}^{N_I}\sum_{J=1}^{N_{JI}} (\xi_I^0)_i(\xi_I^0)_p\Big[-(\xi_I^0)_j\mathcal{H}_{4,kl} - (\xi_I^0)_k\mathcal{H}_{4,jl} \\ &\quad - (\xi_I^0)_l\mathcal{H}_{4,jk} + a_I\mathcal{H}_{4,jkl}\Big] \\ &\quad + (\xi_I^0)_i(\lambda_{JI}^0)_p\mathcal{H}_{5,jkl} + (\xi_I^0)_i(\eta_{JI}^0)_p\mathcal{H}_{6,jkl}\end{aligned} \tag{3.54a}$$

$$\begin{aligned}\Phi_{p,ij} &= \delta_{ip}\Phi_{,j} + \delta_{jp}\Phi_{,i} + x_p\Phi_{,ij} \\ &\quad - \sum_{I=1}^{N_I}\sum_{J=1}^{N_{JI}} (\xi_I^0)_i(\xi_I^0)_p\Big[-(\xi_I^0)_j\mathcal{H}_1 + a_I\mathcal{H}_{1,j}\Big] \\ &\quad + (\xi_I^0)_i(\lambda_{JI}^0)_p\mathcal{H}_{2,j} + (\xi_I^0)_i(\eta_{JI}^0)_p\mathcal{H}_{3,j}\end{aligned} \tag{3.54b}$$

Notice that for Eshelby's tensor of displacement (third-rank tensor), it involves the original form of Φ in Eq. (3.105). Following the same fashion, the quadratic term is derived by splitting the source terms into observing terms and components of a distance vector as $x'_px'_q = (x'_p - x_p)(x'_q - x_q) - x_px_q + x'_px_q + x'_qx_p$. By substituting into the quadratic domain integrals we can write the two potentials as

$$\begin{aligned}\Psi_{pq} &= \int_\Omega |\mathbf{x} - \mathbf{x}'|\big((x'_p - x_p)(x'_q - x_q) - x_px_q + x'_px_q + x_px'_q\big)d\mathbf{x}' \\ &= -x_px_q\Psi + x_p\Psi_q + x_q\Psi_p + \int_\Omega |\mathbf{x} - \mathbf{x}'|(x'_p - x_p)(x'_q - x_q)d\mathbf{x}'\end{aligned} \tag{3.55a}$$

$$\begin{aligned}\Phi_{pq} &= \int_\Omega \frac{(x'_p - x_p)(x'_q - x_q) - x_px_q + x'_px_q + x_px'_q}{|\mathbf{x} - \mathbf{x}'|}d\mathbf{x}' \\ &= -x_px_q\Phi + x_p\Phi_q + x_q\Phi_p + \int_\Omega \frac{(x'_p - x_p)(x'_q - x_q)}{|\mathbf{x} - \mathbf{x}'|}d\mathbf{x}'\end{aligned} \tag{3.55b}$$

where the product of source terms such as x'_px_q can be written as $x_q\Psi_p$. With the closed-form expressions of uniform and linear potentials, the quadratic potentials are obtained once the last integral terms in Eq. (3.55) are solved. Since the derivatives for the two quadratic potentials are similar, only the harmonic potentials are exhibited for illustration,

$$\begin{aligned}&\int_{A_I} \frac{(x'_p - x_p)(x'_q - x_q)}{|\mathbf{x} - \mathbf{x}'|} \\ &= a_I^2(\xi_I^0)_p(\xi_I^0)_q\mathcal{H}_1 + a_I\big((\xi_I^0)_p(\lambda_{JI}^0)_q + (\xi_I^0)_q(\lambda_{JI}^0)_p\big)\mathcal{H}_2\end{aligned}$$

$$
\begin{aligned}
&+ a_I\big((\xi_I^0)_p(\eta_{JI}^0)_q + (\xi_I^0)_q(\eta_{JI}^0)_p\big)\mathcal{H}_3 + (\lambda_{JI}^0)_p(\lambda_{JI}^0)_q\mathcal{H}_7 \\
&+ \big((\lambda_{JI}^0)_p(\eta_{JI}^0)_q + (\lambda_{JI}^0)_q(\eta_{JI}^0)_p\big)\mathcal{H}_8 + (\eta_{JI}^0)_p(\eta_{JI}^0)_q\mathcal{H}_9
\end{aligned} \tag{3.56}
$$

In the following, the fourth-derivative of Ψ_{pq} and second-derivative of Φ_{pq} are provided as Eq. (3.120) in Appendix 3.C. Following the same procedure, this method can be extended to higher-order Eshelby's tensor. Notice that for inhomogeneity problems, although higher-order eigenstrain distribution can provide better accuracy, it is impossible to approach the exact solution by increasing the order of the polynomial eigenstrain [93,94] because the eigenstrain changes at the neighborhood of each vertex, which makes it ineffective to use a single polynomial eigenstrain to describe the whole particle with multiple vertices. However, particle discretization with piecewise continuous polynomial eigenstrain, which is shown in the next section, can achieve the goal. Therefore, for simplicity, polynomial eigenstrain terms with an order higher than quadratic are not considered in the current version. All the components of partial derivatives are provided in Appendix 3.C.

3.3.3 Eshelby's tensor for a uniform eigenstrain on a polygonal inclusion

Consider a plane strain problem containing an infinitely long arbitrary prismatic inclusion Ω embedded in the infinite homogeneous isotropic elastic domain $\mathcal{D}$. The Eshelby-type problem can be simplified as a 2D configuration, which is exactly the same as Eq. (3.43). Based on the assumption for the plane strain problem, the volume integrals Ψ and Φ can be further reduced to the area integrals through the Hadamard regularization [27,54,58], and thus the two potentials can be rewritten as

$$
\begin{aligned}
\Psi &= \int_{\Omega_C} -|\mathbf{x}-\mathbf{x}'|^2 \frac{\ln|\mathbf{x}-\mathbf{x}'|^2 - 1}{2} + C\,dA(\mathbf{x}') \\
\Phi &= \int_{\Omega_C} -\ln|\mathbf{x}-\mathbf{x}'|^2\,dA(\mathbf{x}')
\end{aligned} \tag{3.57}
$$

where C is an integral constant, which can be eliminated by partial differentiation to obtain Eshelby's tensor.

In the following, we derive the integrals over polygonal inclusions. Without loss of generality, consider an N_F-sided cross-section of the cylindrical inclusion in the x_1–x_2 plane. Let $\mathbf{x}$ and $\mathbf{x}'$ be the observing and source points, respectively. Following Rodin [49], we construct a transformed coordinate (TC) on each edge in Fig. 3.5. The base vectors of the fth TC, namely the unit tangent and outward normal vectors of the fth edge on the right-hand basis, are

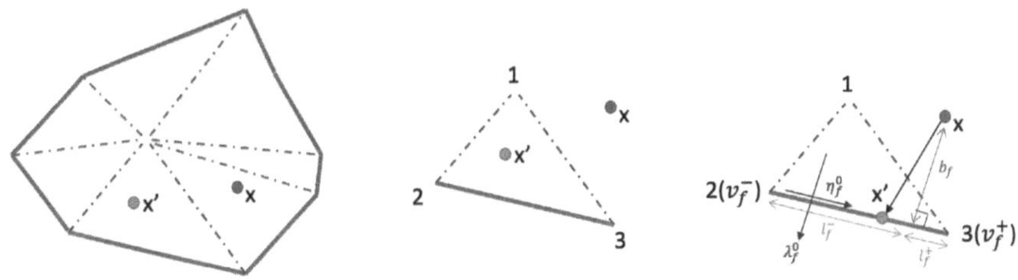

FIGURE 3.5 Schematic plot of the geometry construction of the transformed coordinates.

written as $\boldsymbol{\eta}_f^0$ and $\boldsymbol{\lambda}_f^0$, respectively, and the variables of the TC at the fth edge are given as functions of the observing point $\mathbf{x}$ and two vertices $\mathbf{v}_f^-$, $\mathbf{v}_f^+$ [49,54]:

$$b_f = ((v_f^+)_k - x_k)(\lambda_f^0)_k, \quad l_f^\pm = ((v_f^\pm)_k - x_k)(\eta_f^0)_k \tag{3.58}$$

where b_f is the perpendicular distance between the observing point to the edge. The vector between the observing point $\mathbf{x}$ and the source point $\mathbf{x}'$ is written as $\mathbf{x} - \mathbf{x}' = -b_f\boldsymbol{\lambda}_f^0 - \eta\boldsymbol{\eta}_f^0$, where $\eta = \boldsymbol{\eta}_f^0(\mathbf{x}' - \mathbf{x})$ is the position of the source point on the edge. To derive Eshelby's tensor, partial derivatives of the potentials are required, and therefore Green's theorem can be used to simplify potentials as contour integrals. However, this work uses an alternative scheme to solve the potentials through direct integral over the source domain. As shown in Fig. 3.5b, the integral of any arbitrary piecewise smooth function $\mathcal{F}(|\mathbf{x} - \mathbf{x}'|)$ can be performed as follows:

$$I = \int_{\Omega_c} \mathcal{F}(|\mathbf{x} - \mathbf{x}'|)dA(\mathbf{x}') = \sum_{f=1}^{N_F} \int_{\tan^{-1}(\frac{l_f^-}{b_f})}^{\tan^{-1}(\frac{l_f^+}{b_f})} \int_0^{b_f\sqrt{1+\tan^2[\theta]}} \mathcal{F}(\rho)\rho\, d\rho d\theta \tag{3.59}$$

Let $\mathcal{F}_\Phi = \ln|\mathbf{x} - \mathbf{x}'|^2$ and $\mathcal{F}_\Psi = -|\mathbf{x} - \mathbf{x}'|^2\frac{\ln|\mathbf{x}-\mathbf{x}'|^2-1}{2}$. The analytical solutions of the integral of the harmonic and biharmonic potentials are obtained as Eqs. (3.105) and (3.106), respectively.

3.3.4 Eshelby's tensor for higher-order eigenstrains in a polygonal inclusion

Following Moschovidis and Mura [4] and Mura [1], the continuous eigenstrain field (ε_{ij}^*) can be expanded in the polynomial form of a Taylor series: $\varepsilon_{ij}^*(\mathbf{x}') = \varepsilon_{ij}^{0*} + x_p'\varepsilon_{ijp}^{1*} + x_p'x_q'\varepsilon_{ijpq}^{2*} + \cdots$. The elastic field caused by the eigenstrain can be calculated with the superposition of the contribution of each order of the eigenstrains. Accordingly, Eshelby's tensor can be calculated for each order of the polynomial series. For example, similarly to Eq. (3.44) for constant eigenstrain, the eigenstrain term of $x_m'x_p'x_q'\varepsilon_{ijmpq}^{2*}$ leads to the following integrals:

$$g_{iklmpq...} = \frac{1}{8\pi(1-\nu)}\{\Psi_{mpq...,ikl} - 2\nu\delta_{kl}\Phi_{mpq...,i} - 2(1-\nu)(\delta_{il}\Phi_{mpq...,k} + \delta_{ik}\Phi_{mpq...,l})\} \tag{3.60}$$

$$S_{ijklmpq...} = \frac{1}{8\pi(1-\nu)}\{\Psi_{mpq...,klij} - 2\nu\delta_{kl}\Phi_{mpq...,ij} - (1-\nu)(\delta_{il}\Phi_{mpq...,jk} + \delta_{ik}\Phi_{mpq...,jl} + \delta_{jl}\Phi_{mpq...,ik} + \delta_{jk}\Phi_{mpq...,il})\} \tag{3.61}$$

where $S_{ijklmpq...}$ is the polynomial series of Eshelby's tensor,

$$\Phi_{mpq...} = \int_\Omega -\ln|\mathbf{x}-\mathbf{x}'|^2 x'_m x'_p x'_q ... dA(\mathbf{x}'),$$

and

$$\Psi_{mpq...} = \int_\Omega -|\mathbf{x}-\mathbf{x}'|^2 \frac{\ln|\mathbf{x}-\mathbf{x}'|^2 - 1}{2} x'_m x'_p x'_q ... dA(\mathbf{x}').$$

Since only the partial derivatives of the above potentials are of interest, the area integral can be transferred to the boundary by using Green's theorem. As shown in Fig. 3.5b, the integral variable $\mathbf{x}'$ moves along each edge, and thus the distance between observing and source points, $|\mathbf{x}-\mathbf{x}'|$ can be written as $\sqrt{b_f^2+\eta^2}$ when the source point is on the fth edge. A detailed differentiation process and analytical solution to the uniform Eshelby tensor can be found in the previous work [52,54], which will be used in the following derivation.

3.3.4.1 Linear Eshelby's tensor for the polygonal inclusion

Similarly to Eqs. (3.105) and (3.106), the components of the linear Eshelby tensor can be written as

$$\Phi_p = \int_\Omega (-\ln|\mathbf{x}-\mathbf{x}'|^2)[x_p + x'_p - x_p]dA(\mathbf{x}') = x_p\Phi + \int_\Omega (-\ln|\mathbf{x}-\mathbf{x}'|^2)[x'_p - x_p]dA(\mathbf{x}') \tag{3.62}$$

and

$$\Psi_p = \int_\Omega \left(-|\mathbf{x}-\mathbf{x}'|^2 \frac{\ln|\mathbf{x}-\mathbf{x}'|^2 - 1}{2}\right)[x_p + x'_p - x_p]dA(\mathbf{x}') \tag{3.63}$$

$$= x_p\Psi + \int_\Omega \left(-|\mathbf{x}-\mathbf{x}'|^2 \frac{\ln|\mathbf{x}-\mathbf{x}'|^2 - 1}{2}\right)[x'_p - x_p]dA(\mathbf{x}') \tag{3.64}$$

where Ψ and Φ are defined in Eq. (3.105). The integral limits are determined as $l_f^+ = \boldsymbol{\eta}_f^0.(\mathbf{v}_f^+ - \mathbf{x})$ and $l_f^- = \boldsymbol{\eta}_f^0.(\mathbf{v}_f^- - \mathbf{x})$. Following the chain rule of derivatives, the piecewise smooth function $\mathcal{P}(b_f, l_f^+, l_f^-)$ could be differentiated as follows:

$$\mathcal{P}_{,i} = -(\lambda_f^0)_i \frac{\partial \mathcal{P}}{\partial b_f} - (\eta_f^0)_i [\frac{\partial \mathcal{P}}{\partial l_f^+} + \frac{\partial \mathcal{P}}{\partial l_f^-}] \tag{3.65}$$

and the higher-order differentiation formulas are similarly presented in Appendix 3.B.1. Note that in Fig. 3.5b the source point $\mathbf{x}'$ exists in the area, and to apply Green's theorem, the distance vector at the fth edge $\mathbf{x}' - \mathbf{x}$ is simplified as $b_f \boldsymbol{\lambda}_f^0 + \eta \boldsymbol{\eta}_f^0$. Thus the derivatives of the two above linear potentials, which will be used in the fourth-rank linear Eshelby tensor, are written as

$$\begin{aligned}\Psi_{p,ijkl} = {} & \delta_{pi}\Psi_{,jkl} + \delta_{pj}\Psi_{,ikl} + \delta_{pk}\Psi_{,ijl} + \delta_{pl}\Psi_{,ijk} \\ & + \sum_{f=1}^{N_F} (\lambda_f^0)_i (\lambda_f^0)_p [-(\lambda_f^0)_j (\Psi_0^f)_{,kl} - (\lambda_f^0)_k (\Psi_0^f)_{,jl} \\ & - (\lambda_f^0)_l (\Psi_0^f)_{,jk}] + (\lambda_f^0)_i (\eta_f^0)_p (\Psi_I^f)_{,jkl}\end{aligned} \tag{3.66a}$$

$$\begin{aligned}\Phi_{p,ij} = {} & \delta_{pi}\Phi_{,j} + \delta_{pj}\Phi_{,i} + x_p \Phi_{,ij} \\ & + \sum_{f=1}^{N_F} (\lambda_f^0)_i (\lambda_f^0)_p [-(\lambda_f^0)_j \Phi_0^f + b_f (\Phi_0^f)_{,j}] \\ & + (\lambda_f^0)_i (\eta_f^0)_p (\Phi_I^f)_{,j}\end{aligned} \tag{3.66b}$$

where Ψ_0^f, Ψ_1^f, Φ_0^f, and Φ_1^f denote the contour integrals, which are available in Eqs. (3.108) and (3.109), respectively, in the Appendix. The derivatives of the potentials Ψ and Φ can be derived through Eq. (3.105), i.e., $\Psi_{,ijkl} = \sum_{f=1}^{N_F} (\lambda_f^0)_i (\Psi_0^f)_{,jkl}$ by Green's theorem. Notice that the linear Eshelby tensor for the displacement field contains $\delta_{pi}\Phi$, which requires an analytical solution to the harmonic potential in Eq. (3.105). Finally, the substitution of Φ_p and Ψ_p potentials into Eq. (3.61) yields the linear Eshelby tensor.

3.3.4.2 Quadratic Eshelby's tensor for the polygonal inclusion

As previously shown in Section 3.1, the derivation of the linear Eshelby tensor uses the observing points and the distance vector $x_p + (x'_p - x_p)$ to express the first-order source term x'_p in the contour integrals. Similarly, the second-order term $x'_p x'_q$ can be expressed as $(x'_p - x_p)(x'_q - x_q) - x_p x_q + x'_p x_q + x'_q x_p$, and thus the potential components are written as

$$\Psi_{pq} = \int_{\Omega_c} (-|\mathbf{x}-\mathbf{x}'|^2 \frac{\ln|\mathbf{x}-\mathbf{x}'|^2 - 1}{2}) \begin{bmatrix} (x'_p - x_p)(x'_q - x_q) \\ -x_p x_q + x_p x'_q + x_q x'_p \end{bmatrix} dA(\mathbf{x}')$$

$$= -x_p x_q \Psi + x_p \Psi_q + x_q \Psi_p$$

$$+ \int_{\Omega_c} (-|\mathbf{x}-\mathbf{x}'|^2 \frac{\ln|\mathbf{x}-\mathbf{x}'|^2 - 1}{2})(x'_p - x_p)(x'_q - x_q)\, dA(\mathbf{x}') \quad (3.67a)$$

$$\Phi_{pq} = \int_{\Omega_c} (-\ln|\mathbf{x}-\mathbf{x}'|^2) \begin{bmatrix} (x'_p - x_p)(x'_q - x_q) \\ -x_p x_q + x_p x'_q + x_q x'_p \end{bmatrix} dA(\mathbf{x}')$$

$$= -x_p x_q \Phi + x_p \Phi_q + x_q \Phi_p$$

$$+ \int_{\Omega_c} (-\ln|\mathbf{x}-\mathbf{x}'|^2)(x'_p - x_p)(x'_q - x_q)\, dA(\mathbf{x}') \quad (3.67b)$$

where the terms $x_p x'_q$ and $x_q x'_p$ with the integrals are simplified to $x_p(\Phi/\Psi)_q$ and $x_q(\Phi/\Psi)_p$, respectively. Since the uniform/linear potentials have been derived in the previous sections, the analytical solution to the last term in Eq. (3.67) lead to the closed-form expression of the quadratic Eshelby tensor. The derivatives of the two quadratic potentials, which will be used for the quadratic Eshelby tensor, are presented as Eq. (3.107) in Appendix 3.B.3. The same procedure can be extended to derive Eshelby's tensor for higher-order eigenstrains. All the components of partial derivatives are provided in Appendix 3.B.4.

3.4 Properties of Eshelby's tensor

Originally, Eshelby's tensor [2] was for a uniform eigenstrain in an ellipsoidal domain, which can be straightforwardly extended to elliptical domain using the Hadmard regularization [42,90]. In this chapter, we extend it to polygonal and polyhedral domains with polynomial eigenstrain distribution [42,43,49], which can be generalized to an arbitrarily shaped domain with continuous eigenstrain distribution that can be approximated by domain discretization. Considering a uniform eigenstrain distribution, the classic Eshelby tensor for ellipsoidal inclusion is a uniform tensor for internal points of the ellipsoid, whereas for polygonal inclusions, Eshelby's tensor is not uniform anymore, but exhibits a $\ln r$ singularity on the vertices; for polyhedral inclusions, it exhibits a $1/r$ singularity on the vertices and a $\ln r$ singularity on the edges. For any arbitrary inclusion, Eshelby's tensor exhibits the following properties with the following four lemmas for 2D cases, which are generalized to 3D cases as well with corresponding revisions and will be used in Chapter 4 for two theorems.

Lemma 3.1. *Eshelby's tensor S_{klmn} is dimensionless, and its distribution depends on the shape of the source domain and relative location. The integral of Eshelby's tensor of an inclusion over the inclusion itself is proportional to the*

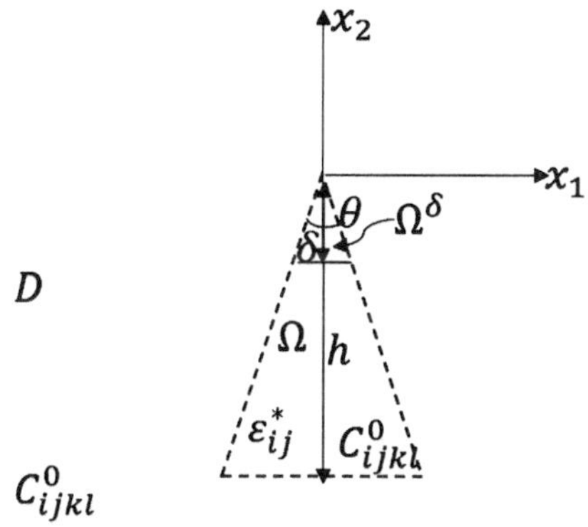

FIGURE 3.6 An isosceles triangular inclusion Ω with height h, top angle θ, and uniform eigenstrain ε_{ij}^* embedded in an infinite domain. The neighborhood of its top vertex is indicated by a similar triangle Ω^δ with height δ.

area or the square of its length scale, referred to the similar shape of inclusion with a unit area.

Therefore, given an isosceles triangle with height h and top angle θ in Fig. 3.6, Eshelby's tensor can be obtained by mapping the value of Eshelby's tensor of a unit triangle Ω^1 with height h_1 and top angle θ (here $h_1^2 \tan(\theta) = 1$ for the unit triangle) to the above-mentioned triangle with length scaling factor h/h_1. Therefore, the stress distribution caused by the same eigenstrain on the similar triangles exhibits the same value on the relative location referred to the unit triangle. This property can be easily demonstrated by elliptical inclusion as well. For an example of a circular integral domain centered at the origin, Eshelby's tensor is written in two zones:

i. For $\mathbf{x}$ in the circle with radius r_0, $|\mathbf{x}| \leq r_0$, it is a constant as in Eq. (3.41).
ii. When $|\mathbf{x}| > r_0$, it depends on the distance and orientation as in Eq. (3.42), where $\rho = a/|\mathbf{x}'|$ and $n_i = \frac{\mathbf{x}'}{|\mathbf{x}'|}$. For the triangular case, Eshelby's tensor is not constant inside the triangle but can still be written in term of the potentials [42] as

$$S_{ijkl} = \frac{1}{8\pi(1-\nu)}[\Psi_{,klij} - 2\nu\delta_{kl}\Phi_{,ij} - (1-\nu)(\delta_{il}\Phi_{,jk} + \delta_{,ik}\Phi_{,jl} + \delta_{jl}\Phi_{,ik} + \delta_{jk}\Phi_{,il})] \tag{3.68}$$

Therefore it is obvious that Eshelby's tensor S_{klmn} is dimensionless.

Through mapping Eshelby's tensor points by points with the area scaling factor for the 2D case, the integral of Eshelby's tensor should be scaled by $(h/h_1)^2$. For example, when a circular inclusion exhibits the constant Eshelby tensor in the circle, the integral is directly proportional to the area. It can be extended to the 3D case with the cubic of length scale. The integral domain can be extended to other parts of integral areas or other shapes of inclusions as long as the mapping with the unit triangle remains.

Lemma 3.2. *Although Eshelby's tensor S_{klmn} of a triangular inclusion Ω with the area A exhibits $\ln r$ singularity at the neighborhood of a vertex, its integral over the inclusion is positive definite and bounded with $\mathbf{0} < \int_{\Omega} \mathbf{S}(\mathbf{x}')d\mathbf{x}' < A\mathbf{I}$ where the unit tensor $I_{klmn} = \frac{1}{2}(\delta_{km}\delta_{ln} + \delta_{kn}\delta_{lm})$.*

This property can be confirmed by the simple fact that if the surrounding material of the triangle is removed, then the final strain will be exactly the eigenstrain, and the integral of any unit eigenstrain component is the area. However, the surrounding material will constrain the strain, which makes the final strain over the inclusion smaller with ratio between 0 and 1. This property can be extended to other shapes of inclusions. For an example of a circular inclusion, any component of Eshelby's tensor in Eq. (3.41) is less than 1, so the integral obviously satisfies the above equation. Define the integral over the unit triangle Ω^1 as

$$S_{klmn}^1 = \int_{\Omega^1} S_{klmn}(\mathbf{x}')d\mathbf{x}' \tag{3.69}$$

We can see that $\mathbf{0} < \mathbf{S}^1 < \mathbf{I}$. For 3D cases, the singularity level at a vertex increases to $1/r$. The integral of Eshelby's tensor exhibits the same property as Eq. (3.69).

Lemma 3.3. *Eshelby's tensor of an area formed by the complement of a large inclusion including a small one is the difference of Eshelby's tensor of the larger inclusion from the smaller one. When they are similar with a shared reference point inside the inclusion, Eshelby's tensor at the reference point is zero.*

This property can be proved by using the superposition of the larger triangle with the same eigenstrain and the smaller triangle with the opposite eigenstrain. This property is applicable to both 2D and 3D cases. Using Lemma 3.1, because of the same values but opposite signs of Eshelby's tensor at the reference point, Eshelby's tensor at that point must be zero. Note that this property is valid when Eshelby's tensor is well defined. When the reference point is at the boundary or vertices, Eshelby's tensor exhibits discontinuity or singularity, so this property cannot be generalized anymore. That case will be addressed in Lemma 3.4.

Particularly, when this property is applied to ellipsoidal/elliptical inclusions, two interesting phenomena can be observed because Eshelby's tensor is constant.

1. When one circle is embedded in another circle, the stress in the small circle caused by the uniform eigenstrain in their complementary area is zero, which means that if the small circle is replaced by any material or void, then the overall stress state will remain the same.

2. Tanaka–Mori theorem [1] immediately shows that the integral of stress or strain over the complementary area caused by any eigenstrain in the small circle is zero.

These phenomena can be straightforwardly generalized to 3D cases for ellipsoidal inclusions that share the same orientation.

Lemma 3.4. *When two triangles share the same vertex and orientation with heights $h^L > h^S$, Eshelby's tensor of the complement of the two triangles at the shared vertex is proportional to* $\ln(h^L/h^S)$.

When two similar triangles share the same vertex and orientation, the complement is an isosceles trapezoid. By Lemma 3.3 it seems that Eshelby's tensor at the shared vertex is zero, but this is not true because the difference between the two singularities cannot be defined. For example, in Fig. 3.6, the complement of the small triangle with top vertex at the origin and height δ is defined as $\overline{\Omega} = \Omega - \Omega^\delta$, where $\overline{\Omega}$ is a quadrilateral area if $\delta \not\to 0$. Eshelby's tensor of $\overline{\Omega}$ at the vertex or origin is written as

$$S_{ijkl}(0) = \sum_{n=1}^{2} \frac{1}{4\pi(1-\nu)} \ln\frac{h}{\delta}\Big[(\lambda_n)_k(\eta_n)_l(\eta_n)_i(\eta_n)_j - 2\nu\delta_{kl}(\lambda_n)_i(\eta_n)_j$$
$$- 4(1-\nu)(\delta_{ik}(\lambda_n)_j(\eta_n)_l + \delta_{il}(\lambda_n)_j(\eta_n)_k + \delta_{jk}(\lambda_n)_i(\eta_n)_l + \delta_{jl}(\lambda_n)_i(\eta_n)_k)\Big] \tag{3.70}$$

where λ_n and η_n are outward and directional vectors of the nth edge (excluding two parallel edges), $\lambda_1 = (-\cos\frac{\theta}{2}, \sin\frac{\theta}{2})$, $\eta_1 = (-\sin\frac{\theta}{2}, -\cos\frac{\theta}{2})$, $\lambda_2 = (\cos\frac{\theta}{2}, \sin\frac{\theta}{2})$, and $\eta_2 = (-\sin\frac{\theta}{2}, \cos\frac{\theta}{2})$. Eq. (3.70) can be obtained by superposition. The contributions of two parallel edges of the trapezoid cancel out themselves, and therefore Eshelby's tensor at the vertex is only related to Poisson's ratio, the angle θ, and the ratio h/δ.

Obviously, because the shared vertex becomes a point outside of the eigenstrain domain of $\overline{\Omega}$, Eshelby's tensor is neither zero nor singular. For 3D cases, the similar property can be obtained for tetrahedrons with a ratio of h^L/h^S instead. The property also shows that the singularity effect of an angular point is localized at a small region. We can use many triangles or tetrahedrons to reproduce Eshelby's tensor of a circle or a sphere [43,74].

*3.5 Extension of the inclusion problem to a bimaterial infinite domain or a semi-infinite domain

For a bimaterial infinite domain, the integrals of harmonic and biharmonic potentials will follow the same way with the Green's function technique. Similarly to Eq. (3.7), the displacement can be written as

$$u_i(\mathbf{x}) = \int_\Omega \frac{\partial G_{ij}}{\partial x'_m} C_{mjkl}\epsilon^*_{kl} d\mathbf{x}' \tag{3.71}$$

where G_{ij} refers to Eqs. (2.15) and (2.16). Considering that $\frac{\partial \phi}{\partial x'_m} = -\frac{\partial \phi}{\partial x_m}$, $\frac{\partial \psi}{\partial x'_m} = -\frac{\partial \psi}{\partial x_m}$, $\frac{\partial \overline{\psi}}{\partial x'_m} = -Q_{mq}\frac{\partial \overline{\psi}}{\partial x_q}$, and $\frac{\partial \overline{\phi}}{\partial x'_m} = -Q_{mq}\frac{\partial \overline{\phi}}{\partial x_q}$ and defining

$$\Gamma^u_{ijm} = -\frac{\partial G_{ij}}{\partial x'_m} \tag{3.72}$$

we can integrate the eigenstrain field over the source domain and the corresponding image domain to obtain the displacement field for the inclusion problem for ellipsoidal, elliptical, polyhedral, and polygonal inclusions. Note that the domain integrals of ϕ, $\overline{\phi}$, ψ, and $\overline{\psi}$ have been obtained in Sections 3.2 and 3.3. The domain integrals of α, $\overline{\alpha}$, β, and $\overline{\beta}$ need to be derived for ellipsoidal, elliptical, polyhedral, and polygonal inclusions. Because they are written in terms of the line integrals of ϕ and $\overline{\phi}$, those domain integrals can be written in terms Φ and $\overline{\Phi}$ as well. Notice that because the source point $\mathbf{x}'$ of the image domain and the field point $\mathbf{x}$ cannot be overlapped except the interface, the integral over an arbitrary shape of the source domain is well defined without any singularity and can be obtained by either analytical or numerical methods. Therefore the EIM can be straightforwardly extended to the bimaterial infinite domain. The interface effect will be induced when the inhomogeneity is close to the interface.

When the stiffness of one material reduces to zero or approaches the infinity, the stress or displacement field in that semi-infinite domain can be immediately obtained, so that the bimaterial problem can be reduced to the elastic analysis of another semi-infinite domain. When the semi-infinite domain exhibits a free surface, it is equivalent to the bimaterial problem containing another zero-stiffness semi-infinite domain; when the semi-infinite exhibits a fixed surface, it is equivalent to the bimaterial problem containing another perfectly rigid semi-infinite domain [86–89]. As $\bar{\mu}^0 \to \infty$, the above Green's function is reduced to Rongved's solution [13,87] for a semi-infinite domain with fixed surface.

*3.5.1 Case study for a semi-infinite domain containing spherical inclusions

Consider a semi-infinite domain with $\mathbf{C}^0$ containing inclusions. The problem can be obtained by reducing the bimaterial infinite domain of $\mathbf{C}^0$ for $x \geq 0$ and

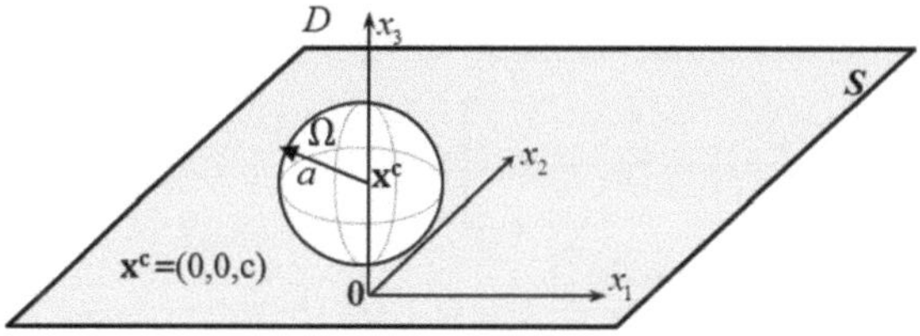

FIGURE 3.7 A semi-infinite domain D ($x_3 >= 0$) containing a subdomain Ω with surface S.

$\bar{\mathbf{C}}^0$ for $x < 0$ with $\bar{\mathbf{C}}^0 = 0$. Using Mindlin's solution of Eq. (2.28), we can write the displacement caused by eigenstrain with the following tensor (see Fig. 3.7):

$$\begin{aligned}
4\pi\mu^0\Gamma^u_{ijm} &= \left[\phi_{,m}\delta_{ij} - \frac{1}{4\left(1-v^0\right)}\psi_{,ijm}\right] \\
&+ Q_{mq}\bar{\phi}_{,q}\delta_{ij} - \frac{\left(3-4v^0\right)}{4\left(1-v^0\right)}Q_{ik}Q_{jl}Q_{mq}\bar{\psi}_{,klq} \\
&- \frac{x_3}{2\left(1-v^0\right)}\left[\begin{array}{c} Q_{jk}Q_{mq}\bar{\psi}_{,ikq3} + 4\left(1-v^0\right)Q_{mq}\delta_{j3}\bar{\phi}_{,iq} \\ +2\left(1-2v^0\right)\delta_{i3}Q_{jk}Q_{mq}\bar{\phi}_{,kq} - Q_{jk}Q_{mq}x_3\bar{\phi}_{,ikq} \end{array}\right] \\
&- 2\left(1-2v^0\right)\left(\delta_{i3}\delta_{jk} - \delta_{ik}\delta_{j3}\right)\overline{\alpha}_{,km'} + \left(1-2v^0\right)Q_{jk}\overline{\beta}_{,ikm'}
\end{aligned} \tag{3.73}$$

Following the previous discussion and assuming a polynomial eigenstrain distribution, we can integrate the displacement as follows:

$$\begin{aligned}
u_i(\mathbf{x}) &= -\int_\Omega \Gamma_{imn}C_{mnkl}\left[E^0_{kl} + E^1_{klp}\left(x'_p - x^c_p\right)\right. \\
&\left. + E^2_{klpq}\left(x'_p - x^c_p\right)\left(x'_q - x^c_q\right)\right]d\mathbf{x}' \\
&= -\frac{1}{4\pi\mu_0}\left(M_{imn}C_{mnkl}E^0_{kl} + N_{pimn}C_{mnkl}E^1_{klp} + T_{pqimn}C_{mnkl}E^2_{klpq}\right)
\end{aligned} \tag{3.74}$$

where M^0_{imn}, N^1_{pimn}, and T^2_{pqimn} are defined as

$$\begin{aligned}
M^0_{imn} &= \frac{1}{2}\left(M^u_{imn} + M^u_{inm}\right) \\
N^1_{pimn} &= \frac{1}{2}\left(N^u_{pimn} + N^u_{pinm}\right) \\
T^2_{pqimn} &= \frac{1}{2}\left(T^u_{pqimn} + T^u_{pqinm}\right)
\end{aligned}$$

and the explicit forms of tensors $\mathbf{M}^u$, $\mathbf{N}^u$, and $\mathbf{T}^u$ are written as

$$
\begin{aligned}
M^u_{imn} = & \left[\Phi_{,n}\delta_{im} - \frac{1}{4\left(1-\upsilon^0\right)}\Psi_{,imn} \right] \\
& + Q_{nq}\bar{\Phi}_{,q}\delta_{im} - \frac{\left(3-4\upsilon^0\right)}{4\left(1-\upsilon^0\right)} Q_{ik}Q_{ml}Q_{nq}\bar{\Psi}_{,klq} \\
& - \frac{x_3}{2\left(1-\upsilon^0\right)} \left[\begin{array}{c} Q_{mk}Q_{nq}\bar{\Psi}_{,ikq3} + 4\left(1-\upsilon^0\right)Q_{nq}\delta_{m3}\bar{\Phi}_{,iq} \\ +2\left(1-2\upsilon^0\right)\delta_{i3}Q_{mk}Q_{nq}\bar{\Phi}_{,kq} - Q_{mk}Q_{nq}x_3\bar{\Phi}_{,ikq} \end{array} \right] \\
& + 2\left(1-2\upsilon^0\right)\left(\delta_{i3}\delta_{mk} - \delta_{ik}\delta_{m3}\right)Q_{nq}A_{,qk} - \left(1-2\upsilon^0\right)Q_{mk}Q_{nq}B_{,qik}
\end{aligned}
\tag{3.75}
$$

$$
\begin{aligned}
N^u_{pimn} = & \left[\Phi_{p,n}\delta_{im} - \frac{1}{4\left(1-\upsilon^0\right)}\Psi_{p,imn} \right] \\
& + Q_{tp}Q_{nq}\bar{\Phi}_{t,q}\delta_{im} - \frac{\left(3-4\upsilon^0\right)}{4\left(1-\upsilon^0\right)} Q_{ik}Q_{ml}Q_{nq}Q_{pt}\bar{\Psi}_{t,klq} \\
& - \frac{x_3}{2\left(1-\upsilon^0\right)} \left[\begin{array}{c} Q_{mk}Q_{nq}Q_{pt}\bar{\Psi}_{t,ikq3} \\ +4\left(1-\upsilon^0\right)Q_{nq}Q_{pt}\delta_{m3}\bar{\Phi}_{t,iq} \\ +2\left(1-2\upsilon^0\right)\delta_{i3}Q_{mk}Q_{nq}Q_{pt}\bar{\Phi}_{t,kq} \\ -Q_{mk}Q_{nq}Q_{pt}x_3\bar{\Phi}_{t,ikq} \end{array} \right] \\
& + 2\left(1-2\upsilon^0\right)\left(\delta_{i3}\delta_{mk} - \delta_{ik}\delta_{m3}\right)Q_{nq}A_{p,qk} - \left(1-2\upsilon^0\right)Q_{mk}Q_{nq}B_{p,qik}
\end{aligned}
\tag{3.76}
$$

and

$$
\begin{aligned}
T^u_{pqimn} = & \left[\Phi_{pq,n}\delta_{im} - \frac{\Psi_{pq,imn}}{4\left(1-\upsilon^0\right)} \right] \\
& + Q_{ng}Q_{ps}Q_{qt}\bar{\Phi}_{st,g}\delta_{im} - \frac{\left(3-4\upsilon^0\right)}{4\left(1-\upsilon^0\right)} Q_{ik}Q_{ml}Q_{ng}Q_{ps}Q_{qt}\bar{\Psi}_{st,klg} \\
& - \frac{x_3}{2\left(1-\upsilon^0\right)} \left[\begin{array}{c} Q_{mk}Q_{ng}Q_{ps}Q_{qt}\bar{\Psi}_{st,ikg3} \\ +4\left(1-\upsilon^0\right)Q_{ng}Q_{ps}Q_{qt}\delta_{m3}\bar{\Phi}_{st,ig} \\ +2\left(1-2\upsilon^0\right)\delta_{i3}Q_{mk}Q_{ng}Q_{ps}Q_{qt}\bar{\Phi}_{st,kg} \\ -Q_{mk}Q_{ng}Q_{ps}Q_{qt}x_3\bar{\Phi}_{st,ikg} \end{array} \right] \\
& + 2\left(1-2\upsilon^0\right)\left(\delta_{i3}\delta_{mk} - \delta_{ik}\delta_{m3}\right)Q_{ng}A_{pq,gk} \\
& - \left(1-2\upsilon^0\right)Q_{mk}Q_{ng}B_{pq,gik}
\end{aligned}
\tag{3.77}
$$

where Φ, Ψ, A, and B denote the integrals of ϕ, ψ, α, and β over the ellipsoidal drop centered at (x_1^c, x_2^c, x_3^c), Φ_p, Ψ_p, A_p, and B_p denote the integrals of $\phi\left(x'_p - x_p^c\right)$, $\psi\left(x'_p - x_p^c\right)$, $\alpha\left(x'_p - x_p^c\right)$, and $\beta\left(x'_p - x_p^c\right)$, and Φ_{pq}, Ψ_{pq}, A_{pq}, and B_{pq} denote the integrals of $\phi\left(x'_p - x_p^c\right)\left(x'_q - x_q^c\right)$, $\psi\left(x'_p - x_p^c\right)\left(x'_q - x_q^c\right)$, $\alpha\left(x'_p - x_p^c\right)\left(x'_q - x_q^c\right)$, and $\beta\left(x'_p - x_p^c\right)\left(x'_q - x_q^c\right)$.

The volume integrals of α and β are given as

$$A(\mathbf{x}, \mathbf{x}^c) = \int_\Omega \left[\ln\left(x_3 + x'_3 + \overline{\psi}\right)\right] d\mathbf{x}' = \frac{4\pi a^3}{3} \ln \widetilde{\psi}(\mathbf{x}, \mathbf{x}^c)$$

and

$$\begin{aligned} B(\mathbf{x}, \mathbf{x}^c) &= \int_\Omega \left[(x_3 + x'_3) \ln\left(x_3 + x'_3 + \overline{\psi}\right) - \overline{\psi}\right] d\mathbf{x}' \\ &= \frac{4\pi a^3}{3} \left[(x_3 + x_3^c) \ln \widetilde{\psi}(\mathbf{x}, \mathbf{x}^c) - \overline{\psi}(\mathbf{x}, \mathbf{x}^c)\right] \end{aligned}$$

where $\widetilde{\psi}(\mathbf{x}, \mathbf{x}^c) = \overline{\psi}(\mathbf{x}, \mathbf{x}^c) + x_3 + x_3^c$.

The volume integrals of $\alpha\left(x'_p - x_p^c\right)$ and $\beta\left(x'_p - x_p^c\right)$ are given as

$$\begin{aligned} A_p(\mathbf{x}, \mathbf{x}^c) &= \int_\Omega \left[\ln\left(x_3 + x'_3 + \overline{\psi}\right)\right]\left(x'_p - x_p^c\right) d\mathbf{x}' \\ &= \begin{cases} -\dfrac{4\pi}{15} a^5 \left(\ln \widetilde{\psi}(\mathbf{x}, \mathbf{x}^c)\right)_{,p} & (p = 1, 2) \\ \dfrac{4\pi}{15} a^5 \widetilde{\phi}(\mathbf{x}, \mathbf{x}^c) & (p = 3) \end{cases} \end{aligned}$$

$$\begin{aligned} B_p(\mathbf{x}, \mathbf{x}^c) &= \int_\Omega \left[(x_3 + x'_3) \ln\left(x_3 + x'_3 + \overline{\psi}\right) - \overline{\psi}\right]\left(x'_p - x_p^c\right) d\mathbf{x}' \\ &= \begin{cases} \dfrac{4\pi a^5}{15} (x_p - x_p^c)\, \widetilde{\phi}(\mathbf{x}, \mathbf{x}^c) & (p = 1, 2) \\ \dfrac{4\pi}{15} a^5 \ln \widetilde{\psi}(x, x^c) & (p = 3) \end{cases} \end{aligned}$$

where $\widetilde{\phi}(\mathbf{x}, \mathbf{x}^c) = \dfrac{1}{\overline{\psi}(\mathbf{x}, \mathbf{x}^c) + x_3 + x_3^c}$. Note that

$$\widetilde{\psi}_{,i} = \overline{\psi}_{,i} + \delta_{3i}, \quad \widetilde{\psi}_{,ij\ldots} = \overline{\psi}_{,ij\ldots}. \tag{3.78}$$

The volume integrals of $\alpha\left(x'_m-x^c_m\right)\left(x'_n-x^c_n\right)$, $\beta\left(x'_m-x^c_m\right)\left(x'_n-x^c_n\right)$ are given as

$$A_{mn}(\mathbf{x},\mathbf{x}^c)=\int_\Omega\left[\ln\left(x_3+x'_3+\overline{\psi}\right)\right]\left(x'_m-x^c_m\right)\left(x'_n-x^c_n\right)d\mathbf{x}'$$

$$=\begin{cases}\dfrac{4\pi a^5}{105}\left\{\begin{array}{c}-a^2\left[\left(\ln\overline{\psi}\right)_{,m}\left(\ln\widetilde{\psi}\right)_{,n}+\left(\ln\widetilde{\psi}\right)_{,m}\left(\ln\overline{\psi}\right)_{,n}\right]\\+\delta_{mn}\left(7\ln\widetilde{\psi}+a^2\widetilde{\phi}\,\overline{\phi}\right)\end{array}\right\}\\(m\neq3;n\neq3)\\-\dfrac{4\pi a^7}{105}\overline{\phi}_{,m}\quad(m\neq3;n=3)\\\dfrac{4\pi a^5}{105}\left(a^2\overline{\phi}_{,3}+7\ln\widetilde{\psi}\right)\quad(m=3;n=3)\end{cases}$$

and

$$B_{mn}(\mathbf{x},\mathbf{x}^c)=\int_\Omega\left[\left(x_3+x'_3\right)\ln\left(x_3+x'_3+\overline{\psi}\right)-\overline{\psi}\right]\left(x'_m-x^c_m\right)\left(x'_n-x^c_n\right)d\mathbf{x}'$$

$$=\begin{cases}\dfrac{4\pi a^5}{105}\left\{\begin{array}{c}-a^2\left(x_m-x^c_m\right)\widetilde{\phi}_{,n}\\+\delta_{mn}\left[7\left(\left(x_3+x^c_3\right)\ln\widetilde{\psi}(\mathbf{x},\mathbf{x}^c)-\overline{\psi}(\mathbf{x},\mathbf{x}^c)\right)-a^2\widetilde{\phi}\right]\end{array}\right\}\\(m\neq3;n\neq3)\\-\dfrac{4\pi a^7}{105}\left(\ln\widetilde{\psi}\right)_{,m}\quad(m\neq3;n=3)\\\dfrac{4\pi a^5}{105}\left\{a^2\overline{\phi}+7\left[\left(x_3+x^c_3\right)\ln\widetilde{\psi}(\mathbf{x},\mathbf{x}^c)-\overline{\psi}(\mathbf{x},\mathbf{x}^c)\right]\right\}\\(m=3;n=3)\end{cases}$$

Also, the strain can be derived as follows:

$$\begin{aligned}\epsilon_{ij}=&-\int_\Omega\Gamma^e_{ijkl}C_{klmn}\Big[E^0_{mn}+E^1_{mnp}\left(x'_p-x^c_p\right)\\&+E^2_{mnst}\left(x'_s-x^c_s\right)\left(x'_t-x^c_t\right)\Big]d\mathbf{x}'\\=&-\frac{1}{4\pi\mu_0}\left(D^0_{ijkl}C_{klmn}E^0_{mn}+D^1_{pijkl}C_{klmn}E^1_{mnp}+D^2_{stijkl}C_{klmn}E^2_{mnst}\right)\end{aligned}\tag{3.79}$$

where D^0_{ijkl}, D^1_{pijkl}, and D^2_{stijkl} are

$$\begin{cases} D^0_{ijkl} = \frac{1}{4}\left(D^e_{ijkl} + D^e_{jikl} + D^e_{ijlk} + D^e_{jilk}\right) \\ D^1_{pijkl} = \frac{1}{4}\left(D^e_{pijkl} + D^e_{pjikl} + D^e_{pijlk} + D^e_{pjilk}\right) \\ D^2_{stijkl} = \frac{1}{4}\left(D^e_{stijkl} + D^e_{stjikl} + D^e_{stijlk} + D^e_{stjilk}\right) \end{cases} \tag{3.80}$$

with D^e_{ijmn}, D^e_{pijkl}, and D^e_{stijkl} given as

$$\begin{aligned} D^e_{ijmn} &= \int_\Omega \Gamma^u_{imn,j} d\mathbf{x}' \\ &= \left[\Phi_{,nj}\delta_{im} - \frac{1}{4\left(1-v^0\right)}\Psi_{,imnj}\right] + Q_{nq}\bar{\Phi}_{,qj}\delta_{im} \\ &- \frac{\left(3-4v^0\right)}{4\left(1-v^0\right)} Q_{ik} Q_{ml} Q_{nq} \bar{\Psi}_{,klqj} \\ &- \frac{x_3}{2\left(1-v^0\right)} \left[\begin{array}{c} Q_{mk} Q_{nq} \bar{\Psi}_{,ikq3j} + 4\left(1-v^0\right) Q_{nq}\delta_{m3}\bar{\Phi}_{,iqj} \\ +2\left(1-2v^0\right)\delta_{i3} Q_{mk} Q_{nq}\bar{\Phi}_{,kqj} \\ -Q_{mk} Q_{nq} x_3 \bar{\Phi}_{,ikqj} - Q_{mk} Q_{nq}\delta_{3j}\bar{\Phi}_{,ikq} \end{array}\right] \\ &- \frac{\delta_{3j}}{2\left(1-v^0\right)} \left[\begin{array}{c} Q_{mk} Q_{nq}\bar{\Psi}_{,ikq3} + 4\left(1-v^0\right) Q_{nq}\delta_{m3}\bar{\Phi}_{,iq} \\ +2\left(1-2v^0\right)\delta_{i3} Q_{mk} Q_{nq}\bar{\Phi}_{,kq} - Q_{mk} Q_{nq} x_3 \bar{\Phi}_{,ikq} \end{array}\right] \\ &+2\left(1-2v^0\right)\left(\delta_{i3}\delta_{mk} - \delta_{ik}\delta_{m3}\right) Q_{nq} A_{,jqk} - \left(1-2v^0\right) Q_{mk} Q_{nq} B_{,qikj} \end{aligned} \tag{3.81}$$

$$\begin{aligned} D^e_{pijmn} &= \int_\Omega \Gamma^u_{imn,j}\left(x'_p - x^c_p\right) d\mathbf{x}' \\ &= \left[\Phi_{p,nj}\delta_{im} - \frac{\Psi_{p,imnj}}{4\left(1-v^0\right)}\Psi_{p,imnj}\right] + Q_{nq} Q_{pt}\bar{\Phi}_{t,qj}\delta_{im} \\ &- \frac{\left(3-4v^0\right)}{4\left(1-v^0\right)} Q_{ik} Q_{ml} Q_{nq} Q_{pt}\bar{\Psi}_{t,klqj} \\ &- \frac{x_3}{2\left(1-v^0\right)} \left[\begin{array}{c} Q_{mk} Q_{nq} Q_{pt}\bar{\Psi}_{t,ikq3j} \\ +4\left(1-v^0\right) Q_{nq}\delta_{m3} Q_{pt}\bar{\Phi}_{t,iqj} \\ +2\left(1-2v^0\right)\delta_{i3} Q_{mk} Q_{nq} Q_{pt}\bar{\Phi}_{t,kqj} \\ -Q_{mk} Q_{nq} x_3 Q_{pt}\bar{\Phi}_{t,ikqj} - Q_{mk} Q_{nq}\delta_{3j} Q_{pt}\bar{\Phi}_{t,ikq} \end{array}\right] \end{aligned}$$

$$
-\frac{\delta_{3j}}{2\left(1-v^0\right)}\left[\begin{array}{c} Q_{mk}Q_{nq}Q_{pt}\bar{\Psi}_{t,ikq3} \\ +4\left(1-v^0\right)Q_{nq}Q_{pt}\delta_{m3}\bar{\Phi}_{t,iq} \\ +2\left(1-2v^0\right)\delta_{i3}Q_{mk}Q_{nq}Q_{pt}\bar{\Phi}_{t,kq} \\ -Q_{mk}Q_{nq}x_3Q_{pt}\bar{\Phi}_{t,ikq} \end{array}\right]
$$
$$
+2\left(1-2v^0\right)\left(\delta_{i3}\delta_{mk}-\delta_{ik}\delta_{m3}\right)Q_{nq}A_{p,jqk}
$$
$$
-\left(1-2v^0\right)Q_{mk}Q_{nq}B_{p,qikj} \tag{3.82}
$$

$$
\begin{aligned}
D^e_{stijmn} &= \int_\Omega \Gamma^u_{imn,j}\left(x'_s-x^c_s\right)\left(x'_t-x^c_t\right)d\mathbf{x}' \\
&= \left[\Phi_{st,nj}\delta_{im}-\frac{\Psi_{st,imnj}}{4\left(1-v^0\right)}\right]+Q_{nq}Q_{ps}Q_{gt}\bar{\Phi}_{pg,qj}\delta_{im} \\
&-\frac{\left(3-4v^0\right)}{4\left(1-v^0\right)}Q_{ik}Q_{ml}Q_{nq}Q_{ps}Q_{gt}\bar{\Psi}_{pg,klqj} \\
&-\frac{x_3}{2\left(1-v^0\right)}\left[\begin{array}{c} Q_{mk}Q_{nq}Q_{ps}Q_{gt}\bar{\Psi}_{pg,ikq3j} \\ +4\left(1-v^0\right)Q_{nq}\delta_{m3}Q_{ps}Q_{gt}\bar{\Phi}_{pg,iqj} \\ +2\left(1-2v^0\right)\delta_{i3}Q_{mk}Q_{nq}Q_{ps}Q_{gt}\bar{\Phi}_{pg,kqj} \\ -Q_{mk}Q_{nq}x_3Q_{ps}Q_{gt}\bar{\Phi}_{pg,ikqj} \\ -Q_{mk}Q_{nq}Q_{ps}Q_{gt}\delta_{3j}\bar{\Phi}_{pg,ikq} \end{array}\right] \\
&-\frac{\delta_{3j}}{2\left(1-v^0\right)}\left[\begin{array}{c} Q_{mk}Q_{nq}Q_{ps}Q_{gt}\bar{\Psi}_{pg,ikq3} \\ +4\left(1-v^0\right)Q_{nq}\delta_{m3}Q_{ps}Q_{gt}\bar{\Phi}_{pg,iq} \\ +2\left(1-2v^0\right)\delta_{i3}Q_{mk}Q_{nq}Q_{ps}Q_{gt}\bar{\Phi}_{pg,kq} \\ -Q_{mk}Q_{nq}x_3Q_{ps}Q_{gt}\bar{\Phi}_{pg,ikq} \end{array}\right] \\
&+2\left(1-2v^0\right)\left(\delta_{i3}\delta_{mk}-\delta_{ik}\delta_{m3}\right)Q_{nq}A_{st,jqk} \\
&-\left(1-2v^0\right)Q_{mk}Q_{nq}B_{st,qikj}
\end{aligned} \tag{3.83}
$$

*3.5.2 Domain integral of fundamental solutions for other shapes of inclusions

In Appendix 2.A, the fundamental solution for the bimaterial infinite domain was presented. Compared to the domain integrals derived above for the harmonic and biharmonic potentials, two additional integrals α and β and their image counterparts of $\overline{\alpha}$ and $\overline{\beta}$ need to be solved. Because the difference of the image counterparts exists only in the x_2 direction, if the integrals of α and β are derived, then the image counterparts can be obtained by simply changing the source positions.

For spherical inclusions, the domain integral was provided in analytical form in the last section, which can be extended to ellipsoidal or elliptical domains in a similar fashion. For polyhedral or polygonal inhomogeneities, the domain integral can be derived analytically but will be of complicated form.

As an exercise, without the loss of generality, consider the polygonal inhomogeneity located in the x_1–x_2 plane and suppose that only the derivatives of α and β are derived. Rewrite α in its first partial derivative form

$$
\begin{aligned}
\frac{\partial}{\partial x_i}\int_{\Omega}\alpha d\mathbf{x}' &= -\delta_{i2}\Phi + \sum_{f=1}^{N_F} -(\lambda_f^0)_i \int_{l_f^-}^{l_f^+} \frac{x_2'-x_2}{|\mathbf{x}-\mathbf{x}'|} d\mathbf{x}' \\
&+ \sum_{f=1}^{N_F} -(\lambda_f^0)_i \int_{l_f^-}^{l_f^+} (x_1'-x_1)\tan^{-1}[\frac{x_1'-x_1}{x_2'-x_2}] d\mathbf{x}' \\
&= -\delta_{i2}\Phi - \Psi_{,i2} + \sum_{f=1}^{N_F} -(\lambda_f^0)_i \int_{l_f^-}^{l_f^+} (x_1'-x_1)\tan^{-1}[\frac{x_1'-x_1}{x_2'-x_2}] d\mathbf{x}'
\end{aligned}
\tag{3.84}
$$

where the first two terms are provided already in Section 3.3.3. The components of the distance vector $x_i' - x_i = b_f(\lambda_f^0)_i + \omega(\eta_f^0)_i$ can be further simplified given the unit vector as follows:

$$
\boldsymbol{\eta}_f^{\mathbf{0}} = (m, n), \quad (\boldsymbol{\lambda}_f^{\mathbf{0}}) = (n, -m) \tag{3.85}
$$

where an edge of the polygon in the counterclockwise numbering sequence is indicated by the direction norm (m, n) with $m^2 + n^2 = 1$. Therefore the last term in Eq. (3.84) can be written as follows:

$$
\begin{aligned}
&\int_{l_f^-}^{l_f^+} (x_1'-x_1)\tan^{-1}[\frac{x_1'-x_1}{x_2'-x_2}] d\mathbf{x}' \\
&= \int_{l_f^-}^{l_f^+} (b_f n + \omega m)\tan^{-1}[\frac{b_f n + \omega m}{-b_f m + \omega n}] d\omega \\
&= \frac{1}{2}\Big[b_f m(l_f^+ - l_f^-) + b_f^2 m(\tan^{-1}[\frac{b_f}{l_f^+}] - \tan^{-1}[\frac{b_f}{l_f^-}]) \\
&+ l_f^-(l_f^- m + 2b_f n)\tan^{-1}\frac{l_f^- m + b_f n}{b_f m - l_f^- n} \\
&- l_f^+(l_f^+ m \tan^{-1} + 2b_f n)\tan^{-1}\frac{l_f^+ m + b_f n}{b_f m - l_f^+ n} \\
&+ b_f^2 n\big(\ln[b_f^2 + (l_f^+)^2] - \ln[b_f^2 + (l_f^-)^2]\big)\Big]
\end{aligned}
\tag{3.86}
$$

Similarly, the domain integral of β can be derived. For the polyhedral inclusion, the integral will be more complicated. However, numerical integral and differential schemes can be implemented to solve the domain integrals and their derivatives.

Appendix 3.A Ellipsoidal/elliptical domain integrals of ψ and ϕ and their derivatives

3.A.1 Coefficients of Eshelby's tensor

In this subsection, we provide all tensor-related coefficients to Eq. (3.22) [91]:

$$S_{IK}^{1} = -\frac{v}{2(1-v)} J_I(\lambda) + \frac{1}{4(1-v)} \left[\frac{a_I^2}{a_I^2 - a_K^2} J_I(\lambda) + \frac{a_K^2}{a_K^2 - a_I^2} J_K(\lambda) \right] \tag{3.87}$$

$$S_{IJ}^{2} = -\frac{1}{4} [J_I(\lambda) + J_I(\lambda)] + \frac{1}{4(1-v)} \left[\frac{a_I^2}{a_I^2 - a_J^2} J_I(\lambda) + \frac{a_J^2}{a_J^2 - a_I^2} J_J(\lambda) \right] \tag{3.88}$$

$$S_I^3 = \frac{\rho^3(\lambda)}{2(1-v)} \left[1 - \rho_I^2(\lambda) \right] \tag{3.89}$$

$$S_K^4 = \frac{\rho^3(\lambda)}{2(1-v)} \left[1 - 2v - \rho_K^2(\lambda) \right] \tag{3.90}$$

$$S_I^5 = \frac{\rho^3(\lambda)}{2(1-v)} \left[v - \rho_I^2(\lambda) \right] \tag{3.91}$$

$$S_J^6 = \frac{\rho^3(\lambda)}{2(1-v)} \left[v - \rho_J^2(\lambda) \right] \tag{3.92}$$

$$S_{IJKL}^{7} = \frac{\rho^3(\lambda)}{2(1-v)} \left[\begin{array}{c} 2\left(\rho_I^2(\lambda) + \rho_J^2(\lambda) + \rho_K^2(\lambda) + \rho_L^2(\lambda)\right) \\ +\rho_m(\lambda)\rho_m(\lambda) - \frac{4\rho_M^2(\lambda)\Theta_m(\lambda)\Theta_m(\lambda)}{\Theta^2(\lambda)} \end{array} \right] \tag{3.93}$$

in which

$$\rho_I(\lambda) = \frac{a_I}{\sqrt{a_I^2 + \lambda}}, \quad \rho(\lambda) = \sqrt[3]{\rho_1(\lambda)\rho_2(\lambda)\rho_3(\lambda)} \tag{3.94}$$

and

$$J_I(\lambda) = -\int_\lambda^\infty \frac{\rho^3(s)}{(a_I^2 + s)}. \tag{3.95}$$

3.A.2 Explicit expression of Φ, Φ_p, Φ_{pq}, Ψ, Ψ_p, and Ψ_{pq} in terms of integrals I and V

For the inclusion problems, given the prescribed eigenstrain, the displacement field is related to the first and third partial derivatives of harmonic and biharmonic potentials, respectively, whereas the strain (stress) field is determined by the second and fourth derivatives. In Chapter 4, solving the inhomogeneity problem with the equivalent inclusion method (EIM) requires higher-order derivatives, which can be directly derived based on Eqs. (3.96)–(3.102).

To simplify the lengthy expressions, following Moschovidis [4], define V functions in terms of I as

$$V(\mathbf{x}) = \frac{1}{2}\left[I(\lambda) - x_r x_r I_R(\lambda)\right] \tag{3.96a}$$

$$V_i(\mathbf{x}) = \frac{1}{2}\left[I_i(\lambda) - x_r x_r I_{Ri}(\lambda)\right] \tag{3.96b}$$

$$V_{ij}(\mathbf{x}) = \frac{1}{2}\left[I_{ij}(\lambda) - x_r x_r I_{Rij}(\lambda)\right] \tag{3.96c}$$

where the I integrals are defined in Eq. (3.19) for the general 3D cases, which can be simplified for special geometric shapes, such as sphere (Eq. (3.21)), circle (Eq. (3.40)), and so on:

$$\Phi_{,i} = V_{,i} \quad \Phi_{,ij} = V_{,ij} \tag{3.97a}$$

$$\Psi_{,ijk} = \delta_{ij}(V - a_I^2 V_I)_{,k} + x_i (V - a_I^2 V_I)_{,j} \tag{3.98a}$$

$$\begin{aligned}\Psi_{,ijkl} = {} & \delta_{ij}(V - a_I^2)_{,kl} + \delta_{ik}(V - a_I^2 V_I)_{,jl} \\ & + \delta_{il}(V - a_I^2 V_I)_{,jk} + x_i (V - a_I^2 V_I)_{,jkl}\end{aligned} \tag{3.98b}$$

$$\Phi_{p,i} = a_P^2(\delta_{ip} V_P + x_p V_{P,i}) \tag{3.99a}$$

$$\Phi_{p,ij} = a_P^2(\delta_{ip} V_{P,j} + \delta_{pj} V_{P,i} + x_p V_{P,ij}) \tag{3.99b}$$

$$\begin{aligned}\Psi_{p,ijk} = a_P^2\Big\{ & \delta_{ip} x_j (V_J - a_P^2 V_{JP}) + (\delta_{pj} x_i + \delta_{ij} x_p)(V_I - a_P^2 V_{IP}) \\ & + x_p x_i (V_I - a_P^2 V_{IP})_{,j}\Big\}\end{aligned} \tag{3.100a}$$

$$\begin{aligned}\Psi_{p,ijkl} = a_P^2\Big\{ & \delta_{ip}\delta_{jk}(V_J - a_P^2 V_{JP})_{,l} + \delta_{ip}\delta_{jl}(V_J - a_P^2 V_{JP})_{,k} \\ & + \delta_{ip} x_j (V_J - a_P^2 V_{JP})_{,kl} + (\delta_{pj}\delta_{ik} + \delta_{ij}\delta_{pk})(V_I - a_P^2 V_{IP})_{,l}\end{aligned}$$

$$+ (\delta_{pj}\delta_{il} + \delta_{ij}\delta_{pl})(V_I - a_P^2 V_{IP})_{,k} + (\delta_{pj}x_i + \delta_{ij}x_p)(V_I - a_P^2 V_{IP})_{,kl}$$
$$+ (\delta_{pk}\delta_{il} + \delta_{ik}\delta_{pl})(V_I - a_P^2 V_{IP})_{,j} + (\delta_{pk}x_i + \delta_{ik}x_p)(V_I - a_P^2 V_{IP})_{,jl}$$
$$+ (\delta_{pl}x_i + \delta_{il}x_p)(V_I - a_P^2 V_{IP})_{,jk} + x_p x_i (V_I - a_P^2 V_{IP})_{,jkl}\Big\} \tag{3.100b}$$

In Eq. (3.27) the eigenstrain is approximated by Taylor's series in a polynomial form. Although Chapter 3 only uses up-to the linear terms in derivation, quadratic terms can provide better approximation, especially for eigenstrain field with larger variations. In the following, we also list quadratic potentials for interested readers:

$$\Phi_{pq,i} = a_P^2 \left\{ a_Q^2 (\delta_{pi}x_q + \delta_{qi}x_m) V_{PQ,i} - \delta_{pq}x_i (V_I - a_P^2 V_{IP}) \right\} \tag{3.101a}$$

$$\Phi_{pq,ij} = a_P^2 \Big\{ a_Q^2 (\delta_{pi}\delta_{qj} + \delta_{qi}\delta_{pj}) V_{PQ} + a_Q^2 (\delta_{pi}x_q + \delta_{qi}x_p) V_{PQ,j}$$
$$+ a_Q^2 (\delta_{pj}x_q + \delta_{qj}x_p) V_{PQ,i} + a_Q^2 x_p x_q V_{PQ,ij}$$
$$- \delta_{pq}\big[\delta_{ij}(V_I - a_P^2 V_{IP}) - x_i (V_{I,j} - a_P^2 V_{IP,j})\big]\Big\} \tag{3.101b}$$

$$\Psi_{pq,ijk} = a_P^2 a_Q^2 \Big\{ (\delta_{pi}\delta_{qj} + \delta_{qi}\delta_{pj}) x_j (V_{PK} - a_Q^2 V_{QPK})$$
$$+ \big[(\delta_{pi}\delta_{qk} + \delta_{qi}\delta_{pk}) x_j + (\delta_{pi}x_q + \delta_{qi}x_p)\delta_{jk}\big](V_{PJ} - a_Q^2 V_{QPJ})$$
$$+ (\delta_{pj}x_q + \delta_{qj}x_p) x_j (V_{PJ} - a_Q^2 V_{QPJ})_{,k}$$
$$+ \big[\delta_{pj}(\delta_{qk}x_i + \delta_{ik}x_q) + \delta_{qj}(\delta_{pk}x_i + \delta_{ik}x_p)$$
$$+ \delta_{ij}(\delta_{pk}x_q + \delta_{qk}x_p)\big](V_{PQ} - a_I^2 V_{IPQ})$$
$$+ (\delta_{pj}x_q x_i + \delta_{qj}x_p x_i + \delta_{ij}x_p x_q)(V_{PQ} - a_I^2 V_{IPQ})_{,k}$$
$$+ (\delta_{pk}x_q x_i + \delta_{qk}x_p x_i + \delta_{ik}x_p x_q)(V_{PQ} - a_I^2 V_{IPQ})_{,j}$$
$$+ x_p x_q x_i (V_{PQ} - a_I^2 V_{IPQ})_{,jk}\Big\}$$
$$+ a_P^2 \delta_{pq} \Big\{ - \delta_{ij} x_k \big[V_K - (a_I^2 + a_P^2) V_{PK} + a_I^4 V_{IPK}\big]$$
$$- (\delta_{ik}x_j + \delta_{jk}x_i)\big[V_J - (a_I^2 + a_P^2) V_{PJ} + a_I^4 V_{IPJ}\big]$$
$$- x_i x_j \big[V_J - (a_I^2 + a_P^2) V_{PJ} + a_I^4 V_{IPJ}\big]\Big\} \tag{3.102a}$$

$$\Psi_{pq,ijkl} = a_P^2 a_Q^2 \Big\{ (\delta_{pi}\delta_{qj} + \delta_{qi}\delta_{pj}) \delta_{kl} (V_{PK} - a_Q^2 V_{QPK})$$
$$+ (\delta_{pi}\delta_{qj} + \delta_{qi}\delta_{pj}) x_k (V_{PK} - a_Q^2 V_{QMK})_{,l}$$
$$+ \big[(\delta_{pi}\delta_{qk} + \delta_{qi}\delta_{pk})\delta_{jl} + (\delta_{pi}\delta_{ql} + \delta_{qi}\delta_{pl})\delta_{jk}\big](V_{PJ} - a_Q^2 V_{NPJ})$$
$$+ \big[(\delta_{pi}\delta_{qk} + \delta_{qi}\delta_{pk}) x_j + (\delta_{pi}x_q + \delta_{qi}x_p)\delta_{jk}\big](V_{PJ} - a_Q^2 V_{QPJ})_{,l}$$

$$
\begin{aligned}
&+\left[(\delta_{pi}\delta_{ql}+\delta_{qi}\delta_{pl})x_j+(\delta_{pi}x_q+\delta_{qi}x_p)\delta_{jl}\right](V_{PJ}-a_Q^2V_{NPJ})_{,k}\\
&+(\delta_{pi}x_q+\delta_{qi}x_m)x_j(V_{PJ}-a_Q^2V_{QPJ})_{,kl}\\
&+\left[\delta_{pj}(\delta_{qk}\delta_{il}+\delta_{ik}\delta_{ql})+\delta_{qj}(\delta_{pk}\delta_{il}+\delta_{ik}\delta_{pl})\right.\\
&\left.+\delta_{ij}(\delta_{pk}\delta_{ql}+\delta_{qk}\delta_{pl})\right](V_{MN}-a_I^2V_{IPQ})\\
&+\left[(\delta_{pj}(\delta_{qk}x_i+\delta_{ik}x_q)+\delta_{qj}(\delta_{pk}x_i+\delta_{ik}x_p)\right.\\
&\left.+\delta_{ij}(\delta_{pk}x_q+\delta_{qk}x_p)\right](V_{PQ}-a_I^2V_{IPQ})_{,l}\\
&+\left[\delta_{pj}(\delta_{ql}x_i+\delta_{il}x_q)+\delta_{qj}(\delta_{pl}x_i+\delta_{il}x_p)\right.\\
&\left.+\delta_{ij}(\delta_{pl}x_q+\delta_{ql}x_p)\right](V_{PQ}-a_I^2V_{IPQ})_{,k}\\
&+(\delta_{pj}x_qx_i+\delta_{qj}x_px_i+\delta_{ij}x_px_q)(V_{PQ}-a_I^2V_{IPQ})_{,kl}\\
&+\left[\delta_{pk}(\delta_{ql}x_i+\delta_{il}x_q)+\delta_{qk}(\delta_{pl}x_i+\delta_{il}x_p)+\delta_{ik}(\delta_{pl}x_q\right.\\
&\left.+\delta_{ql}x_p)(V_{PQ}-a_I^2V_{IPQ})_{,j}\right]\\
&+(\delta_{pk}x_qx_i+\delta_{qk}x_px_i+\delta_{ik}x_px_q)(V_{PQ}-a_I^2V_{IPQ})_{,jl}\\
&+(\delta_{pl}x_qx_i+\delta_{ql}x_px_i+\delta_{il}x_px_q)(V_{PQ}-a_I^2V_{IPQ})_{,jk}\\
&\left.+x_px_qx_i(V_{PQ}-a_I^2V_{IPQ})_{,jkl}\right\}\\
&+a_P^2\delta_{pq}\left\{-\delta_{ij}\delta_{kl}\left[V_K-(a_I^2+a_P^2)V_{PK}+a_I^4V_{IPK}\right]\right.\\
&-\delta_{ij}x_k\left[V_K-(a_I^2+a_P^2)V_{PK}+a_I^4V_{IPK}\right]_{,l}\\
&-(\delta_{ik}\delta_{jl}+\delta_{jk}\delta_{il})\left[V_J-(a_I^2+a_P^2)V_{PJ}+a_I^4V_{IPJ}\right]\\
&-(\delta_{ik}x_j+\delta_{jk}x_i)\left[V_J-(a_I^2+a_P^2)V_{PJ}+a_I^4V_{IPJ}\right]_{,l}\\
&-(\delta_{il}x_j+\delta_{jl}x_i)\left[V_J-(a_I^2+a_P^2)V_{PJ}+a_I^4V_{IPJ}\right]_{,k}\\
&\left.-x_ix_j\left[V_J-(a_I^2+a_P^2)V_{PJ}+a_I^4V_{IPJ}\right]_{,kl}\right\}
\end{aligned}
\tag{3.102b}
$$

Appendix 3.B Closed-form domain integral of polygonal inclusion and their derivatives

3.B.1 Partial differentiation chain rule of functions $\mathcal{P}(b_f, l_f^-, l_f^+)$

Based on Rodin [49] and Liu's [54], the differentiation chain rule uses the geometry construction of the variable b_f, l_f^-, l_f^+ as shown in Eq. (3.65). Here the second/third differentiation procedures are attached:

$$
\begin{aligned}
\mathcal{P}_{,ij}=&\frac{\partial^2\mathcal{P}}{\partial b_f^2}(\lambda_f^0)_i(\lambda_f^0)_j+\left[\frac{\partial^2\mathcal{P}}{\partial(l_f^+)^2}+\frac{\partial^2\mathcal{P}}{\partial(l_f^-)^2}\right](\eta_f^0)_i(\eta_f^0)_j\\
&+\left(\frac{\partial^2\mathcal{P}}{\partial b_f\partial l_f^+}+\frac{\partial^2\mathcal{P}}{\partial b_f\partial l_f^-}\right)\left[(\lambda_f^0)_i(\eta_f^0)_j+(\lambda_f^0)_j(\eta_f^0)_i\right]
\end{aligned}
\tag{3.103}
$$

$$\begin{aligned}
\mathcal{P}_{,ijk} = & -\frac{\partial^3 \mathcal{P}}{\partial b_f^3}(\lambda_f^0)_i(\lambda_f^0)_j(\lambda_f^0)_k \\
& -(\frac{\partial^3 \mathcal{P}}{\partial b_f^2 \partial l_f^+}+\frac{\partial^3 \mathcal{P}}{\partial b_f^2 \partial l_f^-})\begin{bmatrix} (\lambda_f^0)_i(\lambda_f^0)_j(\eta_f^0)_k \\ +(\lambda_f^0)_i(\eta_f^0)_j(\lambda_f^0)_k \\ +(\eta_f^0)_i(\lambda_f^0)_j(\lambda_f^0)_k \end{bmatrix} \\
& -[\frac{\partial^3 \mathcal{P}}{\partial b_f \partial (l_f^+)^2}+\frac{\partial^3 \mathcal{P}}{\partial b_f \partial (l_f^-)^2}]\begin{bmatrix} (\lambda_f^0)_i(\eta_f^0)_j(\eta_f^0)_k \\ +(\eta_f^0)_i(\lambda_f^0)_j(\eta_f^0)_k \\ +(\eta_f^0)_i(\eta_f^0)_j(\lambda_f^0)_k \end{bmatrix} \\
& -[\frac{\partial^3 \mathcal{P}}{\partial (l_f^+)^3}+\frac{\partial^3 \mathcal{P}}{\partial (l_f^-)^3}](\eta_f^0)_i(\eta_f^0)_j(\eta_f^0)_k
\end{aligned} \tag{3.104}$$

Since the upper and lower limits of the integral are l_f^+ and l_f^-, respectively, the derivatives with respect to the combination of l_f^- and l_f^+ ($\frac{\partial^{n+m}}{\partial (l_f^-)^n \partial (l_f^+)^m}$ with $n, m > 0$) are not physical and thus discarded.

3.B.2 Domain integral of Φ and Ψ and their derivatives

The harmonic potential Φ with direct area domain integral:

$$\begin{aligned}
\Phi = & \sum_{f=1}^{N_F} b_f\{-\frac{3}{2}(l_f^- - l_f^+) + b_f(\tan^{-1}(\frac{l_f^-}{b_f}) - \tan^{-1}(\frac{l_f^+}{b_f})) \\
& + \frac{1}{2} l_f^- \ln(b_f^2 + (l_f^-)^2) - \frac{1}{2} l_f^+ \ln(b_f^2 + (l_f^+)^2)\}
\end{aligned} \tag{3.105}$$

The biharmonic potential Ψ with direct area domain integral:

$$\begin{aligned}
\Psi = & \sum_{f=1}^{N_F} \frac{1}{144} b_f\{-51 b_f^2 l_f^- - 13(l_f^-)^3 + 51 b_f^2 l_f^+ + 13(l_f^+)^3 \\
& + 24 b_f^3 \tan^{-1}(\frac{l_f^-}{b_f}) - 24 b_f^3 \tan^{-1}(\frac{l_f^+}{b_f}) + 6 l_f^-(2b_f^2 + (l_f^-)^2)\ln(b_f^2 + (l_f^-)^2) \\
& - 6 l_f^+(2b_f^2 + (l_f^+)^2)\ln(b_f^2 + (l_f^+)^2)\}
\end{aligned} \tag{3.106}$$

3.B.3 Expression of the fourth derivative of Ψ_{pq} and second derivative of Φ_{pq}

$$\begin{aligned}
\Psi_{pq,ijkl} = & [\delta_{pi}\Psi_{q,jkl} + \delta_{pj}\Psi_{q,ikl} + \delta_{pk}\Psi_{q,ijl} + \delta_{pl}\Psi_{q,ijk} + x_p\Psi_{q,ijkl}] \\
& + [\delta_{qi}\Psi_{p,jkl} + \delta_{qj}\Psi_{p,ikl} + \delta_{qk}\Psi_{p,ijl} + \delta_{ql}\Psi_{p,ijk} + x_q\Psi_{p,ijkl}]
\end{aligned}$$

$$
\begin{aligned}
&-[\delta_{pi}\delta_{qj}+\delta_{qi}\delta_{pj}]\Psi_{,kl}-[\delta_{pi}\delta_{qk}+\delta_{qi}\delta_{pk}]\Psi_{,jl}\\
&-[\delta_{pi}\delta_{ql}+\delta_{pl}\delta_{qi}]\Psi_{,jk}-[\delta_{pi}x_q+x_p\delta_{qi}]\Psi_{,jkl}\\
&-[\delta_{pj}\delta_{qk}+\delta_{pk}\delta_{qj}]\Psi_{,il}-[\delta_{pj}\delta_{ql}+\delta_{pl}\delta_{qj}]\Psi_{,ik}\\
&-[\delta_{pj}x_q+x_p\delta_{qj}]\Psi_{,ikl}-[\delta_{pk}\delta_{ql}+\delta_{pl}\delta_{qk}]\Psi_{,ij}\\
&-[\delta_{pk}x_q+x_p\delta_{qk}]\Psi_{,ij}-[\delta_{pl}x_q+x_p\delta_{ql}]\Psi_{,ijk}-x_px_q\Psi_{,ijkl}\\
&+\sum_{f=1}^{N_F}(\lambda_f^0)_i(\lambda_f^0)_p(\lambda_f^0)_q[2(\lambda_f^0)_k(\lambda_f^0)_j\Psi^f_{,l}+2(\lambda_f^0)_l(\lambda_f^0)_j\Psi^f_{,k}\\
&-2b_f(\lambda_f^0)_j\Psi^f_{,kl}+2(\lambda_f^0)_l(\lambda_f^0)_k\Psi^f_{,j}-2b_f(\lambda_f^0)_k\Psi^f_{,jl}-2b_f(\lambda_f^0)_l\Psi^f_{,jk}\\
&+b_f^2\Psi^f_{,jkl}]+(\lambda_f^0)_i((\lambda_f^0)_p(\eta_f^0)_q+(\lambda_f^0)_q(\eta_f^0)_p)[-(\lambda_f^0)_j(\Psi_I^f)_{,kl}\\
&-(\lambda_f^0)_k(\Psi_I^f)_{,jl}-(\lambda_f^0)_l(\Psi_I^f)_{,jk}+b_f(\Psi_I^f)_{,jkl}]\\
&+(\lambda_f^0)_i(\eta_f^0)_p(\eta_f^0)_q(\Psi_{II}^f)_{jkl}
\end{aligned}
\tag{3.107a}
$$

$$
\begin{aligned}
\Phi_{pq,ij}=&[\delta_{pi}\Phi_{q,j}+\delta_{pj}\Phi_{q,i}+x_p\Phi_{q,ij}]+[\delta_{qi}\Phi_{p,j}+\delta_{qj}\Phi_{p,i}+x_q\Phi_{p,ij}]\\
&-\delta_{pi}\delta_{qj}\Phi-[\delta_{pi}x_q+\delta_{qi}x_p]\Phi_{,j}-[\delta_{pj}x_q+x_p\delta_{qj}]\Phi_{,i}-x_px_q\Phi_{,ij}\\
&+\sum_{f=1}^{N_F}(\lambda_f^0)_i(\lambda_f^0)_p(\lambda_f^0)_q[-2b_f(\lambda_f^0)_j\Phi^f+b_f^2\Phi^f_{,j}]\\
&+(\lambda_f^0)_i[(\lambda_f^0)_p(\eta_f^0)_q+(\lambda_f^0)_q(\eta_f^0)_p][-(\lambda_f^0)_j\Phi^f+b_f\Phi^f_{,j}]\\
&+(\lambda_f^0)_i(\eta_f^0)_p(\eta_f^0)_q(\Phi_{II}^f)_{,j}
\end{aligned}
\tag{3.107b}
$$

3.B.4 Domain integral with Green's theorem and the components of contour part

$$
\begin{aligned}
\Psi_0^f=&\int_{l_f^-}^{l_f^+}(b_f^2+\eta^2)\frac{\ln(b_f^2+\eta^2)-1}{2}d\eta\\
=&-\frac{7b_f^2(l_f^+-l_f^-)}{6}-\frac{5\left[(l_f^+)^3-(l_f^-)^3\right]}{18}\\
&-\frac{1}{2}\ln(b_f^2+(l_f^-)^2)[b_f^2l_f^-+\frac{(l_f^-)^3}{3}]\\
&+\frac{2b_f^3}{3}\left[\tan^{-1}(\frac{l_f^+}{b_f})-\tan^{-1}(\frac{l_f^-}{b_f})+\frac{1}{2}\ln(b_f^2+(l_f^+)^2)[b_f^2l_f^++\frac{(l_f^+)^3}{3}\right]
\end{aligned}
\tag{3.108a}
$$

$$\Phi_0^f = \int_{l_f^-}^{l_f^+} \ln(b_f^2+\eta^2)\,d\eta$$
$$= 2b_f[\tan^{-1}(\frac{l_f^+}{b_f}) - \tan^{-1}(\frac{l_f^-}{b_f})] + l_f^+\ln(b_f^2+(l_f^+)^2) - l_f^-\ln(b_f^2+(l_f^-)^2) \tag{3.108b}$$

and

$$\Psi_I^f = \int_{l_f^-}^{l_f^+} \eta(b_f^2+\eta^2)\frac{\ln(b_f^2+\eta^2)-1}{2}\,d\eta$$
$$= \frac{1}{16}[2b_f^4+6b_f^2(l_f^-)^2+3(l_f^-)^4-2(b_f^2+(l_f^-)^2)^2\ln(b_f^2+(l_f^-)^2)]$$
$$-\frac{1}{16}[2b_f^4+6b_f^2(l_f^+)^2+3(l_f^+)^4-2(b_f^2+(l_f^+)^2)^2\ln(b_f^2+(l_f^+)^2)] \tag{3.109a}$$

$$\Phi_1^f = \int_{l_f^-}^{l_f^+} \eta\ln(b_f^2+\eta^2)\,d\eta$$
$$= \frac{1}{2}((l_f^-)^2-(b_f^2+(l_f^-)^2)\ln(b_f^2+(l_f^-)^2))$$
$$-\frac{1}{2}((l_f^+)^2-(b_f^2+(l_f^+)^2)\ln(b_f^2+(l_f^+)^2)) \tag{3.109b}$$

$$\Psi_{II}^f = \int_{l_f^-}^{l_f^+} \eta^2(b_f^2+\eta^2)\frac{\ln(b_f^2+\eta^2)-1}{2}\,d\eta$$
$$= -\frac{1}{2}[\frac{-4b_f^4 l_f^-}{15}+\frac{19b_f^2(l_f^-)^3}{45}+\frac{7(l_f^-)^5}{25}+\frac{4}{15}b_f^5\tan^{-1}(\frac{l_f^-}{b_f})$$
$$-\frac{1}{3}b_f^2(l_f^-)^3\ln(b_f^2+(l_f^-)^2)-\frac{1}{5}(l_f^-)^5\ln(b_f^2+(l_f^-)^2)]$$
$$+\frac{1}{2}[\frac{-4b_f^4 l_f^+}{15}+\frac{19b_f^2(l_f^+)^3}{45}+\frac{7(l_f^+)^5}{25}+\frac{4}{15}b_f^5\tan^{-1}(\frac{l_f^+}{b_f})$$
$$-\frac{1}{3}b_f^2(l_f^+)^3\ln(b_f^2+(l_f^+)^2)-\frac{1}{5}(l_f^+)^5\ln(b_f^2+(l_f^+)^2)] \tag{3.110a}$$

$$\Phi_{II}^f = \int_{l_f^-}^{l_f^+} \eta^2\ln(b_f^2+\eta^2)\,d\eta$$
$$= \frac{1}{9}[-6b_f^2 l_f^- + 2(l_f^-)^3 + 6b_f^2(l_f^+)^2 - 2(l_f^+)^3 + 6b_f^3\tan^{-1}(\frac{l_f^+}{b_f})$$

$$- 6b_f^3 \tan^{-1}(\frac{l_f^-}{b_f}) - 3(l_f^-)^3 \ln(b_f^2 + (l_f^-)^2) + 3(l_f^+)^3 \ln(b_f^2 + (l_f^+)^2)] \tag{3.110b}$$

Derivatives of the Potential Components

Derivatives of Ψ_0^f:

$$\frac{\partial^3 \Psi^f}{\partial (l_f^+)^3} = \frac{2b_f^4(l_f^+)^2 + 4b_f^2(l_f^+)^4 + 2(l_f^+)^6 + (b_f^6 + 3b_f^4(l_f^+)^2 + 3b_f^2(l_f^+)^4 + (l_f^+)^6)\ln[b_f^2 + (l_f^+)^2]}{(b_f^2 + (l_f^+)^2)^3} \tag{3.111a}$$

$$\frac{\partial^3 \Psi^f}{\partial (l_f^-)^3} = -\frac{2b_f^4(l_f^-)^2 + 4b_f^2(l_f^-)^4 + 2(l_f^-)^6 + (b_f^6 + 3b_f^4(l_f^-)^2 + 3b_f^2(l_f^-)^4 + (l_f^-)^6)\ln[b_f^2 + (l_f^-)^2]}{(b_f^2 + (l_f^-)^2)^3} \tag{3.111b}$$

$$\frac{\partial^3 \Psi^f}{\partial (l_f^+)^2 \partial b_f} = \frac{2b_f^5 l_f^+ + 4b_f^3(l_f^+)^3 + 2b_f(l_f^+)^5}{(b_f^2 + (l_f^+)^2)^3} \tag{3.111c}$$

$$\frac{\partial^3 \Psi^f}{\partial (l_f^-)^2 \partial b_f} = -\frac{2b_f^5 l_f^- + 4b_f^3(l_f^-)^3 + 2b_f(l_f^-)^5}{(b_f^2 + (l_f^-)^2)^3} \tag{3.111d}$$

$$\frac{\partial^3 \Psi^f}{\partial l_f^+ \partial b_f^2} = \frac{2b_f^6 + 4b_f^4(l_f^+)^2 + 2b_f^2(l_f^+)^4 + (b_f^6 + 3b_f^4(l_f^+)^2 + 3b_f^2(l_f^+)^4 + (l_f^+)^6)\ln[b_f^2 + (l_f^+)^2]}{(b_f^2 + (l_f^+)^2)^3} \tag{3.111e}$$

$$\frac{\partial^3 \Psi^f}{\partial l_f^- \partial b_f^2} = -\frac{2b_f^6 + 4b_f^4(l_f^-)^2 + 2b_f^2(l_f^-)^4 + (b_f^6 + 3b_f^4(l_f^-)^2 + 3b_f^2(l_f^-)^4 + (l_f^-)^6)\ln[b_f^2 + (l_f^-)^2]}{(b_f^2 + (l_f^-)^2)^3} \tag{3.111f}$$

Derivatives of Φ_0^f:

$$\frac{\partial \Phi^f}{\partial b_f} = 2(\tan^{-1}[\frac{l_f^+}{b_f}] - \tan^{-1}[\frac{l_f^-}{b_f}]) \tag{3.112a}$$

$$\frac{\partial \Phi^f}{\partial l_f^+} = \ln[b_f^2 + (l_f^+)^2] \tag{3.112b}$$

$$\frac{\partial \Phi^f}{\partial l_f^-} = -\ln[b_f^2 + (l_f^-)^2] \tag{3.112c}$$

Derivatives of Ψ_I^f:

$$\frac{\partial^3 \Psi_I^f}{\partial b_f^3} = \frac{2b_f^3[(l_f^-)^2 - (l_f^+)^2]}{(b_f^2 + (l_f^-)^2)(b_f^2 + (l_f^+)^2)} - 3b_f \ln[b_f^2 + (l_f^-)^2] + 3b_f \ln[b_f^2 + (l_f^+)^2] \tag{3.113a}$$

$$\frac{\partial^3 \Psi_I^f}{\partial b_f^2 \partial l_f^+} = l_f^+ [\frac{2b_f^2}{b_f^2 + (l_f^+)^2} + \ln[b_f^2 + (l_f^+)^2]] \tag{3.113b}$$

$$\frac{\partial^3 \Psi_I^f}{\partial b_f^2 \partial l_f^-} = -l_f^- [\frac{2b_f^2}{b_f^2 + (l_f^-)^2} + \ln[b_f^2 + (l_f^-)^2]] \tag{3.113c}$$

$$\frac{\partial^3 \Psi_I^f}{\partial b_f \partial (l_f^+)^2} = b_f [\frac{2(l_f^+)^2}{b_f^2 + (l_f^+)^2} + \ln[b_f^2 + (l_f^+)^2]] \tag{3.113d}$$

$$\frac{\partial^3 \Psi_I^f}{\partial b_f \partial (l_f^-)^2} = -b_f [\frac{2(l_f^-)^2}{b_f^2 + (l_f^-)^2} + \ln[b_f^2 + (l_f^-)^2]] \tag{3.113e}$$

$$\frac{\partial^3 \Psi_I^f}{\partial (l_f^+)^3} = \frac{2(l_f^+)^3}{b_f^2 + (l_f^+)^2} + 3l_f^+ \ln[b_f^2 + (l_f^+)^2] \tag{3.113f}$$

$$\frac{\partial^3 \Psi_I^f}{\partial (l_f^-)^3} = -\frac{2(l_f^-)^3}{b_f^2 + (l_f^-)^2} + 3l_f^+ \ln[b_f^2 + (l_f^-)^2] \tag{3.113g}$$

Derivatives of Φ_I^f:

$$\frac{\partial \Phi_P^f}{\partial b_f} = b_f[-\ln(b_f^2 + (l_f^-)^2) + \ln(b_f^2 + (l_f^+)^2)] \tag{3.114a}$$

$$\frac{\partial \Phi_P^f}{\partial l_f^+} = l_f^+ \ln[b_f^2 + (l_f^+)^2] \tag{3.114b}$$

$$\frac{\partial \Phi_P^f}{\partial l_f^-} = -l_f^- \ln[b_f^2 + (l_f^-)^2] \tag{3.114c}$$

Derivatives of Ψ_{II}^f:

$$\begin{aligned}\frac{\partial^3 \Psi_{II}^f}{\partial b_f^3} = {} & \frac{2b_f}{(b_f^2 + (l_f^-)^2)(b_f^2 + (l_f^+)^2)}[-4b_f^4 l_f^- - 3b_f^2 (l_f^-)^3 + 4b_f^4 l_f^+ \\ & 4b_f^2 (l_f^-)^2 l_f^+ - 4b_f^2 (l_f^+)^2 l_f^- - 3(l_f^-)^3 (l_f^+)^2 \\ & + 3b_f^2 (l_f^+)^3 + 3(l_f^-)^2 (l_f^+)^3 \\ & + 4b_f (b_f^2 + (l_f^-)^2)(b_f^2 + (l_f^+)^2)(\tan^{-1}[\frac{l_f^-}{b_f}] - \tan^{-1}[\frac{l_f^+}{b_f}])]\end{aligned} \tag{3.115a}$$

$$\frac{\partial^3 \Psi_{II}^f}{\partial b_f \partial (l_f^+)^2} = 2b_f l_f^+ (\frac{(l_f^+)^2}{b_f^2 + (l_f^+)^2} + \ln[b_f^2 + (l_f^+)^2]) \tag{3.115b}$$

$$\frac{\partial^3\Psi_{II}^f}{\partial b_f^2\partial l_f^+}=(l_f^+)^2(\frac{2b_f^2}{b_f^2+(l_f^+)^2}+\ln[b_f^2+(l_f^+)^2]) \tag{3.115c}$$

$$\frac{\partial^3\Psi_{II}^f}{\partial (l_f^+)^3}=\frac{-b_f^4-2b_f^2(l_f^+)^2+(l_f^+)^4+(b_f^4+7b_f^2(l_f^+)^2+6(l_f^+)^4)\ln[b_f^2+(l_f^+)^2]}{b_f^2+(l_f^+)^2} \tag{3.115d}$$

The derivatives with respect to l_f^- are similar to Eqs. (3.115b), (3.115c), (3.115d).

Derivatives of Φ_{II}^f:

$$\frac{\partial\Phi_{II}^f}{\partial b_f}=2b_f[-l_f^-+l_f^++b_f(\tan^{-1}[\frac{l_f^-}{b_f}]-\tan^{-1}[\frac{l_f^+}{b_f}])] \tag{3.116a}$$

$$\frac{\partial\Phi_{II}^f}{\partial l_f^+}=(l_f^+)^2\ln[b_f^2+(l_f^+)^2] \tag{3.116b}$$

Appendix 3.C Closed-form domain integral of polyhedral inclusion and their derivatives

Let le denote the variable representing either l_{JI}^+ or l_{JI}^-.

3.C.1 Partial differentiation chain rules in 3D transformed coordinate (TC)

The first-order derivative:

$$\frac{\partial\mathcal{F}(a_I,b_{JI},le)}{\partial x_i}=(-\xi_I^0)_i\frac{\partial\mathcal{F}}{\partial a_I}+(-\lambda_{JI}^0)_i\frac{\partial\mathcal{F}}{b_{JI}}+(-\eta_{JI}^0)\frac{\partial\mathcal{F}}{\partial le} \tag{3.117}$$

The second-order derivative:

$$\begin{aligned}
&\frac{\partial^2\mathcal{F}(a_I,b_{JI},le)}{\partial x_i\partial x_j}\\
&=(\xi_I^0)_i(\xi_I^0)\frac{\partial^2\mathcal{F}}{\partial a_I^2}+(\lambda_{JI})_i(\lambda_{JI})_j\frac{\partial^2\mathcal{F}}{\partial b_{JI}^2}+(\eta_{JI}^0)_i(\eta_{JI}^0)_j\frac{\partial^2\mathcal{F}}{\partial le^2}\\
&+((\xi_I^0)_i(\lambda_{JI}^0)_j+(\xi_I^0)_j(\lambda_{JI}^0)_i)\frac{\partial^2\mathcal{F}}{\partial a_I\partial b_{JI}}\\
&+((\xi_I^0)_i(\eta_{JI}^0)_j+(\xi_I^0)_j(\eta_{JI}^0)_i)\frac{\partial^2\mathcal{F}}{\partial a_I\partial le}\\
&+((\lambda_I^0)_i(\eta_{JI}^0)_j+(\lambda_I^0)_j(\eta_{JI}^0)_i)\frac{\partial^2\mathcal{F}}{\partial b_{JI}\partial le}
\end{aligned} \tag{3.118}$$

The third-order derivative:

$$
\begin{aligned}
&\frac{\partial^3 \mathcal{F}(a_I, b_{JI}, le)}{\partial x_i \partial x_j \partial x_k}\\
&= -(\xi_I^0)_i(\xi_I^0)_j(\xi_I^0)_k \frac{\partial^3 \mathcal{F}}{\partial a_I^3} - (\lambda_{JI}^0)_i(\lambda_{JI}^0)_j(\lambda_{JI}^0)_k \frac{\partial^3 \mathcal{F}}{\partial b_{JI}^3}\\
&- (\eta_{JI}^0)_i(\eta_{JI}^0)_j(\eta_{JI}^0)_k \frac{\partial^3 \mathcal{F}}{\partial le^3}\\
&- \left((\xi_I^0)_i(\xi_I^0)_j(\lambda_{JI}^0)_k + (\xi_I^0)_i(\xi_I^0)_k(\lambda_{JI}^0)_j + (\xi_I^0)_j(\xi_I^0)_k(\lambda_{JI}^0)_i\right) \frac{\partial^3 \mathcal{F}}{\partial a_I^2 \partial b_{JI}}\\
&- \left((\xi_I^0)_i(\xi_I^0)_j(\eta_{JI}^0)_k + (\xi_I^0)_i(\xi_I^0)_k(\eta_{JI}^0)_j + (\xi_I^0)_j(\xi_I^0)_k(\eta_{JI}^0)_i\right) \frac{\partial^3 \mathcal{F}}{\partial a_I^2 \partial le}\\
&- \left((\xi_I^0)_i(\lambda_I^0)_j(\lambda_{JI}^0)_k + (\xi_I^0)_i(\lambda_I^0)_k(\lambda_{JI}^0)_j + (\xi_I^0)_j(\lambda_I^0)_k(\lambda_{JI}^0)_i\right) \frac{\partial^3 \mathcal{F}}{\partial a_I \partial b_{JI}^2}\\
&- \left((\lambda_I^0)_i(\lambda_I^0)_j(\eta_{JI}^0)_k + (\lambda_I^0)_i(\lambda_I^0)_k(\eta_{JI}^0)_j + (\lambda_I^0)_j(\lambda_I^0)_k(\eta_{JI}^0)_i\right) \frac{\partial^3 \mathcal{F}}{\partial b_{JI}^2 \partial le}\\
&- \left((\xi_I^0)_i(\eta_I^0)_j(\eta_{JI}^0)_k + (\xi_I^0)_i(\eta_I^0)_k(\eta_{JI}^0)_j + (\xi_I^0)_j(\eta_I^0)_k(\eta_{JI}^0)_i\right) \frac{\partial^3 \mathcal{F}}{\partial a_I \partial le^2}\\
&- \left((\lambda_I^0)_i(\eta_I^0)_j(\eta_{JI}^0)_k + (\lambda_I^0)_i(\eta_I^0)_k(\eta_{JI}^0)_j + (\lambda_I^0)_j(\eta_I^0)_k(\eta_{JI}^0)_i\right) \frac{\partial^3 \mathcal{F}}{\partial b_{JI} \partial le^2}\\
&- \left((\xi_I^0)_i(\lambda_{JI}^0)_j(\eta_{JI}^0)_k + (\xi_I^0)_i(\lambda_{JI}^0)_k(\eta_{JI}^0)_j + (\xi_I^0)_j(\lambda_{JI}^0)_i(\eta_{JI}^0)_k\right.\\
&\left. + (\xi_I^0)_j(\lambda_{JI}^0)_k(\eta_{JI}^0)_i + (\xi_I^0)_k(\lambda_{JI}^0)_i(\eta_{JI}^0)_j + (\xi_I^0)_k(\lambda_{JI}^0)_j(\eta_{JI}^0)_i\right) \frac{\partial^3 \mathcal{F}}{\partial a_I \partial b_{JI} \partial le}
\end{aligned}
\tag{3.119}
$$

3.C.2 Expressions of the fourth derivative of Ψ_{pq} and second derivative of Φ_{pq}

$$
\begin{aligned}
\Psi_{pq,ij} &= -x_p x_q \Psi_{,ijkl} - (\delta_{pl}x_q + \delta_{ql}x_p)\Psi_{,ijk} - (\delta_{pk}\delta_{ql} + \delta_{qk}\delta_{pl})\Psi_{,ij}\\
&- (\delta_{pk}x_q + \delta_{qk}x_p)\Psi_{,ijl} - (\delta_{pj}\delta_{qk} + \delta_{qj}\delta_{pk})\Psi_{,il} - (\delta_{pj}\delta_{ql} + \delta_{pl}\delta_{qj})\Psi_{,ik}\\
&- (\delta_{pj}x_q + \delta_{qj}x_p)\Psi_{,ikl} - (\delta_{pi}\delta_{qk} + \delta_{pk}\delta_{qi})\Psi_{,jl} - (\delta_{pi}\delta_{ql} + \delta_{qi}\delta_{pl})\Psi_{,jk}\\
&- (\delta_{pi}x_q + \delta_{qi}x_p)\Psi_{,jkl} - (\delta_{pj}\delta_{qi} + \delta_{pi}\delta_{qj})\Psi_{,kl} + \delta_{ip}\Psi_{q,jkl} + \delta_{pj}\Psi_{q,ikl}\\
&+ \delta_{pk}\Psi_{q,ijl} + \delta_{pl}\Psi_{q,ijk} + x_p\Psi_{q,ijkl} + \delta_{qk}\Psi_{p,ijl} + \delta_{ql}\Psi_{p,ijk} + x_q\Psi_{p,ijkl}\\
&- \sum_{I=1}^{N_I}\sum_{J=1}^{N_{JI}} (\xi_I^0)_i(\xi_I^0)_p(\xi_I^0)_q \Big[2(\xi_I^0)_j(\xi_I^0)_k\mathcal{H}_{4,l} + 2(\xi_I^0)_k(\xi_I^0)_l\mathcal{H}_{4,j}\\
&+ 2(\xi_I^0)_j(\xi_I^0)_l\mathcal{H}_{4,k} - 2a_I(\xi_I^0)_j\mathcal{H}_{4,kl} - 2a_I(\xi_I^0)_k\mathcal{H}_{4,jl}
\end{aligned}
$$

$$
\begin{aligned}
&- 2a_I(\xi_I^0)_l\mathcal{H}_{4,jk} + a_I^2\mathcal{H}_{4,jkl}\Big] + (\xi_I^0)_i\big((\lambda_{JI}^0)_p(\xi_I^0)_q + (\lambda_{JI}^0)_q(\xi_I^0)_p\big)\\
&\times\Big[-(\xi_I^0)_j\mathcal{H}_{5,kl} - (\xi_I^0)_k\mathcal{H}_{5,jl} - (\xi_I^0)_l\mathcal{H}_{5,jk} + a_I\mathcal{H}_{5,jkl}\Big]\\
&+ (\xi_I^0)_i\big((\eta_{JI}^0)_p(\xi_I^0)_q + (\eta_{JI}^0)_q(\xi_I^0)_p\big)\\
&\times\Big[-(\xi_I^0)_j\mathcal{H}_{6,kl} - (\xi_I^0)_k\mathcal{H}_{6,jl} - (\xi_I^0)_l\mathcal{H}_{6,jk} + a_I\mathcal{H}_{6,jkl}\Big]\\
&+ (\xi_I^0)_i(\lambda_{JI}^0)_p(\lambda_{JI}^0)_q\mathcal{H}_{10,jkl} + (\xi_I^0)_i\big((\lambda_{JI}^0)_p(\eta_{JI}^0)_q + (\lambda_{JI}^0)_q(\eta_{JI}^0)_p\big)\mathcal{H}_{11,jkl}\\
&+ (\xi_I^0)_i(\eta_{JI}^0)_p(\eta_{JI}^0)_q\mathcal{H}_{12,jkl}
\end{aligned}
\tag{3.120a}
$$

$$
\begin{aligned}
\Phi_{pq,ij} &= -x_px_q\Phi_{,ij} - (\delta_{pj}x_q + \delta_{qj}x_p)\Phi_i - (\delta_{pi}x_q + \delta_{qi}x_p)\Phi_j\\
&- (\delta_{pj}\delta_{qi} + \delta_{pi}\delta_{qj})\Phi + \delta_{pi}\Phi_{q,j} + \delta_{pj}\Phi_{q,i} + \delta_{qi}\Phi_{p,j}\\
&+ \delta_{qj}\Phi_{p,i} + x_p\Phi_{q,ij} + x_q\Phi_{p,ij}\\
&- \sum_{I=1}^{N_I}\sum_{J=1}^{N_{JI}}(\xi_I^0)_i(\xi_I^0)_p(\xi_I^0)_q\Big[-2a_I(\xi_I^0)_j\mathcal{H}_1 + a_I^2\mathcal{H}_{1,j}\Big]\\
&+ (\xi_I)_i\big((\lambda_{JI}^0)_p(\xi_I^0)_q + (\lambda_{JI}^0)_q(\xi_I^0)_p\big)\Big[-(\xi_I^0)_j\mathcal{H}_2 + a_I\mathcal{H}_{2,j}\Big]\\
&+ (\xi_I)_i\big((\eta_{JI}^0)_p(\xi_I^0)_q + (\eta_{JI}^0)_q(\xi_I^0)_p\big)\Big[-(\xi_I^0)_j\mathcal{H}_3 + a_I\mathcal{H}_{3,j}\Big]\\
&+ (\xi_I^0)_i(\lambda_{JI}^0)_p(\lambda_{JI}^0)_q\mathcal{H}_{7,j} + (\xi_I^0)_i\big((\lambda_{JI}^0)_p(\eta_{JI}^0)_q + (\lambda_{JI}^0)_q(\eta_{JI}^0)_p\big)\mathcal{H}_{8,j}\\
&+ (\xi_I^0)_i(\eta_{JI}^0)_p(\eta_{JI}^0)_q\mathcal{H}_{9,j}
\end{aligned}
\tag{3.120b}
$$

3.C.3 Domain integral of Φ and Ψ and their derivatives

The harmonic potential

$$
\begin{aligned}
\Phi &= \sum_{I=1}^{N_I}\sum_{J=1}^{N_{JI}}\frac{a_I}{4}\Big(-2|a_I|(\tan^{-1}[\frac{l_{JI}^+}{b_{JI}}] - \tan^{-1}[\frac{l_{JI}^-}{b_{JI}}])\\
&+ 2b_{JI}(\ln[l_{JI}^+ + \sqrt{a_I^2 + b_{JI}^2 + (l_{JI}^+)^2}] - \ln[l_{JI}^- + \sqrt{a_I^2 + b_{JI}^2 + (l_{JI}^-)^2}])\\
&+ 2a_I(\tan^{-1}[\frac{a_I l_{JI}^+}{b_{JI}\sqrt{a_I^2 + b_{JI}^2 + (l_{JI}^+)^2}}] - \tan^{-1}[\frac{a_I l_{JI}^-}{b_{JI}\sqrt{a_I^2 + b_{JI}^2 + (l_{JI}^-)^2}}])\Big)
\end{aligned}
\tag{3.121}
$$

The biharmonic potential

$$
\begin{aligned}
\Psi &= \sum_{I=1}^{N_I}\sum_{J=1}^{N_{JI}} -\frac{a_I}{24}\Big(-b_{JI}l_{JI}^+\sqrt{a_I^2 + b_{JI}^2 + (l_{JI}^+)^2} + b_{JI}l_{JI}^-\sqrt{a_I^2 + b_{JI}^2 + (l_{JI}^-)^2}\\
&+ 2|a_I|^3(\tan^{-1}[\frac{l_{JI}^+}{b_{JI}}] - \tan^{-1}[\frac{l_{JI}^-}{b_{JI}}])
\end{aligned}
$$

$$
\begin{aligned}
& - b_{JI}(3a_I^2 + b_{JI}^2)(\tanh^{-1}[\frac{l_{JI}^{+}\sqrt{b_{JI}^2 + (l_{JI}^{+})^2}}{\sqrt{a_I^2 + b_{JI}^2 + (l_{JI}^{+})^2}}] \\
& - \tanh^{-1}[\frac{l_{JI}^{-}\sqrt{b_{JI}^2 + (l_{JI}^{-})^2}}{\sqrt{a_I^2 + b_{JI}^2 + (l_{JI}^{-})^2}}]) \\
& - 2a_I^3(\tan^{-1}[\frac{l_{JI}^{+}|a|}{b_{JI}\sqrt{a_I^2 + b_{JI}^2 + (l_{JI}^{+})^2}}] - \tan^{-1}[\frac{l_{JI}^{-}|a|}{b_{JI}\sqrt{a_I^2 + b_{JI}^2 + (l_{JI}^{-})^2}}])\Big)
\end{aligned}
\tag{3.122}
$$

3.C.4 Explicit forms of $\mathcal{F}(a_I, b_{JI}, le)$ functions

$$
\begin{aligned}
\mathcal{F}_1(a_I, b_{JI}, le) = {} & a_I \frac{b_{JI}}{|b_{JI}|} \sin^{-1}[\frac{a_I le}{\sqrt{(a_I^2 + b_{JI}^2)(b_{JI}^2 + le^2)}}] \\
& - |a_I| \tan^{-1}[\frac{le}{b_{JI}}] + b_{JI} \tanh^{-1}[\frac{le}{\sqrt{a_I^2 + b_{JI}^2 + le^2}}]
\end{aligned}
\tag{3.123a}
$$

$$
\begin{aligned}
\mathcal{F}_2(a_I, b_{JI}, le) = {} & \frac{1}{2}\Bigg[\frac{-a_I^2 le \tanh^{-1}[\frac{\sqrt{b_{JI}^2 + le^2}}{\sqrt{a_I^2 + b_{JI}^2 + le^2}}]}{\sqrt{b_{JI}^2 + le^2}} \\
& + (a_I^2 + b_{JI}^2) \tanh^{-1}[\frac{le}{\sqrt{a_I^2 + b_{JI}^2 + le^2}}]\Bigg]
\end{aligned}
\tag{3.123b}
$$

$$
\mathcal{F}_3(a_I, b_{JI}, le) = \frac{b_{JI}}{2}\Bigg[\sqrt{a_I^2 + b_{JI}^2 + le^2} + \frac{a_I^2 \tanh^{-1}[\frac{\sqrt{b_{JI}^2 + le^2}}{\sqrt{a_I^2 + b_{JI}^2 + le^2}}]}{\sqrt{b_{JI}^2 + le^2}}\Bigg]
\tag{3.123c}
$$

$$
\begin{aligned}
\mathcal{F}_4(a_I, b_{JI}, le) = {} & \frac{1}{6}\Big[b_{JI} le \sqrt{a_I^2 + b_{JI}^2 + le^2} \\
& + 2|a_I|^3(-\tan^{-1}[\frac{le}{b_{JI}}] + \tan^{-1}[\frac{le|a_I|}{b_{JI}\sqrt{a_I^2 + b_{JI}^2 + le^2}}]) \\
& + b_{JI}(3a_I^2 + b_{JI}^2) \tanh^{-1}[\frac{le}{\sqrt{a_I^2 + b_{JI}^2 + le^2}}]\Big]
\end{aligned}
\tag{3.123d}
$$

$$\mathcal{F}_5(a_I, b_{JI}, le) = \frac{1}{8\sqrt{b_{JI}^2 + le^2}}\Big[b_{JI}^2 le\sqrt{(b_{JI}^2 + le^2)(a_I^2 + b_{JI}^2 + le^2)}$$

$$- a_I^3 le|a_I| \sinh^{-1}[\frac{\sqrt{b_{JI}^2 + le^2}}{a_I}]$$

$$+ (a_I^2 + b_{JI}^2)^2\sqrt{b_{JI}^2 + le^2} \tanh^{-1}[\frac{le}{a_I^2 + b_{JI}^2 + le^2}]\Big] \quad (3.123e)$$

$$\mathcal{F}_6(a_I, b_{JI}, le) = \frac{b_{JI}}{24\sqrt{a_I^2 + b_{JI}^2 + le^2}}\Big[(a_I^2 + b_{JI}^2 + le^2)(5a_I^2 + 2b_{JI}^2 + 2le^2)$$

$$+ 3a_I^5\sqrt{\frac{a_I^2 + b_{JI}^2 + le^2}{a_I^2(b_{JI}^2 + le^2)}} \sinh^{-1}[\frac{\sqrt{b_{JI}^2 + le^2}}{a_I}]\Big] \quad (3.123f)$$

$$\mathcal{F}_7(a_I, b_{JI}, le) = \frac{1}{3}\Big[\frac{a_I^2 b_{JI} le(|a_I| - \sqrt{a_I^2 + b_{JI}^2 + le^2})}{b_{JI}^2 + le^2}$$

$$+ |a_I|^3 \tan^{-1}[\frac{le}{b_{JI}}] + b_{JI}^3 \tanh^{-1}[\frac{le}{\sqrt{a_I^2 + b_{JI}^2 + le^2}}]$$

$$- a_I^3 \tan^{-1}[\frac{a_I le}{b_{JI}\sqrt{a_I^2 + b_{JI}^2 + le^2}}]\Big] \quad (3.123g)$$

$$\mathcal{F}_8(a_I, b_{JI}, le) = \frac{b_{JI}^2}{3(b_{JI}^2 + le^2)}\Big[-|a_I|^3 + (a_I^2 + b_{JI}^2 + le^2)^{3/2}\Big] \quad (3.123h)$$

$$\mathcal{F}_9(a_I, b_{JI}, le) = \frac{1}{6}\Big[\frac{b_{JI} le}{b_{JI}^2 + le^2}(-2|a_I|^3 + \sqrt{a_I^2 + b_{JI}^2 + le^2}$$

$$\times (2a_I^2 + b_{JI}^2 + le^2)) + 2|a_I|^3 \tan^{-1}[\frac{le}{b}]$$

$$- b_{JI}(3a_I^2 + b_{JI}^2) \tanh^{-1}[\frac{le}{\sqrt{a_I^2 + b_{JI}^2 + le^2}}]$$

$$- 2a_I^3 \tan^{-1}[\frac{a_I le}{b_{JI}\sqrt{a_I^2 + b_{JI}^2 + le^2}}]\Big] \quad (3.123i)$$

$$\mathcal{F}_{10}(a_I, b_{JI}, le) = \frac{1}{15}\Big[|a_I|^5(\frac{b_{JI} le}{b_{JI}^2 + le^2} + \tan^{-1}[\frac{le}{b_{JI}}])$$

$$- \frac{a_I^4 b_{JI} le\sqrt{a_I^2 + b_{JI}^2 + le^2}}{b_{JI}^2 + le^2} + \frac{3b_{JI}^3 le\sqrt{a_I^2 + b_{JI}^2 + le^2}}{2}$$

$$+\frac{b_{JI}^3}{2}(5a_I^2+3b_{JI}^2)\tanh^{-1}[\frac{le}{\sqrt{a_I^2+b_{JI}^2+le^2}}]$$

$$-a_I^5\tan^{-1}[\frac{a_I le}{b_{JI}\sqrt{a_I^2+b_{JI}^2+le^2}}]\Big] \quad (3.123j)$$

$$\mathcal{F}_{11}(a_I,b_{JI},le)=\frac{|a_I|^5(-b_{JI}^2+le^2)+2b_{JI}^2(a_I^2+b_{JI}^2+le^2)^{5/2}}{30(b_{JI}^2+le^2)} \quad (3.123k)$$

$$\mathcal{F}_{12}(a_I,b_{JI},le)=\frac{1}{60}\Big[4|a_I|^5(\frac{-b_{JI}le}{b_{JI}^2+le^2}+\tan^{-1}[\frac{le}{b_{JI}}])$$

$$+\frac{4a_I^4 b_{JI}le\sqrt{a_I^2+b_{JI}^2+le^2}}{b_{JI}^2+le^2}$$

$$-4a_I^5\tan^{-1}[\frac{a_I le}{b_{JI}\sqrt{a_I^2+b_{JI}^2+le^2}}]$$

$$+\frac{b_{JI}le\sqrt{a_I^2+b_{JI}^2+le^2}}{2}(7a_I^2+3b_{JI}^2+6le^2)$$

$$-\frac{b_{JI}}{2}(15a_I^4+10a_I^2b_{JI}^2+3b_{JI}^4)\tanh^{-1}[\frac{le}{\sqrt{a_I^2+b_{JI}^2+le^2}}]\Big] \quad (3.123l)$$

Partial derivatives of the integrand functions $\mathcal{F}_1,\ldots,\mathcal{F}_{12}$

The implementation of Eshelby's tensor for displacement and strain only requires the original and first-order derivatives of harmonic potentials:

$$\frac{\partial F_1}{\partial a_I}=\sin^{-1}[\frac{a_I le|b_{JI}|}{b\sqrt{(a_I^2+b_{JI}^2)(b_{JI}^2+le^2)}}]-\tan^{-1}[\frac{le}{b_{JI}}]\frac{|a_I|}{a_I} \quad (3.124a)$$

$$\frac{\partial F_1}{\partial b_{JI}}=\frac{le(|a_I|-\sqrt{a_I^2+b_{JI}^2+le^2})}{b_{JI}^2+le^2}+\tanh^{-1}[\frac{le}{\sqrt{a_I^2+b_{JI}^2+le^2}}] \quad (3.124b)$$

$$\frac{\partial F_1}{\partial le}=\frac{b_{JI}}{b_{JI}^2+le^2}(-|a_I|+\sqrt{a_I^2+b_{JI}^2+le^2}) \quad (3.124c)$$

$$\frac{\partial F_2}{\partial a_I}=a_I\Big(\tanh^{-1}[\frac{le}{\sqrt{a_I^2+b_{JI}^2+le^2}}]-\frac{le}{\sqrt{b_{JI}^2+le^2}}\tanh^{-1}[\frac{\sqrt{b_{JI}^2+le^2}}{\sqrt{a_I^2+b_{JI}^2}}]\Big) \quad (3.125a)$$

$$\frac{\partial F_2}{\partial b_{JI}} = \frac{b_{JI}}{2}\Big(-\frac{le\sqrt{a_I^2+b_{JI}^2+le^2}}{b_{JI}^2+le^2} + 2\tanh[\frac{le}{\sqrt{a_I^2+b_{JI}^2+le^2}}]$$

$$+ \frac{a_I^2 \tanh^{-1}[\sqrt{\frac{b_{JI}^2+le^2}{a_I^2+b_{JI}^2+le^2}}]}{(b_{JI}^2+le^2)^(3/2)}\Big) \tag{3.125b}$$

$$\frac{\partial F_2}{\partial le} = \frac{b_{JI}^2\sqrt{a_I^2+b_{JI}^2+le^2}\Big(b_{JI}^2+le^2-a_I^2\sqrt{\frac{b_{JI}^2+le^2}{a_I^2+b_{JI}^2+le^2}}\tanh^{-1}[\sqrt{\frac{b_{JI}^2+le^2}{a_I^2+b_{JI}^2+le^2}}]\Big)}{2\left(b_{JI}^2+le^2\right)^2} \tag{3.125c}$$

$$\frac{\partial F_3}{\partial a_I} = \frac{a_I b_{JI}\tanh^{-1}[\sqrt{\frac{b_{JI}^2+le^2}{a_I^2+b_{JI}^2+le^2}}]}{\sqrt{b_{JI}^2+le^2}} \tag{3.126a}$$

$$\frac{\partial F_3}{\partial b_{JI}} = \frac{\sqrt{(b_{JI}^2+le^2)(a_I^2+b_{JI}^2+le^2)}(2b_{JI}^2+le^2)+a_I^2 le^2\tanh^{-1}[\sqrt{\frac{b_{JI}^2+le^2}{a_I^2+b_{JI}^2+le^2}}]}{2\left(b_{JI}^2+le^2\right)^{3/2}} \tag{3.126b}$$

$$\frac{\partial F_3}{\partial le} = \frac{b_{JI}le}{2(b_{JI}^2+le^2)^{3/2}}\Big(\sqrt{(b_{JI}^2+le^2)(a_I^2+b_{JI}^2+le^2)}$$

$$-a_I^2\tanh^{-1}[\sqrt{\frac{b_{JI}^2+le^2}{a_I^2+b_{JI}^2+le^2}}]\Big) \tag{3.126c}$$

$$\frac{\partial F_7}{\partial a_I} = a_I\Big(-\frac{b_{JI}le\sqrt{a_I^2+b_{JI}^2+le^2}}{b_{JI}^2+le^2} + |a_I|\big(\frac{b_{JI}le}{b_{JI}^2+le^2}+\tan^{-1}[\frac{le}{b_{JI}}]\big)$$

$$-a_I\tan^{-1}[\frac{a_I le}{b_{JI}\sqrt{a_I^2+b_{JI}^2+le^2}}]\Big) \tag{3.127a}$$

$$\frac{\partial F_7}{\partial b_{JI}} = \frac{b_{JI}^2}{3(b_{JI}^2+le^2)^2}\Big(-le(-2a_I^2+b_{JI}^2+le^2)\sqrt{a_I^2+b_{JI}^2+le^2}-2le|a_I|^3$$

$$+3(b_{JI}^2+le^2)\tanh^{-1}[\frac{le}{\sqrt{a_I^2+b_{JI}^2+le^2}}]\Big) \tag{3.127b}$$

$$\frac{\partial F_7}{\partial le} = \frac{b_{JI}^3}{3(b_{JI}^2 + le^2)\sqrt{a_I^2 + b_{JI}^2 + le^2}} \times \left(-2a_I^4 - a_I^2(b_{JI}^2 + le^2) + (b_{JI}^2 + le^2)^2 + 2\sqrt{a_I^2 + b_{JI}^2 + le^2}|a_I|^3\right) \tag{3.127c}$$

$$\frac{\partial F_8}{\partial a_I} = \frac{a_I b_{JI}^2}{b_{JI}^2 + le^2}\left(\sqrt{a_I^2 + b_{JI}^2 + le^2} - |a_I|\right) \tag{3.128a}$$

$$\frac{\partial F_8}{\partial b_{JI}} = \frac{b_{JI}}{3(b_{JI}^2 + le^2)} \times \left(\sqrt{a_I^2 + b_{JI}^2 + le^2}(3b_{JI}^4 + 5b_{JI}^2 le^2 + 2le^2(a_I^2 + le^2)) - 2le^2|a_I|^3\right) \tag{3.128b}$$

$$\frac{\partial F_8}{\partial le} = \frac{b_{JI}^2 le}{3(b_{JI}^2 + le^2)}\left(\sqrt{a_I^2 + b_{JI}^2 + le^2}(-2a_I^2 + b_{JI}^2 + le^2) + 2|a_I|^3\right) \tag{3.128c}$$

$$\begin{aligned}\frac{\partial F_9}{\partial a_I} = \frac{-a_I}{b_{JI}^2 + le^2}\Bigg(&-b_{JI} le\sqrt{a_I^2 + b_{JI}^2 + le^2} \\ &+ |a_I|(b_{JI} le - (b_{JI}^2 + le^2)\tan^{-1}[\frac{le}{b_{JI}}]) \\ &+ (b_{JI}^2 + le^2)(a_I \tan^{-1}[\frac{a_I le}{b_{JI}\sqrt{a_I^2 + b_{JI}^2 + le^2}}] \\ &+ b_{JI}\tanh^{-1}[\frac{le}{\sqrt{a_I^2 + b_{JI}^2 + le^2}}])\Bigg)\end{aligned} \tag{3.129a}$$

$$\begin{aligned}\frac{\partial F_9}{\partial b_{JI}} = \frac{1}{6(b_{JI}^2 + le^2)} &\times \Big(le\sqrt{a_I^2 + b_{JI}^2 + le^2}(3b_{JI}^2 + 4(a_I^2 + b_{JI}^2)le^2 + le^4) - 4le^3|a_I|^3 \\ &- 3(a_I^2 + b_{JI}^2)(b_{JI}^2 + le^2)^2\tanh^{-1}[\frac{le}{\sqrt{a_I^2 + b_{JI}^2 + le^2}}]\Big)\end{aligned} \tag{3.129b}$$

$$\frac{\partial F_9}{\partial le} = \frac{b_{JI} le^2}{3(b_{JI}^2 + le^2)\sqrt{a_I^2 + b_{JI}^2 + le^2}}$$

$$\times\left(-2a_I^4-a_I^2(b_{JI}^2+le^2)+(b_{JI}^2+le^2)^2+2\sqrt{a_I^2+b_{JI}^2+le^2}|a_I|^3\right) \tag{3.129c}$$

The implementation of Eshelby's tensor for displacement and strain only requires the second- and third-order derivatives of biharmonic potentials, which are provided as follows.

(1) Second-order partial derivatives:

$$\frac{\partial^2 F_4}{\partial a_I^2}=2a_I\left(\frac{-a_I}{|a_I|}\tan^{-1}[\frac{le}{b_{JI}}]+b_{JI}\tanh^{-1}[\frac{le}{\sqrt{a_I^2+b_{JI}^2+le^2}}]\right) \tag{3.130a}$$

$$\frac{\partial^2 F_4}{\partial a_I\partial b_{JI}}=a_I\left[\tanh^{-1}[\frac{le}{\sqrt{a_I^2+b_{JI}^2+le^2}}]+\frac{le}{a_I+\frac{a_I}{|a_I|}\sqrt{a_I^2+b_{JI}^2+le^2}}\right] \tag{3.130b}$$

$$\frac{\partial^2 F_4}{\partial a_I\partial le}=\frac{a_Ib_{JI}\left(\sqrt{a_I^2+b_{JI}^2+le^2}-|a_I|\right)}{b_{JI}^2+le^2} \tag{3.130c}$$

$$\begin{aligned}\frac{\partial^2 F_4}{\partial b_{JI}^2}&=\frac{b_{JI}}{3(b_{JI}^2+le^2)^2}\\&\quad\times\left[-le(-2a_I^2+b_{JI}^2+le^2)\sqrt{a_I^2+b_{JI}^2+le^2}-2le|a_I|^3\right]\\&\quad+b_{JI}\tanh^{-1}[\frac{le}{\sqrt{a_I^2+b_{JI}^2+le^2}}]\end{aligned} \tag{3.130d}$$

$$\begin{aligned}\frac{\partial^2 F_4}{\partial b_{JI}\partial le}&=\frac{1}{3(b_{JI}^2+le^2)}\Big[\sqrt{a_I^2+b_{JI}^2+le^2}(2b_{JI}^4+3b_{JI}^2le^2+le^4\\&\quad+a_I^2(le^2-b_{JI}^2))+(b_{JI}^2-le^2)|a_I|\Big]\end{aligned} \tag{3.130e}$$

$$\frac{\partial^2 F_4}{\partial le^2}=b_{JI}le\left[\frac{-2a_I^4-a_I^2(b_{JI}^2+le^2)+(b_{JI}^2+le^2)^2+2|a_I|^3\sqrt{a_I^2+b_{JI}^2+le^2}}{3(b_{JI}^2+le^2)^2\sqrt{a_I^2+b_{JI}^2+le^2}}\right] \tag{3.130f}$$

$$\begin{aligned}\frac{\partial^2 F_5}{\partial a_I^2}&=\frac{1}{2}\Bigg(-\frac{3a_Ile|a_I|\sinh^{-1}[\frac{\sqrt{b_{JI}^2+le^2}}{a_I}]}{\sqrt{b_{JI}^2+le^2}}\\&\quad+(3a_I^2+b_{JI}^2)\tanh^{-1}[\frac{le}{\sqrt{a_I^2+b_{JI}^2+le^2}}]\Bigg)\end{aligned} \tag{3.131a}$$

$$\frac{\partial^2 F_5}{\partial a_I \partial b_{JI}} = \frac{a_I b_{JI}}{2(b_{JI}^2 + le^2)^{3/2}} \Big(-le\sqrt{(b_{JI}^2 + le^2)(a_I^2 + b_{JI}^2 + le^2)} + a_I le |a_I| \sinh^{-1}[\frac{b_{JI}}{a_I}] \Big) + a_I b_{JI} \tanh^{-1}[\frac{le}{\sqrt{a_I^2 + b_{JI}^2 + le^2}}] \tag{3.131b}$$

$$\frac{\partial^2 F_5}{\partial a_I \partial le} = \frac{a_I b_{JI}^2 \Big(\sqrt{(b_{JI}^2 + le^2)(a_I^2 + b_{JI}^2 + le^2)} - a_I |a_I| \sinh^{-1}[\frac{\sqrt{b_{JI}^2 + le^2}}{a_I}] \Big)}{2\left(b_{JI}^2 + le^2\right)^{3/2}} \tag{3.131c}$$

$$\frac{\partial^2 F_5}{\partial b_{JI}^2} = \frac{1}{8} \Big(\frac{le\sqrt{a_I^2 + b_{JI}^2 + le^2}(a_I^2(2b_{JI}^2 - le^2) + 2le^2(b_{JI}^2 + le^2))}{(b_{JI}^2 + le^2)^2} + \frac{a_I^3 le(-2b_{JI}^2 + le^2)|a_I| \sinh^{-1}[\frac{a_I}{\sqrt{b_{JI}^2 + le^2}}]}{(b_{JI}^2 + le^2)^{5/2}} + 4(a_I^2 + 3b_{JI}^2) \tanh^{-1}[\frac{le}{\sqrt{a_I^2 + b_{JI}^2 + le^2}}] \Big) \tag{3.131d}$$

$$\frac{\partial^2 F_5}{\partial b_{JI} \partial le} = \frac{b_{JI}}{8(b_{JI}^2 + le^2)^{5/2}} \Big(\sqrt{(b_{JI}^2 + le^2)(a_I^2 + b_{JI}^2 + le^2)} \times (-a_I^2 b_{JI}^2 + 6b_{JI}^4 + 2(a_I^2 + 5b_{JI}^2)le^2 + 4b_{JI}^4) + a_I^3 (b_{JI}^2 - 2le^2)|a_I| \sinh^{-1}[\frac{a_I}{\sqrt{b_{JI}^2 + le^2}}] \Big) \tag{3.131e}$$

$$\frac{\partial^2 F_5}{\partial le^2} = \frac{b_{JI}^2 le}{8(b_{JI}^2 + le^2)^{5/2}} \Big(\sqrt{(b_{JI}^2 + le^2)(a_I^2 + b_{JI}^2 + le^2)} \times (-3a_I^2 + 2(b_{JI}^2 + le^2)) + 3a_I^3 |a_I| \sinh^{-1}[\frac{\sqrt{b_{JI}^2 + le^2}}{a_I}] \Big) \tag{3.131f}$$

$$\frac{\partial^2 F_6}{\partial a_I^2} = \frac{b_{JI}}{2\sqrt{a_I^2 + b_{JI}^2 + le^2}} \Big(a_I^2 + b_{JI}^2 + le^2 + 3a_I \sqrt{1 + \frac{a_I^2}{b_{JI}^2 + le^2}} |a_I| \sinh^{-1}[\frac{\sqrt{b_{JI}^2 + le^2}}{a_I}] \Big) \tag{3.132a}$$

$$\frac{\partial^2 F_6}{\partial a_I \partial b_{JI}} = \frac{a_I\sqrt{a_I^2+b_{JI}^2+le^2}(2b_{JI}^2+le^2)}{2(b_{JI}^2+le^2)} + \frac{a_I^4 le^2 \sinh^{-1}[\frac{\sqrt{b_{JI}^2+le^2}}{a_I}]}{2(b_{JI}^2+le^2)^{3/2}|a_I|} \tag{3.132b}$$

$$\frac{\partial^2 F_6}{\partial a_I \partial le} = \frac{a_I b_{JI} le}{2(b_{JI}^2+le^2)} - \frac{a_I^2|a_I|b_{JI}le}{2(b_{JI}^2+le^2)^{3/2}}\sinh^{-1}[\frac{\sqrt{b_{JI}^2+le^2}}{a_I}] \tag{3.132c}$$

$$\frac{\partial^2 F_6}{\partial b_{JI}^2} = \frac{b_{JI}(a_I^2+b_{JI}^2+le^2)}{8(b_{JI}^2+le^2)^2}\left(8b_{JI}^4+(3a_I^2+14b_{JI}^2)le^2+6le^4\right)$$
$$- \frac{3a_I^3|a_I|b_{JI}le^2}{8(b_{JI}^2+le^2)^{5/2}}\sinh^{-1}[\frac{\sqrt{b_{JI}^2+le^2}}{a_I}] \tag{3.132d}$$

$$\frac{\partial^2 F_6}{\partial b_{JI} \partial le} = \frac{le\sqrt{a_I^2+b_{JI}^2+le^2}}{8(b_{JI}^2+le^2)^2}(a_I^2(-2b_{JI}^2+le^2)+2(b_{JI}^2+le^2)(2b_{JI}^2+le^2))$$
$$- \frac{le a_I^3|a_I|(-2b_{JI}^2+le^2)}{8(b_{JI}^2+le^2)^{5/2}}\sinh^{-1}[\frac{\sqrt{b_{JI}^2+le^2}}{a_I}] \tag{3.132e}$$

$$\frac{\partial^2 F_6}{\partial le^2} = \frac{b_{JI}\sqrt{a_I^2+b_{JI}^2+le^2}}{8(b_{JI}^2+le^2)^2}(a_I^2(b_{JI}^2-2le^2)+2(b_{JI}^2+le^2)(b_{JI}^2+2le^2))$$
$$- \frac{b_{JI}a_I^3|a_I|(b_{JI}^2-2le^2)}{8(b_{JI}^2+le^2)^{5/2}}\sinh^{-1}[\frac{\sqrt{b_{JI}^2+le^2}}{a_I}] \tag{3.132f}$$

(2) Third-order partial derivatives:

$$\frac{\partial^3 F_4}{\partial a_I^3} = \frac{a_I b_{JI} le}{(a_I^2+b_{JI}^2)\sqrt{a_I^2+b_{JI}^2+le^2}} - \frac{2a_I}{|a_I|}\tan^{-1}[\frac{le}{b_{JI}}]$$
$$+ 2\tan^{-1}[\frac{a_I le}{b_{JI}\sqrt{a_I^2+b_{JI}^2+le^2}}] \tag{3.133a}$$

$$\frac{\partial^3 F_4}{\partial a_I^2 \partial b_{JI}} = \frac{-le\left(2a_I^2+b_{JI}^2+\frac{a_I b_{JI}^2}{\sqrt{a_I^2+b_{JI}^2+le^2}}\right)}{(a_I^2+b_{JI}^2)(-a_I+\sqrt{a_I^2+b_{JI}^2+le^2})}$$
$$+ \tanh^{-1}[\frac{le}{\sqrt{a_I^2+b_{JI}^2+le^2}}] \tag{3.133b}$$

$$\frac{\partial^3 F_4}{\partial a_I^2 \partial le} = \frac{b_{JI}}{b_{JI}^2 + le^2}\left(\frac{2a_I^2 + b_{JI}^2 + le^2}{\sqrt{a_I^2 + b_{JI}^2 + le^2}} - 2|a_I|\right) \tag{3.133c}$$

$$\frac{\partial^3 F_4}{\partial a_I \partial b_{JI}^2} = \frac{a_I b_{JI} le\left(2a_I^4 - le^2(b_{JI}^2 + le^2) + a_I^2(3b_{JI}^2 + le^2) - 2(a_I^2 + b_{JI}^2)\sqrt{a_I^2 + b_{JI}^2 + le^2}\right)}{(a_I^2 + b_{JI}^2)(b_{JI}^2 + le^2)\sqrt{a_I^2 + b_{JI}^2 + le^2}} \tag{3.133d}$$

$$\frac{\partial^3 F_4}{\partial a_I \partial b_{JI} \partial le} = \frac{a_I\left[-a_I^2 b_{JI}^2 + (a_I^2 + b_{JI}^2)le^2 + le^4 + (b_{JI}^2 + le^2)\sqrt{a_I^2 + b_{JI}^2 + le^2}|a_I|\right]}{(b_{JI}^2 + le^2)\sqrt{a_I^2 + b_{JI}^2 + le^2}} \tag{3.133e}$$

$$\begin{aligned}\frac{\partial^3 F_4}{\partial a_I \partial le^2} &= \frac{-a_I b_{JI} le}{(b_{JI}^2 + le^2)^2\sqrt{a_I^2 + b_{JI}^2 + le^2}} \\ &\quad \times \left[2a_I^2 + b_{JI}^2 + le^2 - 2\sqrt{a_I^2 + b_{JI}^2 + le^2}|a_I|\right]\end{aligned} \tag{3.133f}$$

$$\begin{aligned}\frac{\partial^3 F_4}{\partial b_{JI}^3} &= \frac{2a_I^6 le(-3a_I^2 + le^2) - b_{JI}^2 le(b_{JI}^2 + le^2)^2(3b_{JI}^2 + 4le^2) - a_I^2 le(b_{JI}^2 + le^2)(3b_{JI}^4 + le^4)}{3(a_I^2 + b_{JI}^2)(b_{JI}^2 + le^2)^3\sqrt{a_I^2 + b_{JI}^2 + le^2}} \\ &\quad + \frac{a_I^4(-9b_{JI}^4 + le^5)}{3(a_I^2 + b_{JI}^2)(b_{JI}^2 + le^2)^3\sqrt{a_I^2 + b_{JI}^2 + le^2}} \\ &\quad - \frac{2le(-3b_{JI}^2 + le^2)|a_I|^3}{3(b_{JI}^2 + le^2)^3} + \tanh^{-1}\left[\frac{le}{\sqrt{a_I^2 + b_{JI}^2 + le^2}}\right]\end{aligned} \tag{3.133g}$$

$$\begin{aligned}\frac{\partial^3 F_4}{\partial b_{JI}^2 \partial le} &= \frac{b_{JI}}{3(a_I^2 + b_{JI}^2)(b_{JI}^2 + le^2)^3\sqrt{a_I^2 + b_{JI}^2 + le^2}} \\ &\quad \times \left[(b_{JI}^2 - 3le^2)(2a_I^4 + a_I^2 b_{JI}^2 + a_I^2 le^2)\right. \\ &\quad + (b_{JI}^2 + le^2)^2(2b_{JI}^2 + 3le^2) \\ &\quad \left. - 2(b_{JI}^2 - 3le^2)\sqrt{a_I^2 + b_{JI}^2 + le^2}|a_I|^3\right]\end{aligned} \tag{3.133h}$$

$$\begin{aligned}\frac{\partial^3 F_4}{\partial b_{JI} \partial le^2} &= \frac{b_{JI}}{3(a_I^2 + b_{JI}^2)(b_{JI}^2 + le^2)^3\sqrt{a_I^2 + b_{JI}^2 + le^2}} \\ &\quad \times \left[-(2a_I^4 + a_I^2 b_{JI}^2 + a_I^2 le^2)(b_{JI}^2 - 3le^2)\right. \\ &\quad \left. + b_{JI}^2(b_{JI}^2 + le^2)^2 + 2(b_{JI}^2 - 3le^2)\sqrt{a_I^2 + b_{JI}^2 + le^2}|a_I|^3\right]\end{aligned} \tag{3.133i}$$

$$\frac{\partial^3 F_4}{\partial le^3} = \frac{-a_I b_{JI} le}{(b_{JI}^2 + le^2)^2\sqrt{a_I^2 + b_{JI}^2 + le^2}} \times \left[2a_I^2 + b_{JI}^2 + le^2 - 2\sqrt{a_I^2 + b_{JI}^2 + le^2}|a_I|\right] \tag{3.133j}$$

$$\frac{\partial^3 F_5}{\partial a_I^3} = -\frac{a_I le(4a_I^4 + (5a_I^2 + b_{JI}^2)(b_{JI}^2 + le^2))}{4(a_I^2 + b_{JI}^2)(a_I^2 + b_{JI}^2 + le^2)^{3/2}} - \frac{3le|a_I|\sinh^{-1}[\frac{\sqrt{b_{JI}^2+le^2}}{a_I}]}{2\sqrt{b_{JI}^2 + le^2}} + 3a_I\tanh^{-1}[\frac{le}{\sqrt{a_I^2 + b_{JI}^2 + le^2}}] \tag{3.134a}$$

$$\frac{\partial^3 F_5}{\partial a_I^2 \partial b_{JI}} = \frac{b_{JI}}{4}\Big(\frac{a_I^2 le}{(a_I^2 + b_{JI}^2 + le^2)^{3/2}} - \frac{2le(3a_I^2 + b_{JI}^2)}{(a_I^2 + b_{JI}^2)\sqrt{a_I^2 + b_{JI}^2 + le^2}} - \frac{3a_I^2 le}{\sqrt{a_I^2 + b_{JI}^2 + le^2}(b_{JI}^2 + le^2)} + \frac{3a_I le|a_I|\sinh^{-1}[\frac{\sqrt{b_{JI}^2+le^2}}{a_I}]}{(b_{JI}^2 + le^2)^{3/2}} + 4\tanh^{-1}[\frac{le}{\sqrt{a_I^2 + b_{JI}^2 + le^2}}]\Big) \tag{3.134b}$$

$$\frac{\partial^3 F_5}{\partial a_I^2 \partial le} = \frac{1}{8((b_{JI}^2 + le^2)(a_I^2 + b_{JI}^2 + le^2))^{3/2}} \times \Big(2\sqrt{b_{JI}^2 + le^2}\big(2b_{JI}^2(b_{JI}^2 + le^2)^2 + a_I^4(5b_{JI}^2 + 2le^2) + a_I^2(b_{JI}^2 + le^2)(7b_{JI}^2 + 3le^2)\big) - 6a_I b_{JI}^2(a_I^2 + b_{JI}^2 + le^2)^{3/2}|a_I|\sinh^{-1}[\frac{\sqrt{b_{JI}^2 + le^2}}{a_I}]\Big) \tag{3.134c}$$

$$\frac{\partial^3 F_5}{\partial a_I \partial b_{JI}^2} = \frac{a_I b_{JI}}{2}\Big(\frac{2b_{JI}^4 + 5b_{JI}^2 le^2 + 3le^2(a_I^2 + le^2)}{(b_{JI}^2 + le^2)^2\sqrt{a_I^2 + b_{JI}^2 + le^2}} - \frac{3a_I le^2|a_I|\sinh^{-1}[\frac{\sqrt{b_{JI}^2+le^2}}{a_I}]}{(b_{JI}^2 + le^2)^{5/2}}\Big) \tag{3.134d}$$

$$\frac{\partial^3 F_5}{\partial a_I \partial b_{JI} \partial le} = \frac{a_I le(-2a_I^2 + b_{JI}^2) + (a_I^2 + b_{JI}^2 le^2 + le^4)}{(b_{JI}^2 + le^2)^2\sqrt{a_I^2 + b_{JI}^2 + le^2}}$$

$$-\frac{a_I^2|a_I|le(-2b_{JI}^2+le^2)}{(b_{JI}^2+le^2)^{5/2}}\sinh^{-1}[\frac{\sqrt{b_{JI}^2+le^2}}{a_I}] \tag{3.134e}$$

$$\frac{\partial^3 F_5}{\partial a_I \partial le^2}=\frac{-a_I b_{JI}^2 le\sqrt{\frac{b_{JI}^2+le^2}{a_I^2+b_{JI}^2+le^2}}(3a_I^2+b_{JI}^2+le^2)-3a_I|a_I|\sinh^{-1}[\frac{\sqrt{b_{JI}^2+le^2}}{a_I}]}{2(b_{JI}^2+le^2)^{5/2}} \tag{3.134f}$$

$$\begin{aligned}\frac{\partial^3 F_5}{\partial b_{JI}^3}=&\frac{b_{JI}}{8}\Big(\frac{3a_I^3 le(2b_{JI}^2-3le^2)|a_I|\sinh^{-1}[\frac{\sqrt{b_{JI}^2+le^2}}{a_I}]}{(b_{JI}^2+le^2)^{7/2}}\\&+24\tanh^{-1}[\frac{le}{\sqrt{a_I^2+b_{JI}^2+le^2}}]\\&-\frac{le}{(a_I^2+b_{JI}^2)(b_{JI}^2+le^2)^3\sqrt{a_I^2+b_{JI}^2+le^2}}\\&\times\big[a_I^6(6b_{JI}^2-9le^2)+a_I^4(2b_{JI}^2-le^4)(4b_{JI}^2+le^2)\\&+2b_{JI}^2(b_{JI}^2+le^2)^2(6b_{JI}^2+7le^2)\\&+a_I^2(b_{JI}^2+le^2)(6b_{JI}^4+7b_{JI}^2le^2+6le^4)\big]\Big)\end{aligned} \tag{3.134g}$$

$$\begin{aligned}\frac{\partial^3 F_5}{\partial b_{JI}^2\partial le}=&\frac{1}{8(b_{JI}^2+le^2)^{7/2}}\\&\times\Big(\sqrt{\frac{b_{JI}^2+le^2}{a_I^2+b_{JI}^2+le^2}}(a_I^4(2b_{JI}^4-11b_{JI}^2le^2+2le^4)\\&+2(b_{JI}^2+le^2)^2(6b_{JI}^4+9b_{JI}^2le^2+2le^4)\\&+a_I^2(b_{JI}^2+le^2)(6b_{JI}^4+7b_{JI}^2le^2+6le^4))\\&+a_I^3(-2b_{JI}^4+11b_{JI}^2le^2-2le^4)|a_I|\sinh^{-1}[\frac{\sqrt{b_{JI}^2+le^2}}{a_I}]\Big)\end{aligned} \tag{3.134h}$$

$$\begin{aligned}\frac{\partial^3 F_5}{\partial b_{JI}\partial le^2}=&\frac{b_{JI}le}{8}\Big(\frac{a_I^4(9b_{JI}^2-6b_{JI}^2)+2(b_{JI}^2+le^2)^2(b_{JI}^2+2le^2)+a_I^2(3b_{JI}^4+b_{JI}^2le^2-2le^4)}{(b_{JI}^2+le^2)^3\sqrt{a_I^2+b_{JI}^2+le^2}}\\&+\frac{3a_I^3(-3b_{JI}^2+2le^2)|a_I|\sinh^{-1}[\frac{\sqrt{b_{JI}^2+le^2}}{a_I}]}{(b_{JI}^2+le^2)^{7/2}}\Big)\end{aligned} \tag{3.134i}$$

$$\frac{\partial^3 F_5}{\partial le^3}=\frac{b_{JI}^2}{8}\Big(\frac{-3a_I^4(b_{JI}^2-4le^2)+2b_{JI}^2(b_{JI}^2+le^2)^2+a_I^2(-b_{JI}^4+3b_{JI}^2le^2+4le^4)}{(b_{JI}^2+le^2)^3\sqrt{a_I^2+b_{JI}^2+le^2}}$$

$$+\frac{3a_I^3(b_I^2-4le^2)|a_I|\sinh^{-1}[\frac{\sqrt{b_{JI}^2+le^2}}{a_I}]}{(b_{JI}^2+le^2)^{7/2}}\Bigg) \tag{3.134j}$$

$$\frac{\partial^3 F_6}{\partial a_I^3}=-\frac{a_I b_{JI}}{(b_{JI}^2+le^2)\sqrt{a_I^2+b_{JI}^2+le^2}}(b_{JI}^2+le^2)$$
$$-3\frac{|a_I|b_{JI}}{\sqrt{b_{JI}^2+le^2}}\sinh^{-1}[\frac{\sqrt{b_{JI}^2+le^2}}{a_I}] \tag{3.135a}$$

$$\frac{\partial^3 F_6}{\partial a_I^2\partial b_{JI}}=\frac{2b_{JI}^4+3b_{JI}^2le^2+le^4+a_I^2(4b_{JI}^2+le^2)}{2(b_{JI}^2+le^2)\sqrt{a_I^2+b_{JI}^2+le^2}}$$
$$+\frac{3a_I|a_I|le^2\sinh^{-1}[\frac{\sqrt{b_{JI}^2+le^2}}{a_I}]}{2(b_{JI}^2+le^2)^{3/2}} \tag{3.135b}$$

$$\frac{\partial^3 F_6}{\partial a_I\partial b_{JI}\partial le}=\frac{1}{2(b_{JI}^2+le^2)^3\sqrt{a_I^2+b_{JI}^2+le^2}}$$
$$\times\Big(ale(b_{JI}^2+le^2)(-2a_I^2b_{JI}^2+(a_I^2+b_{JI}^2)le^2+le^4)$$
$$-a_I^2|a_I|le(-2b_{JI}^2+le^2)$$
$$\times\sqrt{(b_{JI}^2+le^2)(a_I^2+b_{JI}^2+le^2)}\sinh^{-1}[\frac{\sqrt{b_{JI}^2+le^2}}{a_I}]\Big) \tag{3.135c}$$

$$\frac{\partial^3 F_6}{\partial a_I\partial le^2}=\frac{a_I b_{JI}}{2(b_{JI}^2+le^2)^3\sqrt{a_I^2+b_{JI}^2+le^2}}$$
$$\times\Big((b_{JI}^2+le^2)(a_I^2(b_{JI}^2-2le^2)+b_{JI}^2(b_{JI}^2+le^2))$$
$$-a_I|a_I|(b_{JI}^2-2le^2)$$
$$\times\sqrt{(b_{JI}^2+le^2)(a_I^2+b_{JI}^2+le^2)}\sinh^{-1}[\frac{\sqrt{b_{JI}^2+le^2}}{a_I}]\Big) \tag{3.135d}$$

$$\frac{\partial^3 F_6}{\partial b_{JI}^3}=\frac{1}{8(b_{JI}^2+le^2)^4\sqrt{a_I^2+b_{JI}^2+le^2}}$$
$$\times\Big((b_{JI}^2+le^2)(3a_I^4le^2(-4b_{JI}^2+le^2)$$

$$+2(b_{JI}^2+le^2)^2(8b_{JI}^4+12b_{JI}^2le^2+3le^4))$$
$$+3a_I^3|a_I|le^2(-4b_{JI}^2+le^2)$$
$$\times\sqrt{(b_{JI}^2+le^2)(a_I^2+b_{JI}^2+le^2)}\sinh^{-1}[\frac{\sqrt{b_{JI}^2+le^2}}{a_I}]\Big) \quad (3.135e)$$

$$\frac{\partial^3 F_6}{\partial b_{JI}^2\partial le}=\frac{b_{JI}le}{8(b_{JI}^2+le^2)^4\sqrt{a_I^2+b_{JI}^2+le^2}}\Big((b_{JI}^2+le^2)(a_I^4(6b_{JI}^2-9le^2)$$
$$+a_I^2(2b_{JI}^2-3le^2)(b_{JI}^2+le^2)+2(b_{JI}^2+le^2)^2(2b_{JI}^2+3le^2))$$
$$+3a_I^3|a_I|\sqrt{(b_{JI}^2+le^2)(a_I^2+b_{JI}^2+le^2)}$$
$$\times(-2b_{JI}^2+3le^2)\sinh^{-1}[\frac{\sqrt{b_{JI}^2+le^2}}{a_I}]\Big) \quad (3.135f)$$

$$\frac{\partial^3 F_6}{\partial b_{JI}\partial le^2}=\frac{1}{8(b_{JI}^2+le^2)^4\sqrt{a_I^2+b_{JI}^2+le^2}}$$
$$\times\Big((b_{JI}^2+le^2)(a_I^4(6b_{JI}^2-9le^2)$$
$$+a_I^2(2b_{JI}^2-3le^2)(b_{JI}^2+le^2)+2(b_{JI}^2+le^2)(2b_{JI}^2+3le^2))$$
$$3a_I^3|a_I|\sqrt{(b_{JI}^2+le^2)(a_I^2+b_{JI}^2+le^2)}$$
$$\times(-2b_{JI}^2+3le^2)\sinh^{-1}[\frac{\sqrt{b_{JI}^2+le^2}}{a_I}]\Big) \quad (3.135g)$$

$$\frac{\partial^3 F_6}{\partial le^3}=\frac{b_{JI}le}{8(b_{JI}^2+le^2)^4\sqrt{a_I^2+b_{JI}^2+le^2}}$$
$$\times\Big((b_{JI}^2+le^2)(2(b_{JI}^2+le^2)^2(3b_{JI}^2+2le^2)$$
$$+a_I^4(-9b_{JI}^2+6le^2)-a_I^2(3b_{JI}^4+b_{JI}^2le^2-2le^4))$$
$$+3a_I^3|a_I|\sqrt{(b_{JI}^2+le^2)(a_I^2+b_{JI}^2+le^2)}\sinh^{-1}[\frac{\sqrt{b_{JI}^2+le^2}}{a_I}]\Big)$$
$$(3.135h)$$

$$\frac{\partial^3 F_{10}}{\partial a_I^3}=a_I\Big(-\frac{b_{JI}le(4a_I^4+3b_{JI}^2(b_{JI}^2+le^2)+4a_I^2(2b_{JI}^2+le^2))}{(a_I^2+b_{JI}^2)(b_{JI}^2+le^2)\sqrt{a_I^2+b_{JI}^2+le^2}}$$
$$-4a_I\tan^{-1}[\frac{a_Ile}{b_{JI}\sqrt{a_I^2+b_{JI}^2+le^2}}]$$

$$+4|a_I|(\frac{b_{JI}le}{b_{JI}^2+le^2}+\tan^{-1}[\frac{le}{b_{JI}}])\Big) \tag{3.136a}$$

$$\frac{\partial^3 F_{10}}{\partial a_I^2 \partial b_{JI}} = \frac{b_{JI}^2}{3(b_{JI}^2+le^2)^2} \times \Big(\frac{le}{(a_I^2+b_{JI}^2)\sqrt{a_I^2+b_{JI}^2+le^2}}(8a_I^6-4a_I^2le^2(b_{JI}^2+le^2) - b_{JI}^2(b_{JI}^2+le^2)^2+4a_I^4(3b_{JI}^2+le^2))-8le|a_I|^3 + 3(b_{JI}^2+le^2)^2\tanh^{-1}[\frac{le}{\sqrt{a_I^2+b_{JI}^2+le^2}}]\Big) \tag{3.136b}$$

$$\frac{\partial^3 F_{10}}{\partial a_I^2 \partial le} = \frac{b_{JI}^3}{3(b_{JI}^2+le^2)^2\sqrt{a_I^2+b_{JI}^2+le^2}} \times \Big(-8a_I^4-4a_I^2(b_{JI}^2+le^2)+(b_{JI}^2+le^2)^2 + 8\sqrt{a_I^2+b_{JI}^2+le^2}|a_I|^3\Big) \tag{3.136c}$$

$$\frac{\partial^3 F_{10}}{\partial a_I \partial b_{JI}^2} = \frac{a_I b_{JI}}{3} \times \Big(\frac{le}{(a_I^2+b_{JI}^2)(b_{JI}^2+le^2)^3\sqrt{a_I^2+b_{JI}^2+le^2}}(4a_I^6(-b_{JI}^2+le^2) - b_{JI}^2(b_{JI}^2+le^2)^2(4b_{JI}^2+5le^2)+2a_I^4(-3b_{JI}^4+2b_{JI}^2le^2+le^4) - a_I^2(b_{JI}^2+le^2)(3b_{JI}^4+b_{JI}^2le^2+2le^4)) + \frac{4(b_{JI}^2-le^2)le|a_I|^3}{(b_{JI}^2+le^2)^3}+6\tanh^{-1}[\frac{le}{\sqrt{a_I^2+b_{JI}^2+le^2}}]\Big) \tag{3.136d}$$

$$\frac{\partial^3 F_{10}}{\partial a_I \partial b_{JI} \partial le} = \frac{a_I b_{JI}^2}{3(b_{JI}^2+le^2)^3\sqrt{a_I^2+b_{JI}^2+le^2}} \times \Big((2a_I^4+a_I^2b_{JI}^2+a_I^2le^2)(b_{JI}^2-3le^2) + (b_{JI}^2+le^2)^2(2b_{JI}^2+3le^2) - 2(b_{JI}^2-3le^2)\sqrt{a_I^2+b_{JI}^2+le^2}|a_I|^3\Big) \tag{3.136e}$$

$$\frac{\partial^3 F_{10}}{\partial a_I \partial le^2} = \frac{-a_I b_{JI}^3 le}{3(b_{JI}^2+le^2)^3\sqrt{a_I^2+b_{JI}^2+le^2}}$$

$$\times\Big(-8a_I^4-4a_I^2(b_{JI}^2+le^2)+(b_{JI}^2+le^2)^2$$
$$+8\sqrt{a_I^2+b_{JI}^2+le^2}|a_I|^3\Big) \tag{3.136f}$$

$$\frac{\partial^3 F_{10}}{\partial b_{JI}^3}=\frac{le}{15(a_I^2+b_{JI}^2)(b_{JI}^2+le^2)^4\sqrt{a_I^2+b_{JI}^2+le^2}}$$
$$\times\Big(-3b_{JI}^2(b_{JI}^2+le^2)^3(6b_{JI}^4+4b_{JI}^2le^2-3le^4)$$
$$+4a_I^8(3b_{JI}^4-8b_{JI}^2le^2+le^4)$$
$$+2a_I^6(9b_{JI}^6-21b_{JI}^4le^2-5b_{JI}^2le^4+le^6)$$
$$+a_I^2(b_{JI}^2+le^2)^2(3b_{JI}^6+19b_{JI}^2le^2+19b_{JI}^2le^4+9le^6)$$
$$+a_I^4(12b_{JI}^8+21b_{JI}^6le^2+37b_{JI}^4le^4+35b_{JI}^2le^4+7le^8)\Big)$$
$$-\frac{4a_I^4le(3b_{JI}^4-8b_{JI}^2le^2+le^4)|a_I|}{15(b_{JI}^2+le^2)^4}$$
$$+(a_I^2+6b_{JI}^2)\tanh^{-1}[\frac{le}{\sqrt{a_I^2+b_{JI}^2+le^2}}] \tag{3.136g}$$

$$\frac{\partial^3 F_{10}}{\partial b_{JI}^2\partial le}=\frac{b_{JI}}{15(b_{JI}^2+le^2)\sqrt{a_I^2+b_{JI}^2+le^2}}$$
$$\times\Big(-4a_I^6(b_{JI}^4-8b_{JI}^2le^2+3le^4)$$
$$+3(b_{JI}^2+le^2)^3(12b_{JI}^4+19b_{JI}^2le^2+6le^4)$$
$$+a_I^2(b_{JI}^2+le^2)^2(23b_{JI}^4+41b_{JI}^2le^2+24le^4)$$
$$-2a_I^4(b_{JI}^4-8b_{JI}^2le^2+3le^4)(b_{JI}^2+le^2$$
$$-2\sqrt{a_I^2+b_{JI}^2+le^2}|a_I|)\Big) \tag{3.136h}$$

$$\frac{\partial^3 F_{10}}{\partial b_{JI}\partial le^2}=\frac{b_{JI}^2le}{5(b_{JI}^2+le^2)\sqrt{a_I^2+b_{JI}^2+le^2}}$$
$$\times\Big(8a_I^6(-b_{JI}^2+le^2)+a_I^2(b_{JI}^2-le^2)(b_{JI}^2+le^2)^2$$
$$+(b_{JI}^2+le^2)^3(2b_{JI}^2+3le^2)$$
$$-4a_I^4(b_{JI}^2-le^2)(b_{JI}^2+le^2-2\sqrt{a_I^2+b_{JI}^2+le^2}|a_I|)\Big)$$
$$\tag{3.136i}$$

$$\frac{\partial^3 F_{10}}{\partial le^3} = \frac{b_{JI}^3}{15(b_{JI}^2+le^2)^4\sqrt{a_I^2+b_{JI}^2+le^2}}$$
$$\times \Big(8a_I^6(b_{JI}^2-5le^2) - a_I^2(b_{JI}^2-5le^2)(b_{JI}^2+le^2)^2$$
$$+ 3b_{JI}^2(b_{JI}^2+le^2)^3 + 4a_I^4(b_{JI}^2-5le^2)(b_{JI}^2+le^2$$
$$- 2\sqrt{a_I^2+b_{JI}^2+le^2}|a_I|)\Big) \tag{3.136j}$$

$$\frac{\partial^3 F_{11}}{\partial a_I^3} = \frac{a_I}{b_{JI}^2+le^2}\left(\frac{b_{JI}^2(4a_I^2+3(b_{JI}^2+le^2))}{\sqrt{a_I^2+b_{JI}^2+le^2}} + 2|a_I|(-b_{JI}^2+le^2)\right) \tag{3.137a}$$

$$\frac{\partial^3 F_{11}}{\partial a_I^2 \partial b_{JI}} = \frac{b_{JI}}{3(b_{JI}^2+le^2)^2\sqrt{a_I^2+b_{JI}^2+le^2}}$$
$$\times \Big((3b_{JI}^4(2a_I^2+b_{JI}^2) + 8(a_I^2+b_{JI}^2)^2 le^2$$
$$+ (10a_I^2+7b_{JI}^2)le^4 + 2le^6 - 8le^2\sqrt{a_I^2+b_{JI}^2+le^2}|a_I|^3)\Big) \tag{3.137b}$$

$$\frac{\partial^3 F_{11}}{\partial a_I^2 \partial le} = \frac{b_{JI}^2 le}{3(b_{JI}^2+le^2)\sqrt{a_I^2+b_{JI}^2+le^2}}$$
$$\times \Big(-8a_I^4 - 4a_I^2(b_{JI}^2+le^2) + (b_{JI}^2+le^2)^2$$
$$+ 8\sqrt{a_I^2+b_{JI}^2+le^2}|a_I|^3\Big) \tag{3.137c}$$

$$\frac{\partial^3 F_{11}}{\partial a_I \partial b_{JI}^2} = \frac{a_I}{3(b_{JI}^2+le^2)\sqrt{a_I^2+b_{JI}^2+le^2}}$$
$$\times \Big(2a_I^4 le^2(-3b_{JI}^2+le^2) + (b_{JI}^{+}le^2)^2(6b_{JI}^4+9b_{JI}^2 le^2+2le^4)$$
$$+ a_I^2(b_{JI}^2+le^2)(3b_{JI}^4+3b_{JI}^2 le^2+4le^4)$$
$$- 2le^2(-3b_{JI}^2+le^2)\sqrt{a_I^2+b_{JI}^2+le^2}|a_I|^3\Big) \tag{3.137d}$$

$$\frac{\partial^3 F_{11}}{\partial a_I \partial b_{JI} \partial le} = \frac{a_I b_{JI} le}{3(b_{JI}^2+le^2)^3\sqrt{a_I^2+b_{JI}^2+le^2}}$$
$$\times \Big(4a_I^4(b_{JI}^2-le^2) + (b_{JI}^2+le^2)^2(b_{JI}^2+2le^2)$$

$$+ 2a_I^2(b_{JI}^4 - le^4) + 4(-b_{JI}^2 + le^2)\sqrt{a_I^2 + b_{JI}^2 + le^2}|a_I|^3\Big)$$
(3.137e)

$$\frac{\partial^3 F_{11}}{\partial a_I \partial le^2} = \frac{a_I b_{JI}^2}{3(b_{JI}^2 + le^2)^3\sqrt{a_I^2 + b_{JI}^2 + le^2}}$$
$$\times \Big(-2a_I^4(b_{JI}^2 - 3le^2) - a_I^2(b_{JI}^2 - 3le^2)(b_{JI}^2 + le^2)$$
$$+ b_{JI}^2(b_{JI}^2 + le^2)^2 + 2(b_{JI}^2 - 3le^2)\sqrt{a_I^2 + b_{JI}^2 + le^2}|a_I|^3\Big)$$
(3.137f)

$$\frac{\partial^3 F_{11}}{\partial b_{JI}^3} = \frac{b_{JI}}{5(b_{JI}^2 + le^2)^4\sqrt{a_I^2 + b_{JI}^2 + le^2}}\Big(8a_I^6(b_{JI}^2 - le^2)le^2$$
$$+ (b_{JI}^2 + le^2)^3(20b_{JI}^4 + 33b_{JI}^2 le^2 + 12le^4)$$
$$\times + a_I^2(b_{JI}^2 + le^2)^2(15b_{JI}^4 + 29b_{JI}^2 le^2 + 16le^4) \quad (3.137g)$$
$$+ 4a_I^4 le^2(b_{JI}^2 - le^2)(b_{JI}^2 + le^2 - 2\sqrt{a_I^2 + b_{JI}^2 + le^2}|a_I|)\Big)$$
(3.137h)

$$\frac{\partial^3 F_{11}}{\partial b_{JI}^2 \partial le} = \frac{le}{15(b_{JI}^2 + le^2)^4\sqrt{a_I^2 + b_{JI}^2 + le^2}}\Big(-4a_I^6(3b_{JI}^4 - 8b_{JI}^2 le^2 + le^4)$$
$$+ 3(b_{JI}^2 + le^2)^3(6b_{JI}^4 + 9b_{JI}^2 le^2 + 2le^4)$$
$$+ a_I^2(b_{JI}^2 + le^2)^2(9b_{JI}^4 + 11b_{JI}^2 le^2 + 8le^4)$$
$$- 2a_I^4(3b_{JI}^4 - 8b_{JI}^2 l62 + le^4)(b_{JI}^2 + le^2$$
$$- 2\sqrt{a_I^2 + b_{JI}^2 + le^2}|a_I|)\Big) \quad (3.137i)$$

$$\frac{\partial^3 F_{11}}{\partial b_{JI} \partial le^2} = \frac{b_{JI}}{15(b_{JI}^2 + le^2)^4\sqrt{a_I^2 + b_{JI}^2 + le^2}}\Big(4a_I^6(b_{JI}^4 - 8b_{JI}^2 le^2 + 3le^4)$$
$$+ 3(b_{JI}^2 + le^2)^3(3b_{JI}^4 + 6b_{JI}^2 le^2 + 4le^4)$$
$$+ a_I^2(b_{JI}^2 + le^2)^2(7b_{JI}^4 + 19b_{JI}^2 le^2 + 6le^4)$$
$$+ 2a_I^4(b_{JI}^4 - 8b_{JI}^2 le^2 + 3le^4)(b_{JI}^2 + le^2$$
$$- 2\sqrt{a_I^2 + b_{JI}^2 + le^2}|a_I|)\Big) \quad (3.137j)$$

$$\frac{\partial^3 F_{11}}{\partial le^3} = \frac{b_{JI}^2 le}{5(b_{JI}^2 + le^2)^4\sqrt{a_I^2 + b_{JI}^2 + le^2}}$$

$$\times \Big(8a_I^6(b_{JI}^2 - le^2) - a_I^2(b_{JI}^2 - le^2)(b_{JI}^2 + le^2)^2 + (b_{JI}^2 + le^2)^3(3b_{JI}^2 + 2le^2) + 4a_I^4(b_{JI}^2 - le^2)(b_{JI}^2 + le^2 - 2\sqrt{a_I^2 + b_{JI}^2 + le^2}|a_I|)\Big) \tag{3.137k}$$

$$\frac{\partial^3 F_{12}}{\partial a_I^3} = \frac{a_I}{b_{JI}^2 + le^2}\Big(\frac{b_{JI}le(4a_I^2 + 3b_{JI}^2 + 3le^2)}{\sqrt{a_I^2 + b_{JI}^2 + le^2}} + 4|a_I|(-b_{JI}le + (b_{JI}^2 + le^2)\tan^{-1}[\frac{le}{b_{JI}}]) - (b_{JI}^2 + le^2)(4a_I \tan^{-1}[\frac{a_I le}{b_{JI}\sqrt{a_I^2 + b_{JI}^2 + le^2}}] + 3b_{JI}\tanh^{-1}[\frac{le}{\sqrt{a_I^2 + b_{JI}^2 + le^2}}])\Big) \tag{3.138a}$$

$$\frac{\partial^3 F_{12}}{\partial a_I^2 \partial b_{JI}} = \frac{1}{6(b_{JI}^2 + le^2)\sqrt{a_I^2 + b_{JI}^2 + le^2}} \times \Big(16a_I^4 le^3 + le(b_{JI}^2 + le^2)^2(3b_{JI}^2 + le^2) + a_I^2 le(b_{JI}^2 + le^2)(9b_{JI}^2 + 17le^2) - \sqrt{a_I^2 + b_{JI}^2 + le^2}(16le^3|a_I|^3 + 3(3a_I^2 + b_{JI}^2)(b_{JI}^2 + le^2)^2)\tanh^{-1}[\frac{le}{\sqrt{a_I^2 + b_{JI}^2 + le^2}}]\Big) \tag{3.138b}$$

$$\frac{\partial^3 F_{12}}{\partial a_I^2 \partial le} = \frac{b_{JI}le^2}{3(b_{JI}^2 + le^2)^2\sqrt{a_I^2 + b_{JI}^2 + le^2}} \times \Big(-8a_I^4 - 4a_I^2(b_{JI}^2 + le^2) + (b_{JI}^2 + le^2)^2 + 8\sqrt{a_I^2 + b_{JI}^2 + le^2}|a_I|^3\Big) \tag{3.138c}$$

$$\frac{\partial^3 F_{12}}{\partial a_I \partial b_{JI}^2} = -\frac{a_I b_{JI}}{3(b_{JI}^2 + le^2)^3\sqrt{a_I^2 + b_{JI}^2 + le^2}} \times \Big(-3b_{JI}^6 le + 2(4a_I^4 + 2a_I^2 b_{JI}^2 - 5b_{JI}^4)le^3$$

$$+(4a_I^2-11b_{JI}^2)le^5-4le^7+\sqrt{a_I^2+b_{JI}^2+le^2}(-8le^3|a_I|^3$$
$$+3(b_{JI}^2+le^2)^3\tanh^{-1}[\frac{le}{\sqrt{a_I^2+b_{JI}^2+le^2}}])\Big) \quad (3.138d)$$

$$\frac{\partial^3 F_{12}}{\partial a_I \partial b_{JI} \partial le}=\frac{a_I le^2}{3(b_{JI}^2+le^2)^3\sqrt{a_I^2+b_{JI}^2+le^2}}$$
$$\times\Big(a_I^4(6b_{JI}^2-2le^2)+a_I^2(3b_{JI}^2-le^2)(b_{JI}^2+le^2)$$
$$+le^2(b_{JI}^2+le^2)^2+2(-3b_{JI}^4+le^2)\sqrt{a_I^2+b_{JI}^2+le^2}|a_I|^3\Big) \quad (3.138e)$$

$$\frac{\partial^3 F_{12}}{\partial a_I \partial le^2}=\frac{a_I b_{JI} le}{3(b_{JI}^2+le^2)^3\sqrt{a_I^2+b_{JI}^2+le^2}}$$
$$\times\Big(4a_I^4(-b_{JI}^2+le^2)+(b_{JI}^2+le^2)^2(2b_{JI}^2+le^2)$$
$$+2a_I^2(-b_{JI}^4+le^4)+4(b_{JI}^2-le^2)\sqrt{a_I^2+b_{JI}^2+le^2}|a_I|^3\Big) \quad (3.138f)$$

$$\frac{\partial^3 F_{12}}{\partial b_{JI}^3}=\frac{1}{30}\Big(\frac{le}{(b_{JI}^2+le^2)^4\sqrt{a_I^2+b_{JI}^2+le^2}}(16a_I^6(5b_{JI}^2le^2-le^4)$$
$$+3(b_{JI}^2+le^2)^3(15b_{JI}^4+20b_{JI}^2le^2+3le^4)$$
$$+a_I^2(b_{JI}^2+le^2)^2(15b_{JI}^4+20b_{JI}^2le^2+17b_{JI}^4)$$
$$+8a_I^4(5b_{JI}^4le^2+4b_{JI}^2le^4-le^6))+\frac{16a_I^4le^3(-5b_{JI}^2+le^2)|a_I|}{(b_{JI}^2+le^2)^4}$$
$$-15(a_I^2+3b_{JI}^2)\tanh^{-1}[\frac{le}{\sqrt{a_I^2+b_{JI}^2+le^2}}]\Big) \quad (3.138g)$$

$$\frac{\partial^3 F_{12}}{\partial b_{JI}^2 \partial le}=\frac{b_{JI} le^2}{5(b_{JI}^2+le^2)^4\sqrt{a_I^2+b_{JI}^2+le^2}}$$
$$\times\Big(8a_I^6(-b_{JI}^2+le^2)+a_I^2(b_{JI}^2-le^2)(b_{JI}^2+le^2)^2$$
$$+(b_{JI}^2+le^2)^3(2b_{JI}^2+3le^2)$$
$$-4a_I^4(b_{JI}^2-le^2)(b_{JI}^2+le^2-2\sqrt{a_I^2+b_{JI}^2+le^2}|a_I|)\Big) \quad (3.138h)$$

$$\frac{\partial^3 F_{12}}{\partial b_{JI}\partial le^2} = \frac{le}{15(b_{JI}^2+le^2)^4\sqrt{a_I^2+b_{JI}^2+le^2}}\Big(4a_I^6(3b_{JI}^4-8b_{JI}^2le^2+le^4)$$
$$+3(b_{JI}^2+le^2)^3(4b_{JI}^4+6b_{JI}^2le^2+3le^4)$$
$$+a_I^2(b_{JI}^2+le^2)^2(6b_{JI}^4+19b_{JI}^2le^2+7b_{JI}^4)$$
$$+2a_I^4(3b_{JI}^4-8b_{JI}^2le^2+le^4)(b_{JI}^2+le^2$$
$$-2\sqrt{a_I^2+b_{JI}^2+le^2}|a_I|)\Big) \tag{3.138i}$$

$$\frac{\partial^3 F_{12}}{\partial le^3} = \frac{b_{JI}}{15(b_{JI}^2+le^2)^4\sqrt{a_I^2+b_{JI}^2+le^2}}\Big(-4a_I^6(b_{JI}^4-8b_{JI}^2le^2+3le^4)$$
$$+3(b_{JI}^2+le^2)^3(2b_{JI}^4+9b_{JI}^2le^2+6le^4)$$
$$+a_I^2(b_{JI}^2+le^2)^2(8b_{JI}^4+11b_{JI}^2le^2+9le^4)$$
$$-2a_I^4(b_{JI}^4-8b_{JI}^2le^2+3le^4)(b_{JI}^2+le^2$$
$$-\sqrt{a_I^2+b_{JI}^2+le^2}|a_I|)\Big) \tag{3.138j}$$

Chapter 4

The equivalent inclusion method

Inhomogeneity problems in an unbounded domain

In this chapter, we introduce Eshelby's equivalent inclusion method (EIM) for an ellipsoidal inhomogeneity in an infinite domain and generalizes it for polyhedral inhomogeneities. It provides the exact solution for a single ellipsoidal particle in an infinite domain but generates an approximate one for the case of multiple particles. They can be extended to their 2D counterparts, elliptical and polygonal inhomogeneities, respectively. Particularly, for polyhedral particles, the analytical solution with polynomial eigenstrains is not effective considering the singularity effects in the neighborhood of the vertices. Particle discretization is introduced to obtain higher accuracy. To show the boundary effect, the Green's function for a semi-infinite domain is used in the EIM to study the local field for particles at different distances to the surface.

4.1 Introduction to Eshelby's equivalent inclusion method

The last chapter solves the inclusion problem, in which the particle exhibits the same stiffness as the matrix. This chapter will solve the inhomogeneity problem, in which the materials exhibit different elastic properties. Eshelby [2] proposed the equivalent inclusion method (EIM) to solve the ellipsoidal inhomogeneity problem for an infinite domain. We have extended it to 2D and 3D problems in a finite domain. The key idea of the EIM is introducing an eigenstrain ε_{ij}^* on the particle domain to simulate the material mismatch on the stress field.

Consider an inhomogeneity Ω embedded in an infinite domain $\mathcal{D}$ subjected to a far-field stress σ_{ij}^0 as shown in Fig. 1.2 in the Cartesian coordinate. The matrix domain is denoted by $\mathcal{D} - \Omega$. A distributed body force b_i and an inelastic strain ε_{ij}^p may exist on Ω as well. Here ε_{ij}^p can, for example, be a misfit strain and is a type of prescribed eigenstrain [1]. For the isotropic elastic medium, let $C_{ijkl}^I = \lambda^I \delta_{ij}\delta_{kl} + \mu^I(\delta_{ik}\delta_{jl} + \delta_{il}\delta_{jk})$, where λ^I and μ^I denote the Lame constants of the Ith phase of the composite, in which $I = 0$ or 1 represents the matrix and inhomogeneity, respectively. The governing equations can be written as

$$C_{ijkl}^I(u_{k,li} - \varepsilon_{kl,i}^p) + b_j = 0 \quad \text{for } \mathbf{x} \in \Omega \tag{4.1}$$

The Inclusion-Based Boundary Element Method (iBEM). https://doi.org/10.1016/B978-0-12-819384-6.00012-7

which is the Navier equation in each material domain of $I = 0, 1$. Here the displacement field u_k is continuous and differentiable, whereas the eigenstrain ε_{kl}^{p} is continuous on Ω and zero in the matrix. For simplicity, the body force b_j is taken as a constant on Ω, which can be in polynomial form as well.

Assuming no debonding on the interfaces, the boundary value problem of inhomogeneities can be enacted with the following continuity equations of displacements and tractions:

$$\sigma_{ij}^{+}(\mathbf{x})n_j^{+} = \sigma_{ij}^{-}(\mathbf{x})n_j^{-}, \quad u_i^{+}(\mathbf{x}) = u_i^{-}(\mathbf{x}) \quad \text{on } \partial\Omega \tag{4.2}$$

where $\partial\Omega$ represents the interface between the subdomain and matrix, superscripts $+$ and $-$ denote the outer and inner surfaces of the interface, respectively, and n_j is the unit normal vector of the surface. Considering the far-field stress σ_{ij}^{0}, prescribed eigenstrain ε_{ij}^{p}, and body force b_i on the $\mathcal{D}$ domain, we can establish the boundary value problem. Accordingly, the inhomogeneity problem is reformulated as an equivalent inclusion problem, where the eigenstrain ε_{ij}^{*} is different from the prescribed eigenstrain ε_{ij}^{p}. Although ε_{ij}^{p} contributes to ε_{ij}^{*}, the mismatch eigenstrain at each point on the inhomogeneity is required to satisfy the stress equivalent condition

$$C_{ijkl}^{0} : (\varepsilon_{kl}^{0} + \varepsilon_{kl}' - \varepsilon_{kl}^{p} - \varepsilon_{kl}^{*}) = C_{ijkl}^{I} : (\varepsilon_{kl}^{0} + \varepsilon_{kl}' - \varepsilon_{kl}^{p}) \tag{4.3}$$

where ε_{kl}^{0} is the strain caused by the far-field stress as follows:

$$\varepsilon_{kl}^{0} = C_{klij}^{0^{-1}} \sigma_{ij}^{0} \tag{4.4}$$

and ε_{kl}' is the strain caused by the sources on the particle as follows:

$$\begin{aligned} \varepsilon_{kl}'(\mathbf{x}) = \frac{1}{2}\int_{\Omega} & [G_{kj,l}(\mathbf{x},\mathbf{x}') + G_{lj,k}(\mathbf{x},\mathbf{x}')]b_j \\ & - [G_{ki,jl}(\mathbf{x},\mathbf{x}') + G_{li,jk}(\mathbf{x},\mathbf{x}')]C_{ijmn}^{0}[\varepsilon_{mn}^{*}(\mathbf{x}') + \varepsilon_{mn}^{p}(\mathbf{x}')]d\mathbf{x}' \end{aligned} \tag{4.5}$$

If the equation is satisfied at every point in the inhomogeneity, then the original problem can be exactly solved by an equivalent inclusion problem. However, the mismatch eigenstrain should be carefully calculated. For a single particle in an infinite domain, when Eshelby's tensor is a constant for a certain shape of particles, such as ellipsoidal or elliptical particles, a uniform mismatch eigenstrain field [40] can satisfy the stress equivalent condition (4.3). Therefore the elastic field for the inhomogeneity problem can be immediately obtained by the equivalent inclusion problem. If a uniform body force is applied to the particle, then a linear mismatch eigenstrain is derived [95]. When there are multiple elliptical particles or the domain is bounded, a polynomial eigenstrain based on the Taylor expansion on the centroid of the particle can provide a tailorable accuracy [42]. The stress equivalent condition can be satisfied by each order of the eigenstrain variation. However, for the polyhedral or polygonal inhomogeneity, due to the singularity on the vertex, the eigenstrain distribution will be much more complex. In such case a single polynomial form of eigenstrain based on the Taylor expansion on a reference point for the whole particle may not catch the singularity effect at all vertices. Therefore particle discretization can be used to approximate the eigenstrain distribution and approach the exact solution [43]. In the following, we consider different particle shapes through the equivalent inclusion method. For simplicity, in general, we consider only the far-field stress by assuming that $\mathbf{b} = \mathbf{0}$ and $\boldsymbol{\varepsilon}^p = \mathbf{0}$, unless we specifically mention the consideration of the effect of the prescribed eigenstrain and body force on the particle.

4.2 Ellipsoidal and elliptical inhomogeneities

Eshelby's original work on the equivalent inclusion method was for one ellipsoidal particle embedded in an infinite domain under a uniform far-field stress. Since Eshelby's tensor is constant for a uniform eigenstrain on an ellipsoidal particle, the disturbed strain (4.5) is constant on the particle as well. Therefore a constant eigenstrain can satisfy the stress equivalent condition (4.3), although in the neighborhood of the inhomogeneity, the stress field is not uniform anymore. A disturbed elastic field is assumed by material mismatch written as u', ε', and σ'.

The equivalent inclusion method can be illustrated as follows. For the disturbed field, u_i', ε_{ij}' and σ_{ij}' are caused by the material mismatch of the inhomogeneity Ω, and Hooke's law becomes

$$\begin{cases} \sigma_{ij}^0 + \sigma_{ij}' = C_{ijkl}^* \left(\varepsilon_{kl}^0 + u_{k,l}' \right) & x \in \Omega \\ \sigma_{ij}^0 + \sigma_{ij}' = C_{ijkl} \left(\varepsilon_{kl}^0 + u_{k,l}' \right) & x \in D - \Omega \end{cases} \tag{4.6}$$

Consider an eigenstrain in inclusion Ω. The Hooke's law reads

$$\begin{cases} \sigma_{ij}^0 + \sigma_{ij}' = C_{ijkl} \left(\varepsilon_{kl}^0 + u_{k,l}' - \varepsilon_{kl}^* \right) & x \in \Omega \\ \sigma_{ij}^0 + \sigma_{ij}' = C_{ijkl} \left(\varepsilon_{kl}^0 + u_{k,l}' \right) & x \in D - \Omega \end{cases} \tag{4.7}$$

The disturbed fields caused by ε^* are determined as

$$\begin{cases} u_k' = -\int_\Omega \frac{\partial G_{ki}}{\partial x_j} C_{ijmn} \varepsilon_{mn}^* \\ \varepsilon_{ij}' = S_{ijmn} \varepsilon_{mn}^* \end{cases} \tag{4.8}$$

where Eshelby's tensor S_{ijmn} is a constant on the Ω domain and reduces rapidly toward zero for the field far from the particle.

A necessary and sufficient condition for the two problems in figure to be equivalent is the equivalence of ε'_{ij} and σ'_{ij}:

$$C^*_{ijkl}\left(\varepsilon^0_{kl} + u'_{k,l}\right) = C_{ijkl}\left(\varepsilon^0_{kl} + u'_{k,l} - \varepsilon^*_{kl}\right) \tag{4.9}$$

or

$$C^*_{ijkl}\left(\varepsilon^0_{kl} + \varepsilon'_{kl}\right) = C_{ijkl}\left(\varepsilon^0_{kl} + \varepsilon'_{kl} - \varepsilon^*_{kl}\right) \tag{4.10}$$

Thus

$$\Delta \mathbf{C} : \left(\boldsymbol{\varepsilon}^0 + \boldsymbol{\varepsilon}'\right) = -\mathbf{C} : \boldsymbol{\varepsilon}^* \tag{4.11}$$

where $\Delta \mathbf{C} = \mathbf{C}^* - \mathbf{C}$.

From Eq. (4.8) the disturbed strain field ε' is known as

$$\boldsymbol{\varepsilon}' = -\mathbf{D}^{\Omega} : \mathbf{C} : \boldsymbol{\varepsilon}^* \tag{4.12}$$

where $\mathbf{D}^{\Omega}$ is $\mathbf{D}(\mathbf{x})$ for $\mathbf{x} \in \Omega$ as follows:

$$D_{ijkl} = \begin{cases} \frac{\rho^3}{60\mu(1-v)} \left[\begin{array}{c} \left(5 - 3\rho^2\right)\delta_{ij}\delta_{kl} - \left(5 - 10v + 3\rho^2\right)\left(\delta_{ik}\delta_{jl} + \delta_{il}\delta_{jk}\right) \\ -15(1-\rho^2)(\delta_{ij}n_k n_l + \delta_{kl}n_i n_j) \\ -15(v-\rho^2)(\delta_{ik}n_j n_l + \delta_{il}n_j n_k + \delta_{jk}n_i n_l + \delta_{jl}n_i n_k) \\ +15(5-7\rho^2)n_i n_j n_k n_l \end{array} \right] \\ \text{for } r > a \\ \frac{1}{30\mu(1-v)}\left[\delta_{ij}\delta_{kl} - (4-5v)\left(\delta_{ik}\delta_{jl} + \delta_{il}\delta_{jk}\right)\right] \quad \text{for } r \le a \end{cases} \tag{4.13}$$

which means that $\mathbf{S} = -\mathbf{D}^{\Omega} : \mathbf{C}$. Multiply $\Delta \mathbf{C}^{-1}$ on both sides and substitute Eq. (4.12) into Eq. (4.10). The eigenstrain can be determined as

$$\boldsymbol{\varepsilon}^* = \mathbf{C}^{-1} : \left(\mathbf{D}^{\Omega} - \Delta \mathbf{C}^{-1}\right)^{-1} : \boldsymbol{\varepsilon}^0 \tag{4.14}$$

Thus the disturbed strain field $\boldsymbol{\varepsilon}'$ becomes

$$\boldsymbol{\varepsilon}' = -\mathbf{D}(\mathbf{x}) : \left(\mathbf{D}^{\Omega} - \Delta \mathbf{C}^{-1}\right) : \boldsymbol{\varepsilon}^0 \tag{4.15}$$

Therefore all elastic fields can be written in terms of $\boldsymbol{\varepsilon}^0$.

4.2.1 General cases of inhomogeneity with a prescribed eigenstrain

For the inhomogeneity (Fig. 1.2), which already exists with a prescribed eigenstrain $\boldsymbol{\varepsilon}^p$, such as a thermal strain, the material is still subjected to a far-field strain. The disturbed elastic fields u_i', ε_{ij}', and σ_{ij}' are caused by the material mismatch of the inhomogeneity Ω and the eigenstrain $\boldsymbol{\varepsilon}^p$. Hooke's law becomes

$$\begin{cases} \sigma_{ij}^0 + \sigma_{ij}' = C_{ijkl}^*\left(\varepsilon_{kl}^0 + u_{k,l}' - \varepsilon_{kl}^p\right) & x \in \Omega \\ \sigma_{ij}^0 + \sigma_{ij}' = C_{ijkl}\left(\varepsilon_{kl}^0 + u_{k,l}'\right) & x \in D - \Omega \end{cases} \tag{4.16}$$

Consider an eigenstrain in inclusion Ω. Hooke's law reads

$$\begin{cases} \sigma_{ij}^0 + \sigma_{ij}' = C_{ijkl}\left(\varepsilon_{kl}^0 + u_{k,l}' - \varepsilon_{kl}^* - \varepsilon_{kl}^p\right) & x \in \Omega \\ \sigma_{ij}^0 + \sigma_{ij}' = C_{ijkl}\left(\varepsilon_{kl}^0 + u_{k,l}'\right) & x \in D - \Omega \end{cases} \tag{4.17}$$

where the total eigenstrain can be written as

$$\varepsilon_{kl}^{**} = \varepsilon_{kl}^* + \varepsilon_{kl}^p \tag{4.18}$$

The disturbed fields caused by ε^{**} are determined as

$$\begin{cases} u_k' = -\int \frac{\partial G_{ki}}{\partial x_j} C_{ijmn}\varepsilon_{mn}^{**} \\ \varepsilon_{ij}' = -D_{ijkl}C_{klmn}\varepsilon_{mn}^{**} \end{cases} \tag{4.19}$$

A necessary and sufficient condition for the two problems in Fig. 1.2 to be equivalent is the equivalence of ε_{ij}' and σ_{ij}'

$$C_{ijkl}^*\left(\varepsilon_{kl}^0 + \varepsilon_{kl}' - \varepsilon_{kl}^p\right) = C_{ijkl}\left(\varepsilon_{kl}^0 + \varepsilon_{kl}' - \varepsilon_{kl}^{**}\right) \tag{4.20}$$

which can be rewritten as

$$\Delta\mathbf{C} : \left(\boldsymbol{\varepsilon}^0 + \boldsymbol{\varepsilon}'\right) = -\mathbf{C} : \boldsymbol{\varepsilon}^{**} + \mathbf{C}^* : \boldsymbol{\varepsilon}^p \tag{4.21}$$

Similarly, the disturbed strain field ε' is known as

$$\boldsymbol{\varepsilon}' = -\mathbf{D}^{\Omega} : \mathbf{C} : \boldsymbol{\varepsilon}^{**} \tag{4.22}$$

The total eigenstrain can be determined as

$$\boldsymbol{\varepsilon}^{**} = \mathbf{C}^{-1} : \left(\mathbf{D}^{\Omega} - \Delta\mathbf{C}^{-1}\right)^{-1} : \left[\boldsymbol{\varepsilon}^0 - \boldsymbol{\varepsilon}^p - \Delta\mathbf{C}^{-1} : \mathbf{C} : \boldsymbol{\varepsilon}^p\right] \tag{4.23}$$

Therefore the disturbed strain field $\boldsymbol{\varepsilon}'$ becomes

$$\boldsymbol{\varepsilon}' = -\mathbf{D}(\mathbf{x}) : \left(\mathbf{D}^{\Omega} - \Delta\mathbf{C}^{-1}\right) : \left[\boldsymbol{\varepsilon}^0 - \boldsymbol{\varepsilon}^p - \Delta\mathbf{C}^{-1} : \mathbf{C} : \boldsymbol{\varepsilon}^p\right] \tag{4.24}$$

4.2.2 Interface condition and the uniqueness of solution

For $x \in \partial\Omega$, the continuity of displacement and normal stress are described as

$$\begin{cases} u_i^+ = u_i^- \\ n_i \sigma_{ij}^+ = n_i \sigma_{ij}^- \end{cases} \tag{4.25}$$

The displacement field is obtained as

$$u_k = u_k^0 - \int_\Omega G_{ki,j} C_{ijmn} \varepsilon_{mn}^{**} dx' \tag{4.26}$$

where both u_k^0 and $\int_\Omega G_{ki,j} dx'$ are continuous across the interface. Therefore $u_i^+ = u_i^-$. As for the continuity of the stress field, the interior and exterior stress fields are

$$\begin{cases} \boldsymbol{\sigma}^+ = \mathbf{C} : \left(\boldsymbol{\varepsilon}^0 + \boldsymbol{\varepsilon}'\right) \\ \boldsymbol{\sigma}^- = \mathbf{C} : \left(\boldsymbol{\varepsilon}^0 + \boldsymbol{\varepsilon}' - \boldsymbol{\varepsilon}^{**}\right) \end{cases} \tag{4.27}$$

where the corresponding disturbed strain fields are written as

$$\begin{cases} \boldsymbol{\varepsilon}' = -\mathbf{D}^+ : \mathbf{C} : \boldsymbol{\varepsilon}^{**}, & x \in D - \Omega \\ \boldsymbol{\varepsilon}' = -\mathbf{D}^\Omega : \mathbf{C} : \boldsymbol{\varepsilon}^{**}, & \mathbf{x} \in \Omega \end{cases} \tag{4.28}$$

Using the constitutive law in Eq. (4.27), we can rewrite the stress along the interface as

$$\begin{cases} \boldsymbol{\sigma}^+ = \mathbf{C}\left(\boldsymbol{\varepsilon}^0 - \mathbf{D}\right) : \mathbf{C} : \boldsymbol{\varepsilon}^{**} \\ \boldsymbol{\sigma}^- = \mathbf{C}\left(\boldsymbol{\varepsilon}^0 - \mathbf{D}^\Omega : \mathbf{C} : \boldsymbol{\varepsilon}^{**} - \boldsymbol{\varepsilon}^{**}\right) \end{cases} \tag{4.29}$$

Therefore we obtain the difference of the stress tensor across the interface as

$$\begin{aligned} \boldsymbol{\sigma}^+ - \boldsymbol{\sigma}^- &= \mathbf{C} : \left[\left(\mathbf{D}^\Omega - \mathbf{D}^+\right) : \mathbf{C} + \mathbf{I}\right] : \boldsymbol{\varepsilon}^{**} \\ &= \mu \begin{bmatrix} \frac{2\nu}{1-\nu} \delta_{ij}\delta_{kl} + \left(\delta_{ik}\delta_{jl} + \delta_{il}\delta_{jl}\right) \\ -\frac{2\nu}{1-\nu}\left(\delta_{ij} n_k n_l + \delta_{kl} n_i n_j\right) \\ -\left(\delta_{ik} n_j n_l + \delta_{il} n_j n_k + \delta_{jk} n_i n_l + \delta_{jl} n_i n_k\right) \\ +\frac{2}{1-\nu} n_i n_j n_k n_l \end{bmatrix} \end{aligned} \tag{4.30}$$

We can write the jump of traction along the interface as

$$n_i\left(\sigma_{ij}^+ - \sigma_{ij}^-\right) = \mu \begin{bmatrix} \frac{2\nu}{1-\nu} n_j \delta_{kl} + \left(\delta_{ik} n_k + \delta_{il} n_l\right) \\ -\frac{2\nu}{1-\nu}\left(n_j n_k n_l + \delta_{kl} n_j\right) \\ -\left(2 n_j n_k n_l + \delta_{jk} n_l + \delta_{jl} n_k\right) \\ +\frac{2}{1-\nu} n_j n_k n_l \end{bmatrix} = 0 \tag{4.31}$$

Therefore the interfacial continuity conditions of both the displacement and the surface stress vector or traction are satisfied. Based on the uniqueness of the boundary value problem, the present solution is the exact solution.

4.2.3 Elastic solution for a pair of ellipsoidal inhomogeneities in the infinite domain

Let Ω and $\bar{\Omega}$ be a pair of ellipsoidal inhomogeneities with stiffness C^*_{ijkl} embedded in an infinite medium with stiffness C^0_{ijkl}, which is subjected to a uniform far-field strain ε^0_{ij}. Set two coordinate systems $\mathbf{x}$ and $\bar{\mathbf{x}}$ with origins $\mathbf{0}$ and $\bar{\mathbf{0}}$, respectively, as shown in Fig. 4.1. The coordinate of $\bar{\mathbf{0}}$ in the $\mathbf{x}$ coordinate system is written as $\mathbf{x}^0$. For an arbitrary point P in the domain, x_i^P and $\bar{x}_i^P$ stand for the corresponding coordinates in the two systems, and their relationship is expressed as

$$x_i^P = x_i^0 + Q_{ij}\bar{x}_j^P \text{ or } \bar{x}_j^P = Q_{ij}\left(x_i^P - x_i^0\right) \tag{4.32}$$

where Q_{ij} is the coordination rotational matrix with the direction cosine between the x_i and $\bar{x}_j$ axes. Obviously, the eigenstrain over each particle is not uniform because of the particle interaction. Expanding it over the corresponding local coordinate provides

$$\begin{aligned} \varepsilon^*_{ij}(\mathbf{x}) &= E^I_{ij} + E^I_{ijk}x_k + \cdots \\ \bar{\varepsilon}^*_{ij}(\bar{\mathbf{x}}) &= E^{II}_{ij} + E^{II}_{ijk}\bar{x}_k + \cdots \end{aligned} \tag{4.33}$$

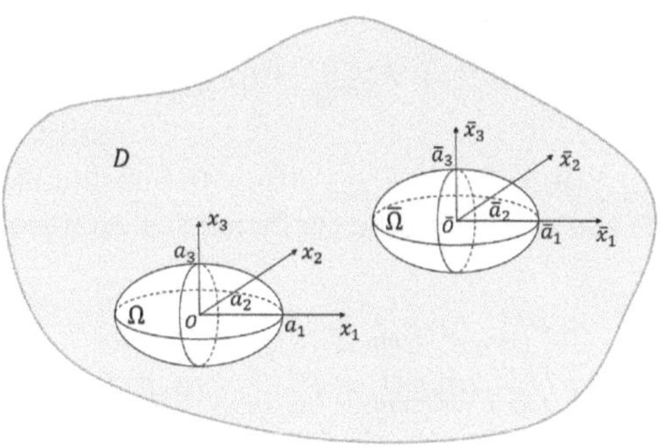

FIGURE 4.1 Interaction between two ellipsoidal particles with local coordinates in parallel.

The strain field of any point $\mathbf{x}$ is the superposition of the field induced by the sources of eigenstrains of the two inhomogeneities as follows:

$$\varepsilon_{ij}(\mathbf{x}) = \varepsilon^0_{ij} + \int_{\Omega+\bar{\Omega}} \Gamma_{ijkl}\left(\mathbf{x}, \mathbf{x}'\right) C_{klmn}\varepsilon^*_{mn}(\mathbf{x}')d\mathbf{x}' \tag{4.34}$$

Referring to the $\mathbf{x}$ coordinate, we can rewrite the induced strain as

$$\begin{aligned}\varepsilon_{ij}(\mathbf{x}) = \varepsilon^0_{ij} + F^I_{ijmn}(\mathbf{x})E^I_{mn} + F^I_{ijmnp}(\mathbf{x})E^I_{mnp} + F^{II}_{ijmn}(\mathbf{x})E^{II}_{mn}\\ + F^{II}_{ijmnp}(\mathbf{x})E^{II}_{mnp} + \cdots\end{aligned} \tag{4.35}$$

where F^I_{ijmn} and F^I_{ijmnp} are the integrals over the first particle, and F^{II}_{ijmn} and F^{II}_{ijmnp} over the second particle.

Similarly, referring to the $\bar{\mathbf{x}}$ coordinate, we can rewrite the induced strain as

$$\begin{aligned}\varepsilon_{ij}(\bar{\mathbf{x}}) = \varepsilon^0_{ij} + F^I_{ijmn}(\bar{\mathbf{x}})E^I_{mn} + F^I_{ijmnp}(\bar{\mathbf{x}})E^I_{mnp} + F^{II}_{ijmn}(\bar{\mathbf{x}})E^{II}_{mn}\\ + F^{II}_{ijmnp}(\bar{\mathbf{x}})E^{II}_{mnp} + \cdots\end{aligned} \tag{4.36}$$

Based on the equivalent inclusion condition, on each inhomogeneity, we can write

$$C^*_{ijkl}\varepsilon_{kl} = C^0_{ijkl}\left(\varepsilon_{kl} - \varepsilon^*_{kl}\right) \tag{4.37}$$

or

$$\boldsymbol{\varepsilon} = -\Delta\mathbf{C}^{-1} : \mathbf{C}^0 : \varepsilon^* \tag{4.38}$$

with $\Delta\mathbf{C} = \mathbf{C}^* - \mathbf{C}^0$.

For the first inhomogeneity, we can expand the strain in Eq. (4.36) in the Taylor series as

$$\begin{aligned}\varepsilon_{ij}(\mathbf{x}) = \varepsilon^0_{ij} + F^I_{ijmn}(\mathbf{0})E^I_{mn} + F^I_{ijmnp}(\mathbf{0})E^I_{mnp} + F^{II}_{ijmn}(\mathbf{0})E^{II}_{mn}\\ + F^{II}_{ijmnp}(\mathbf{0})E^{II}_{mnp} + \Big[F^I_{ijmn,r}(\mathbf{0})E^I_{mn} + F^I_{ijmnp,r}(\mathbf{0})E^I_{mnp}\\ + F^{II}_{ijmn,r}(\mathbf{0})E^{II}_{mn} + F^{II}_{ijmnp,r}(\mathbf{0})E^{II}_{mnp}\Big]x_r + \ldots\end{aligned} \tag{4.39}$$

Notice that $F^I_{ijmnp}(\mathbf{0}) = \mathbf{0}$ and $F^I_{ijmn,r}(\mathbf{0}) = \mathbf{0}$. Substituting the above equation into Eq. (4.38) and comparing the coefficients at each order, we obtain

$$\begin{cases}\varepsilon^0_{ij} + F^I_{ijmn}(\mathbf{0})E^I_{mn} + F^{II}_{ijmn}(\mathbf{0})E^{II}_{mn} + F^{II}_{ijmnp}(\mathbf{0})E^{II}_{mnp} = \Delta C^{-1}_{ijkl}C^0_{klmn}E^I_{mn}\\ F^I_{ijmnp,r}(\mathbf{0})E^I_{mnp} + F^{II}_{ijmn,r}(\mathbf{0})E^{II}_{mn} + F^{II}_{ijmnp,r}(\mathbf{0})E^{II}_{mnp} = \Delta C^{-1}_{ijkl}C^0_{klmn}E^I_{mnr}\end{cases} \tag{4.40}$$

Similarly, for the second inhomogeneity, we can also write two equations as

$$\begin{cases}\varepsilon^0_{ij} + F^{II}_{ijmn}(\bar{\mathbf{0}})E^{II}_{mn} + F^I_{ijmn}(\bar{\mathbf{0}})E^I_{mn} = \Delta C^{-1}_{ijkl}C^0_{klmn}E^{II}_{mn}\\ F^{II}_{ijmnp,r}(\mathbf{0})E^{II}_{mnp} + F^I_{ijmn,r}(\mathbf{0})E^{II}_{mn} + F^I_{ijmnp,r}(\mathbf{0})E^{II}_{mnp} = \Delta C^{-1}_{ijkl}C^0_{klmn}E^{II}_{mnr}\end{cases} \tag{4.41}$$

Solving the linear equation system in the above four sets of equations, we can determine the four unknown tensors E^I_{ij}, E^I_{ijk}, E^{II}_{ij}, and E^{II}_{ijk}. Then we can derive the displacement, strain, and stress fields. The formulation can be extended to multiple particles with multiple material phases. The formulation can be applicable to plane strain problems when 2D harmonic and biharmonic potentials are used.

To exhibit the differences and tailorable accuracy of the polynomial eigenstrain, in Fig. 4.2, we introduce two case studies: 1) top-down interactions and 2) side-by-side interactions in an infinite space. In Chapter 3, we demonstrated domain integrals of the Green's function over various shapes of inclusions and listed some properties, one of which is that the uniform Eshelby tensor is dimensionless. Hence the intensity of the interaction is evaluated with ratio $\frac{d}{a}$, where d and a are the distance between two particles and characteristic length of the particle, respectively. Without the loss of generality, we consider two equal-sized spherical particles with radius $a = 0.05$ m, and their Young's modulus and Poisson's ratio are 2 MPa and 0.25, respectively, whereas Young's modulus and Poisson's ratio of the matrix are 1 MPa and 0.25. Through reducing the ratio $\frac{d}{a}$, the interactions become more intensive, and the differences between uniform, linear, and quadratic terms start to exhibit and increase.

When the ratio is 1, the differences between uniform, linear, and quadratic terms are negligible. However, as the ratio decreases to 0.1, which implies more intensive interactions, Figs. 4.2(a) and 4.2(b) both show obvious differences.

4.3 Polyhedral and polygonal Inhomogeneities with a single polynomial eigenstrain

As in Section 4.2, the polyhedral inhomogeneities could be treated as equivalent inclusions with polynomial form eigenstrain to be determined. For the case of ellipsoidal/elliptical shapes, it is straightforward to employ the center of the inhomogeneity as the point to set up the equivalent stress equations, which captures the inhomogeneity geometric features. As for the polyhedral inhomogeneities with regular shapes, i.e., regular tetrahedron or cube, using the centroid can provide fairly good results. However, when the particle shapes become irregular, the polynomial form of eigenstrain expanded at the centroid of the polyhedron may not be sufficient to solve for the elastic field due to the complexity of the stress field with singularity at the neighborhood of the vertices. To illustrate the complexity of elastic fields with the singularity effects, we first consider two case studies of isosceles triangles (2D polygonal) inhomogeneity in an infinite domain with two different material properties: (1) a void inhomogeneity ($E_1 = 0$) and (2) a structural steel inhomogeneity ($E_s = 210$ GPa, $\nu_s = 0.30$) embedded in an infinite aluminum alloy matrix ($E_a = 70$ GPa, $\nu_a = 0.30$) under a far-field uniform stress. We consider two loading conditions: (1) $\sigma_{11}^0 = -1$ MPa; (2) $\sigma_{22}^0 = -1$ MPa.

Fig. 4.3 shows the comparison of the stress distribution for both void and structural steel [42]. The accuracy of uniform, linear, and quadratic cases of

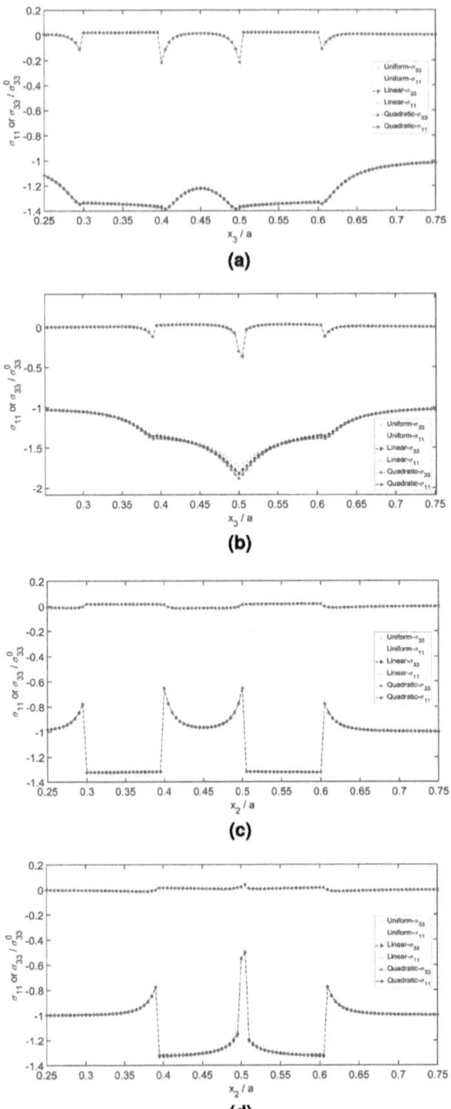

FIGURE 4.2 Variation of normal stresses σ_{11} and σ_{33} of particle interactions of top-down and side-by-side cases: (a), (b) are top-down cases with distance ratios $\frac{d}{a} = 1, 0.1$, respectively; (c), (d) are top-down cases with distance ratios $\frac{d}{a} = 1, 0.1$, respectively.

eigenstrain is investigated, and the quadratic case performs the best among them. We can observe the following features:

1) Unlike the void case with zero stress, the stress on the stiffer inhomogeneity exhibits the same sign as the far-field load, i.e., the compressive stress caused

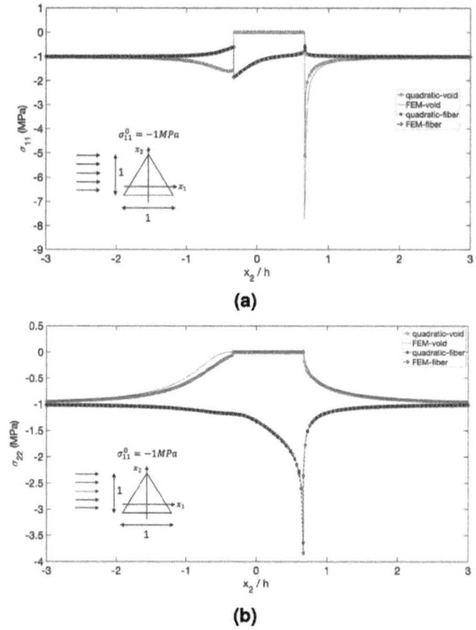

FIGURE 4.3 Variation of stresses (a) σ_{11} under $\sigma_{11}^0 = -1$ MPa and (b) σ_{22} under $\sigma_{22}^0 = -1$ MPa along x_2 for triangular void/stiffer fiber with width and height at 1.

by compressive load for both σ_{11}^0 and σ_{22}^0. Moreover, the stress value on the inhomogeneity is higher than on the far-field load.

2) Although stress singularity still exists at the top vertex, the EIM uses two points to approximate the stress field, and the results are close to that of FEM. The agreement between EIM and FEM is better for the case of a stiffer inhomogeneity because the deformation of the inhomogeneity can be approximated better with a quadratic eigenstrain.

3) In both cases of the uniform horizontal and vertical far-field load conditions, the stress distribution along x_2 reverses the trend of the void case, which can be well explained by Eq. (4.9). Changing the material properties of the inhomogeneity affect the sign of $C_{ijkl}^1 - C_{ijkl}^0$, which produces series of eigenstrains of opposite signs.

Even though the quadratic order of eigenstrain is applied, the obvious difference from the FEM results still exists. Another example of the microstructural effect can be illustrated by aspect ratio (AR). When the AR (width/height) of the triangle changes, the stress distribution can also behave significantly differently. Figs. 4.4(a) and 4.4(b) show the variation of the stress comparison along the x_2 direction with different widths (say 0.5, 1, and 2 m) of a triangular void, whereas Figs. 4.4(c) and 4.4(d) show the case for structural steel inhomogeneity (fiber). We can observe the following features:

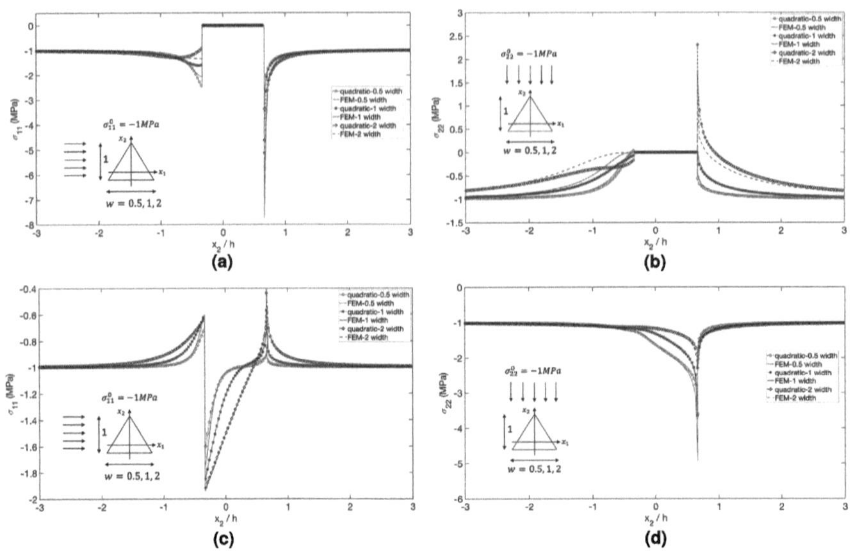

FIGURE 4.4 Variation of stresses (a) σ_{11} under $\sigma^0_{11} = -1$ MPa and (b) σ_{22} under $\sigma^0_{22} = -1$ MPa along x_2 for triangular void; (c) σ_{11} under $\sigma^0_{11} = -1$ MPa and (d) σ_{22} under $\sigma^0_{22} = -1$ MPa along x_2 for triangular stiffer fiber with different widths 0.5, 1, 2 and height at 1.

1) Under a horizontal compressive far-field stress, σ_{11} at the mid of the bottom edge increases with shorter width, which can be caused by the higher influence by two bottom vertices.
2) For the air void case, when a vertical compressive far-field stress is applied, the stress σ_{22} around the top vertex increases significantly with aspect ratio. It even causes a tensile stress for AR = 2.
3) When AR is far from 1, the difference between the FEM and EIM results becomes larger. Because the selected equivalent stress point is the centroid of the triangle, the irregularity of the triangle may lead to the loss of the accuracy of eigenstrain approximated by the Taylor expansion. Higher-order terms of eigenstrain may provide better results.
4) The stiffer inhomogeneity cases still exhibit a reverse trend of the void cases.

Overall, the EIM with quadratic eigenstrain on a single inhomogeneity provides accurate results in comparison with the FEM, which uses a large number of elements. However, the larger AR exhibits a larger difference between EIM and FEM, and we can imagine that when AR = 10 (a flat isosceles triangle), the difference can be significant. As there may exist more microstructural effects, interactions among inhomogeneities and boundary effects as well, to better predict the elastic field, instead of applying EIM over the entire polyhedral/polygonal inhomogeneity domain, particle discretization discussed in the following section can describe a sudden/dramatic change of the continuous eigenstrain field.

In the above polygonal case studies, an initial conclusion was obtained that when the geometry of the particles changes, i.e., the AR increases, a single polynomial eigenstrain can barely represent the overall elastic solutions. Notice that the singularity level of Eshelby's tensor for a polyhedral at vertices is $1/r^2$, which is one order higher than in the polygonal cases. Hence a more rapid fluctuation is expected in the neighborhood of the vertices. Moreover, there exist singularities with $\ln r$ order on the edges. To demonstrate the above singularity effects, we use a tetrahedral inhomogeneity.

Consider a structural steel ($E_s = 200\,\text{GPa}$, $\nu_s = 0.3$) tetrahedron with height $h = 1\,\text{m}$ (along the x_3 axis) and equilateral (edge $l = \sqrt{3}\,\text{m}$) triangular bottom surface embedded in the infinite aluminum alloy (E_a, ν_a) under the uniform far-field load $\sigma^0_{33} = -1\,\text{MPa}$. For the polynomial-form EIM, the stress equivalence is satisfied at the centroid $(0, 0, 0)$ of the tetrahedron, and the accuracy decreases with the distance from the centroid. As demonstrated in the polygonal case [42], the usage of uniform or even linear terms leads to compromised accuracy, since Eshelby's tensor in a tetrahedron is nonuniform. FEM is applied to provide reference of an accurate solution of normal stresses along the x_3 axis. Due to the effect of the singular vertex, the size of elements for the tetrahedron is 0.03 m, and the number of nodes and elements are 4,309,991 and 3,184,212, respectively.

To compare and illustrate the stress distributions, 1201 observing points are evenly distributed in the range of $[-3, 3]\,\text{m}$. Notice that when the observing points are on the surface and vertex, Eshelby's tensor cannot be evaluated due to the jump condition and strong singularity. Thus, to exhibit the two phenomenons, observing points are placed close to them as $\pm 10^{-4}\,\text{m}$.

As shown in Figs. 4.5a and 4.5b, the stress variation mainly concentrates in the neighborhood of the vertices of the tetrahedron, and it rapidly decreases with distance. Though the uniform and linear terms could asymptotically predict the stress distribution, the quadratic term provides the best comparison among them. However, an apparent difference from FEM is observed for σ_{33}. In such a case, higher-order eigenstrains can provide better but not surely accurate results, which is caused by the assumption of distribution of the eigenstrain field. Observing the case studies in Section 3.2.2, for the polyhedral inhomogeneity problem, the strong singularity effect makes it difficult to represent the eigenstrain with a single polynomial. Actually, the initial conclusion can be extended to the polygonal vertices. Although the solution to several triangular cylindrical fibers can be approximated with polynomial-form EIM, when the material filled in the subdomain becomes stiffer or softer or the aspect ratio increases, it is barely possible and challenging to apply the EIM of such a form. Therefore in the following we propose the particle domain discretization method to capture the rapid variation of eigenstrain effect caused by several microstructural effects.

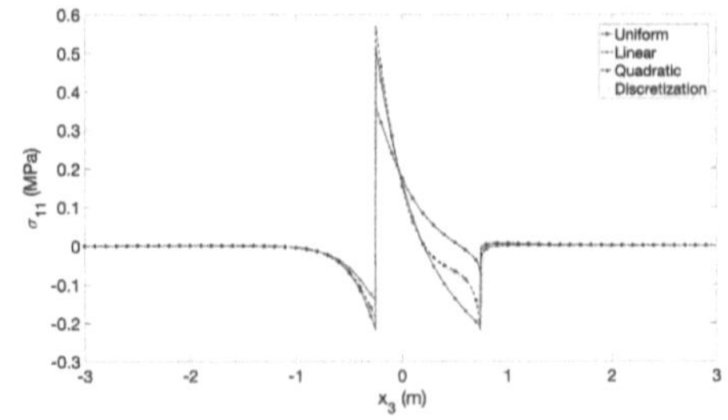

(a)

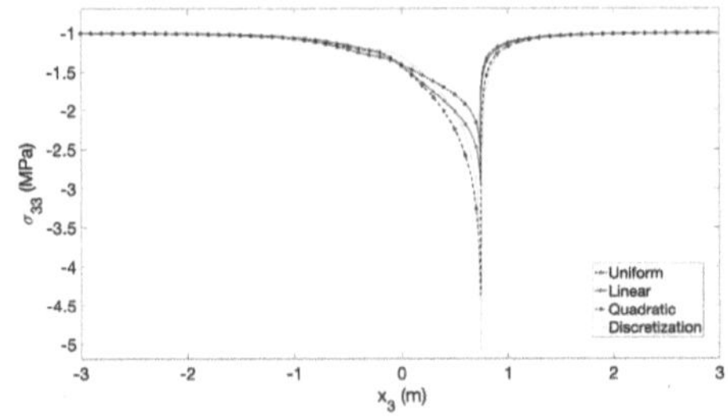

(b)

FIGURE 4.5 Variation of stresses (a) σ_{11} and (b) σ_{33} along x_3 in the range $[-3, 3]$ m under far-field stress $\sigma_{33}^0 = -1$ MPa through uniform, linear, and quadratic order EIM.

4.4 Discretization of the polyhedral/polygonal inhomogeneities

In Section 4.3, due to the microstructural effects, singularity effects, and particle interactions, assuming a simple linear/quadratic eigenstrain distribution over the entire polyhedral/polygonal inhomogeneity domain cannot approximate the actual distribution of the eigenstrain field. Learning from other volume/area discretization methods, such as FEM, finite difference method (FDM), and finite volume method (FVM), provides some insights into the change of fields. Taking the physical vertex singularity as one example, FEM uses very refined mesh in the neighborhood of the vertices, which can describe the dramatic change of displacement and then calculate the strain/stress field. Following the same fashion as other volume discretization methods, the EIM itself can be extended to capture the change of the eigenstrain field around the vertices as well. Thanks to the closed-form domain integrals of the fundamental solutions, the numerical approximation is only dependent on the shape functions and the number of elements.

4.4.1 Polyhedral part

In this section, we illustrate the discretization approach with the simplest element, a 4-node pyramid element, which contains the linear distribution of quantities on the vertices. In comparison to the FEM, which has constant strain over the element, the eigenstrain with C^0 continuity can provide a linear distribution of the strain field over the element. On one hand, to describe the same change of strain, FEM can only refine the mesh or use elements with quadratic shape functions, which requires more computational resources and has an overall worse convergence rate.

The eigenstrain of one element can be written in terms of the eigenstrain assigned at each node by the shape functions, similarly to the FEM, and thus the eigenstrain field inside an element is entirely dependent on the shape functions, which ensures the C^0 continuity. Let E^I and N^I denote the number of elements and nodes of the Ith polyhedral subdomain. For one element with four nodes, say (x_1^1, x_2^1, x_3^1), (x_1^2, x_2^2, x_3^2), (x_1^3, x_2^3, x_3^3), and (x_1^4, x_2^4, x_3^4), we can write the eigenstrain distribution in the element in terms of eigenstrain on the nodes as

$$\varepsilon_{ij}^{*I}(\mathbf{x}') = \sum_{n=1}^{4} (\alpha^n + \beta^n x_1' + \gamma^n x_2' + \chi^n x_3')\,(\varepsilon_{ij}^{*I})^n \tag{4.42}$$

where the superscript n represents the quantity associated with node n, and no index summation rule is applied with it; α^I, β^I, γ^I, and χ^I are the components of the shape functions of the Ith local node as (let the other three local nodes be J, K, L) the matrix operation

$$\begin{bmatrix} \alpha^I & \alpha^J & \alpha^K & \alpha^L \\ \beta^I & \beta^J & \beta^K & \beta^L \\ \gamma^I & \gamma^J & \gamma^K & \gamma^L \\ \chi^I & \chi^J & \chi^K & \chi^L \end{bmatrix} = \begin{bmatrix} 1 & x_1^I & x_2^I & x_3^I \\ 1 & x_1^J & x_2^J & x_3^J \\ 1 & x_1^K & x_2^K & x_3^K \\ 1 & x_1^L & x_2^L & x_3^L \end{bmatrix}^{-1} \tag{4.43}$$

and

$$\alpha^I = \frac{-x_1^L x_2^K x_3^J + x_1^K x_2^L x_3^J + x_1^L x_2^J x_3^K - x_1^J x_2^L x_3^K - x_1^K x_2^J x_3^L + x_1^J x_2^K x_3^L}{6V} \tag{4.44}$$

$$\beta^I = \frac{x_2^K x_3^L - x_2^L x_3^J - x_2^J x_3^K + x_2^L x_3^K + x_2^J x_3^L - x_2^K x_3^L}{6V} \tag{4.45}$$

$$\gamma^I = \frac{-x_1^K x_3^J + x_1^L x_3^J + x_1^J x_3^K - x_1^L x_3^K - x_1^J x_3^L + x_1^K x_3^L}{6V} \tag{4.46}$$

$$\chi^I = \frac{x_1^K x_2^J - x_1^L x_2^J - x_1^J x_2^K + x_1^L x_2^K + x_1^J x_2^L - x_1^K x_2^L}{6V} \tag{4.47}$$

where V is the volume of the element. Obviously, the shape function contains linear source terms, and thus to perform the domain integral of the fundamental solution, we need to separate the source terms into field points and distance vectors, where the closed-form domain integrals have been provided in Section 3.2.

Then we can analytically evaluate the domain integral as

$$-\int_D G_{ij,m}(\mathbf{x},\mathbf{x}')C^0_{mjkl}\varepsilon^*_{kl}(\mathbf{x}')d\mathbf{x}' = \frac{1}{8\pi(1-\nu)}\left[\mathcal{A}_{kl,kli} - 2\nu\mathcal{B}_{kk,i} - 4(1-\nu)\mathcal{B}_{ik,k}\right] \tag{4.48}$$

where

$$\mathcal{A}_{ij} = \int_\Omega \varepsilon_{ij}(\mathbf{x'})L^I(\mathbf{x}')|\mathbf{x}-\mathbf{x}'|d\mathbf{x}' = \sum_{e=1}^{E^I}\sum_{I=1}^{4}(\varepsilon^*_{ij})^I\left[\alpha^I\Psi + \beta^I\Psi_1 + \gamma^I\Psi_2 + \chi^I\Psi_3\right] \tag{4.49a}$$

$$\mathcal{B}_{ij} = \int_\Omega \varepsilon_{ij}(\mathbf{x'})L^I(\mathbf{x}')\frac{1}{|\mathbf{x}-\mathbf{x}'|}d\mathbf{x}' = \sum_{e=1}^{E^I}\sum_{I=1}^{4}(\varepsilon^*_{ij})^I\left[\alpha^I\Phi + \beta^I\Phi_1 + \gamma^I\Phi_2 + \chi^I\Phi_3\right] \tag{4.49b}$$

With the implementation of the above expressions, we take into account the eigenstrain effect analytically, which maintains the advantage of the Green's function.

4.4.2 Polygonal part

Similarly to the 3D problem, in this section, we use a 3-node linear triangular element for demonstration. Consider a 3-node quadratic triangular element in which (x_1^1, x_2^1), (x_1^2, x_2^2), and (x_1^3, x_2^3) are the corner nodes that can be written in terms of the first three nodes. The eigenstrain distribution in the element can be written in terms of eigenstrain on the nodes through their shape functions as

$$\varepsilon^*_{ij}(\mathbf{x}') = \sum_{n=1}^{3}(\alpha^n + \beta^n x_1' + \gamma^n x_2')(\varepsilon^*_{ij})^n \tag{4.50}$$

where the shape function at the nth node $L^n = \alpha^n + \beta^n x_1 + \gamma x_2$ is the area coordinates of the triangular element, α^I, β^I, and γ^I are the components of shape functions of the Ith local node, in which the superscripts J and K represent the other two corner nodes:

$$\alpha^I = \frac{x_1^J x_2^K - x_1^K x_2^J}{2A}, \quad \beta^I = \frac{x_2^J - x_2^K}{2A}, \quad \gamma^I = \frac{x_1^K - x_1^J}{2A}$$

$$2A = det \begin{bmatrix} 1 & x_1^1 & x_2^1 \\ 1 & x_1^2 & x_2^2 \\ 1 & x_1^3 & x_2^3 \end{bmatrix} \tag{4.51}$$

Following the same way as Eq. (4.48) in 3D problem, we take into account the effects of eigenstrain. We provide the integral components $\mathcal{A}_{ij}$ and $\mathcal{B}_{ij}$ for 2D as

$$\mathcal{A}_{ij} = \int_{\Omega} L^I(\mathbf{x}')|\mathbf{x}-\mathbf{x}'|^2 \frac{ln|\mathbf{x}-\mathbf{x}'|^2 - 1}{2} d\mathbf{x}' = \alpha^I \Psi + \beta^I \Psi_1 + \gamma^I \Psi_2^I \tag{4.52}$$

$$\mathcal{B}_{ij} = \int_{\Omega} L^I(\mathbf{x}') ln|\mathbf{x}-\mathbf{x}'|^2 d\mathbf{x}' = \alpha^I \Phi + \beta^I \Phi_1 + \gamma^I \Phi_2^I \tag{4.53}$$

4.4.3 Implementation of particle discretization method

The domain discretization method proposed in Section 4.4 is a subdomain collocation method with the stress equivalent equations established on the particle domain. Because the stress field has been represented by the eigenstrain on all nodes in the particle discretization, this problem includes $3 \times N_I$ (2D) or $6 \times N_I$ (3D) unknowns in total to be solved. If we can establish the same number of linearly independent equations, then the eigenstrain field can be solved, and the elastic field is obtained through the domain integral with above shape functions. It is natural to establish the equations exactly at the nodes themselves. However, there exists discontinuity of edges (2D), surfaces (3D), and strong singularities of vertices (2D & 3D) and edges (3D), which requires proper handling to obtain a practical solution.

In the conventional boundary element method (BEM), there exist strong singularities of the potential/displacement kernel functions when the source coincides with field points, which are recognized as Cauchy principal value (CPV). Besides the well-established rigid body motion (creating a uniform potential/displacement) method in the literature, Li et al. [96] proposed a triangle-polar coordinate, which maps the 4-node linear tetrahedral element into a cubic element. Nakasone et al. [8] later extended this treatment to approximate the singularity of polygonal vertices. Although using the triangle polar coordinate can approximately solve the 2D problem, it downgrades the merit of analytical domain integrals of Eshelby's tensor and is challenging to be extended to 3D problems. Therefore, in this section, we propose an alternative scheme to set up interior stress equivalent equations instead of exactly on the nodes themselves, which avoids the singularity issue.

Notice that the choice of interior stress equivalent points is not arbitrary; close attention should be paid to numerical instability as follows:

(1) Avoid choosing points close to its original boundary nodes. Since the domain integral tends to infinity at the boundary nodes on the edges, if the distance of

interior points to the original node is close, then a small change could result in significance increase, leading to numerical instability.
(2) A number of points close to the original boundary nodes must be included to consider the geometry effect. Because the effect of Green's function decays rapidly with the distance, if all stress equivalent points are selected far from the boundary, then less contribution from the boundary nodes may lead to inaccurate results.
(3) Avoid choosing the stress equivalent points close to each other, which leads to an ill-conditioned matrix as well by reducing the linear independence of the equations.

There are several methods to set up the linear equation system to solve for eigenstrain as follows:

(1) Set up the stress equivalent equation on the Gauss integral points within the element with weight of volume and combine the weighted equations to the corresponding nodes to form a closed system of linear equations.
(2) Use the continuity of eigenstrain distribution, express the eigenstrain of the interior points in terms of three or four nodes for 2D or 3D cases, respectively, by shape function interpolation, to avoid the equation on the boundary nodes.
(3) Set up the stress equivalent equation on a new set of control points instead of the original nodes.

For the first option, because the number of elements can be higher or lower than the number of nodes, extra cautions should be taken to balance the accuracy and computational cost by using more Gauss integral points. For the second option, the results may vary with interpolation methods, i.e., the rational range of shape functions. Particularly, a small error of eigenstrain on the edge may lead to a large change of stress in its neighborhood, which results in the loss of stability of its solution. The third one can be the simplest and robust option, but there exist many ways to construct the system of linear equations. The uniqueness and convergence of the numerical solution are questionable.

In this chapter, we propose the first choice as an alternative method, which sets up stress equivalent equations on the Gauss integral points instead of one node itself, which avoids the singularity issue as follows:

1. For each element, derive the stress residual on each Gauss integral point from Eq. (4.10) as

$$\sigma_{ij}^{d} = \Delta C_{ijkl} : (\varepsilon_{kl}^{0} + \varepsilon_{kl}' - \varepsilon_{kl}^{p}) + C_{ijkl}^{0} : \varepsilon_{kl}^{*} \tag{4.54}$$

where $\Delta C_{ijkl} = C_{ijkl}^{1} - C_{ijkl}^{0}$; ε_{kl}^{p} is the prescribed inelastic strain or eigenstrain. The residual σ_{ij}^{d} is supposed to be zero. However, if the equation is exactly executed, then the number of equations will be much larger than the number of unknowns, and no solution can be obtained.
2. Apply the weight of each Gauss point on its stress residual σ_{ij}^{d} and distribute it on the nodes of the element. For simplicity, this method uses the same number

of Gauss points as the nodes in the element and assigns the residual to the closest node. However, more Gauss points and other distribution methods can be used.
3. Collect the weighted stress residuals on each node and make the sum equal to zero. If one node only appears on one element, then no summation is necessary. Otherwise, the contribution from each element should be considered.

Therefore we can set up the closed linear equation system. Notice that steps 2 and 3 can be implemented by the collection of stress residuals on a node by integrating the weighted residual on the neighboring Gauss points by a kernel function.

Notice that the integral weight on the stress residual on a vertex node is trivial when it is associated with only one element or Gauss point but it is necessary for other nodes located at multiple elements as the element areas (2D) and volumes (3D) can be different. Solving the system of linear equations for eigenstrain on the elements, we can use them to calculate the displacement field by the domain integral in Section 4.4. Therefore we can obtain the strain and stress fields. Although the particle discretization leads to numerical errors with quadratic function of eigenstrain distribution, because the influence zone of errors is small and the domain integral is based on the exact solution, the algorithm can provide high-fidelity results to the elastic field with low mesh sensitivity.

In this section, for demonstration, we applied only two types of linear elements, which could map most of the geometries of subdomains. However, this particle discretization method can be extended for other shape functions, such as bi/trilinear, quadratic, and combined with numerical integral as well, which will be elaborated in the appendix of this chapter.

4.4.4 Evaluation of domain integral at interior nodes

Eshelby's tensor is singular at the edge points (3D) and vertices (2D & 3D), and discontinuous on the other boundary surface points, which can be illustrated numerically by the stress in the neighborhood. Since the 2D and 3D problems are similar, in this section, we only propose the handling of 3D problem. For the interior nodes, we suppose Eshelby's tensor to be continuous and bounded, but for one single element, its domain integral on each node is not well defined. To treat this problem, the singularity isolation method can be used by creating an infinitesimal spherical region around the interior nodes [21,97]. We consider the domain integral of the infinitesimal sphere as the free term, which is also recognized as the interior uniform spherical Eshelby tensor. Since the singular terms cancel out with a spherical shell, the singularity of $1/r$ is reduced to $\ln r$ by application of Gauss' theorem, which could be numerically evaluated.

In this section, with the obtained uniform and polynomial components of potentials e derived from the corresponding order of eigenstrains, we propose a simple technique to avoid any unnecessary numerical integral. Firstly, we obtain the linear and quadratic potentials by dividing them into field point-related and distance vector-related components. There exists no singularity issue at edges and vertices for linear and quadratic potential parts with distance vector components, which implies that the singularity is caused by the uniform eigenstrain only. Secondly, in Eqs. (4.53) and (4.49), when the field point coincides with a vertex, the shape functions for other nodes yield zero, including the field point-related components introduced by linear and quadratic potentials, and thus the coefficients for uniform potentials become 1, which means handling Ψ and Φ together on the coinciding vertex. When the uniform potential or uniform Eshelby's tensors are mentioned, such potentials can be directly written in terms of the eigenstrain on the boundary node without a volume integral. Therefore no numerical integral is necessary, and the analytical integrals can exactly evaluate the stresses at interior nodes when the eigenstrains on all the nodes are given.

4.5 Singularity of stress and eigenstrain in angular particles and its influence zone

For a smooth inhomogeneity, the stress is continuous and bounded in both particle and matrix domains although discontinuities on the stress components exist. However, for angular particles, Eshelby's tensor exhibits singularity at the vertex. The exact integral of Eshelby's tensor for a uniform eigenstrain over a polygonal area [49] provides a straightforward and robust description of stress singularity on the vertex. In our recent work [42], we extended the exact integral to polynomial eigenstrains, which makes it feasible to solve the inhomogeneity problems. However, due to the singularity effect, the solution of eigenstrain field and its effect on the elastic field are not clear. When a certain load, such as a body force or inelastic strain (prescribed eigenstrain), is distributed on an angular domain in the matrix, it can be considered as an inclusion problem, and the singularity directly follows Eshelby's tensor. We call it a load singularity, in which the eigenstrain is prescribed. However, for the inhomogeneity problem, the eigenstrain is unknown, and thus the stress singularity may be different. We call it a material singularity.

Due to the singularity problem, some questions are brought forth as follows:

1) Whether is eigenstrain field solvable, and what does it look like?
2) What is the singularity level for stress distribution in the inhomogeneity problem, and whether it is the same as an inclusion problem with load singularity?
3) When a singularity of eigenstrain exists in material singularity, how does it influence the overall elastic field?

This section addresses these questions.

Without loss of generality, we investigate the singularity at the top vertex of an isosceles triangular inhomogeneity with top angle θ and height h embedded in an infinite domain D subjected to a far-field stress σ_{ij}^0 as Fig. 4.6(a) in the Cartesian coordinate of the x_1–x_2 plane with the origin at the top vertex. The equivalent inclusion problem is also shown in parallel in Fig. 4.6(b), in which an eigenstrain on the particle is introduced to simulate the stiffness mismatch between the inhomogeneity and matrix. If the stress equivalent condition

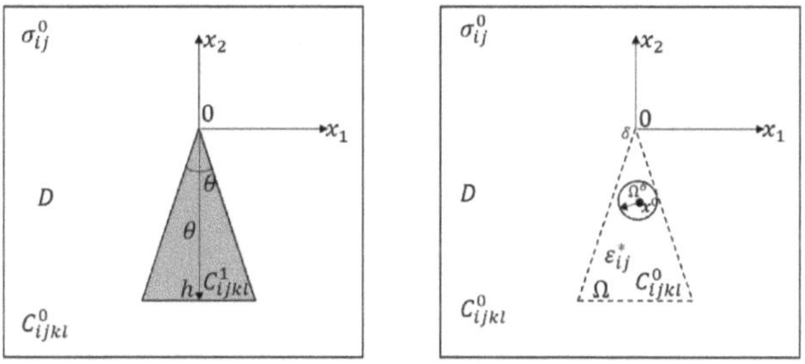

(a) An inhomogeneity embedded in an Infinite domain under a far field stress σ_{ij}^0 (b) An equivalent inclusion with an eigenstrain ε_{ij}^* under a far field stress σ_{ij}^0

FIGURE 4.6 Equivalent inclusion method for an inhomogeneity embedded in an infinite domain: (a) an inhomogeneity embedded in a matrix is simulated by (b) an equivalent inclusion with eigenstrain.

in Eq. (4.3) is satisfied at each position on the inhomogeneity, then the equivalent inclusion problem provides an exact solution to the inhomogeneity problem. Therefore the material singularity in a triangular inhomogeneity is equivalent to the inclusion problem with distributed eigenstrain loading on the subdomain.

Although the continuity of the eigenstrain inside the particle is guaranteed by the elastic equations in the continuum, due to the singularity of Eshelby's tensor [49], the distribution of eigenstrain in the neighborhood of the vertex is open for investigation. Here we use the uniform far-field stress σ_{ij}^0 for a 2D problem as an example of the distribution of eigenstrain on the vertex. The proof can be generalized to other loading situations and 3D cases as well. The stress equivalent condition in Eq. (4.3) can be reduced to

$$C_{ijkl}^0 : (\varepsilon_{kl}^0 + \varepsilon_{kl}' - \varepsilon_{kl}^*) = C_{ijkl}^1 : (\varepsilon_{kl}^0 + \varepsilon_{kl}') \tag{4.55}$$

Because ε_{kl}' is singular at the vertex, the equation cannot be solved exactly at the vertex, which may produce a misconception that the eigenstrain must approach the infinite at the vertex as well. In the following, we can investigate the distribution of mismatch eigenstrain in the particle. In Fig. 4.6(b), we introduce a tiny circular area Ω^δ with radius δ and center at $\mathbf{x}^0$. The integral of ε_{kl}' over Ω^δ can be written as

$$\begin{aligned}\int_{\Omega^\delta} \varepsilon_{kl}'(\mathbf{x})d\mathbf{x} &= -\frac{1}{2}\int_{\Omega^\delta}\int_{\Omega}[G_{ki,jl}(\mathbf{x},\mathbf{x}') + G_{li,jk}(\mathbf{x},\mathbf{x}')]C_{ijmn}^0\varepsilon_{mn}^*(\mathbf{x}')d\mathbf{x}'d\mathbf{x}\\ &= \int_{\Omega}\int_{\Omega^\delta}\overline{G}_{klmn}(\mathbf{x},\mathbf{x}')d\mathbf{x}\varepsilon_{mn}^*(\mathbf{x}')d\mathbf{x}'\\ &= \int_{\Omega} S_{klmn}^\delta(\mathbf{x}' - \mathbf{x}^0)\varepsilon_{mn}^*(\mathbf{x}')d\mathbf{x}' \end{aligned} \tag{4.56}$$

where

$$\overline{G}_{klmn}(\mathbf{x},\mathbf{x}') = -\frac{1}{2}[G_{ki,jl}(\mathbf{x},\mathbf{x}') + G_{li,jk}(\mathbf{x},\mathbf{x}')]C^0_{ijmn}$$
$$= \frac{1}{4\pi(1-\nu)|\mathbf{x}-\mathbf{x}'|^2}[-(1-2\nu)\delta_{kl}\delta_{mn}$$
$$+ (1-2\nu)(\delta_{km}\delta_{ln}+\delta_{kn}\delta_{lm}) + 2\delta_{kl}n_m n_n + 2(1-2\nu)\delta_{mn}n_k n_l$$
$$+ 2\nu(\delta_{km}n_n n_l + \delta_{lm}n_n n_k + \delta_{kn}n_m n_l + \delta_{ln}n_m n_k) - 8n_k n_l n_m n_n] \tag{4.57}$$

in which $\mathbf{n} = \frac{\mathbf{x}'-\mathbf{x}}{|\mathbf{x}'-\mathbf{x}|}$ and $S^\delta_{klmn}(\mathbf{x}'-\mathbf{x}^0)$ is Eshelby's tensor for the small domain Ω^δ,

$$S^\delta_{klmn}(\mathbf{x}'-\mathbf{x}^0) = \begin{cases} \frac{4\nu-1}{8(1-\nu)}\delta_{kl}\delta_{mn} + \frac{3-4\nu}{8(1-\nu)}(\delta_{km}\delta_{ln}+\delta_{kn}\delta_{lm}) \quad \text{for } x' \in (\Omega^\delta) \\ \frac{\rho^2}{8(1-\nu)}[(\rho^2+4\nu-2)\delta_{kl}\delta_{mn} \\ +(\rho^2-4\nu+2)(\delta_{km}\delta_{ln}+\delta_{kn}\delta_{lm}) \\ +4(1-\rho^2)\delta_{kl}n_m n_n + 4(1-2\nu-\rho^2)\delta_{mn}n_k n_l \\ +4(\nu-\rho^2)(\delta_{km}n_n n_l + \delta_{lm}n_n n_k + \delta_{kn}n_m n_l + \delta_{ln}n_m n_k) \\ +8(3\rho^2-2)n_k n_l n_m n_n], \quad \text{for } x' \notin (\Omega^\delta) \end{cases} \tag{4.58}$$

where $\rho = \frac{\delta}{|\mathbf{x}'-\mathbf{x}^0|}$ and $n_i = \frac{x'_i - x^0_i}{|\mathbf{x}'-\mathbf{x}^0|}$ with $\mathbf{x}$ replaced by the circle center at $\mathbf{x}^0$ in Eq. (4.60). For $\mathbf{x}' \in \Omega^\delta$, Eshelby's tensor is a constant, namely S^c_{klmn}. For $\mathbf{x}' \notin \Omega^\delta$ or $|\mathbf{x}'-\mathbf{x}^0| > \delta$, comparing Eqs. (4.60) and (4.61), we can rewrite Eshelby's tensor as follows:

$$S^\delta_{klmn}(\mathbf{x}'-\mathbf{x}^0) = \pi\delta^2\overline{G}_{klmn}(\mathbf{x}',\mathbf{x}^0)$$
$$+ \frac{\delta^4}{8(1-\nu)(|\mathbf{x}'-\mathbf{x}^0|)^4}[\delta_{kl}\delta_{mn} + (\delta_{km}\delta_{ln}+\delta_{kn}\delta_{lm}) - 4\delta_{kl}n_m n_n$$
$$- 4\delta_{mn}n_k n_l - 4(\delta_{km}n_n n_l + \delta_{lm}n_n n_k + \delta_{kn}n_m n_l + \delta_{ln}n_m n_k)$$
$$+ 24 n_k n_l n_m n_n] \tag{4.59}$$

Similarly, for 3D cases, Eqs. (4.57)–(4.59) can be written as

$$\overline{G}_{klmn}(\mathbf{x},\mathbf{x}') = -\frac{1}{2}[G_{ki,jl}(\mathbf{x},\mathbf{x}') + G_{li,jk}(\mathbf{x},\mathbf{x}')]C^0_{ijmn}$$
$$= \frac{1}{8\pi(1-\nu)|\mathbf{x}-\mathbf{x}'|^3}[-(1-2\nu)\delta_{kl}\delta_{mn}$$
$$+ (1-2\nu)(\delta_{km}\delta_{ln}+\delta_{kn}\delta_{lm}) + 3\delta_{kl}n_m n_n + 3(1-2\nu)\delta_{mn}n_k n_l$$
$$+ 3\nu(\delta_{km}n_n n_l + \delta_{lm}n_n n_k + \delta_{kn}n_m n_l + \delta_{ln}n_m n_k) - 15n_k n_l n_m n_n] \tag{4.60}$$

which is obtained from the 3D elastic Green's function,

$$S^{\delta}_{klmn}(\mathbf{x}'-\mathbf{x}^0)=\begin{cases}\frac{5\nu-1}{15(1-\nu)}\delta_{kl}\delta_{mn}+\frac{4-5\nu}{15(1-\nu)}(\delta_{km}\delta_{ln}+\delta_{kn}\delta_{lm}) & \text{for } \mathrm{x}'\in(\Omega^{\delta})\\ \frac{\rho^3}{30(1-\nu)}[(3\rho^2+10\nu-5)\delta_{kl}\delta_{mn} \\ +(3\rho^2-10\nu+5)(\delta_{km}\delta_{ln}+\delta_{kn}\delta_{lm}) \\ +15(1-\rho^2)\delta_{kl}n_m n_n+15(1-2\nu-\rho^2)\delta_{mn}n_k n_l \\ +15(\nu-\rho^2)(\delta_{km}n_n n_l+\delta_{lm}n_n n_k+\delta_{kn}n_m n_l+\delta_{ln}n_m n_k) \\ +15(7\rho^2-5)n_k n_l n_m n_n] & \text{for } \mathrm{x}'\notin(\Omega^{\delta})\end{cases} \tag{4.61}$$

which is obtained from the volume integral on the sphere Ω^{δ}. For $\mathbf{x}'\notin\Omega^{\delta}$ or $|\mathbf{x}'-\mathbf{x}^0|>\delta$, comparing Eqs. (4.60) and (4.61), we can rewrite Eshelby's tensor as follows:

$$\begin{aligned}S^{\delta}_{klmn}(\mathbf{x}'-\mathbf{x}^0)=&\frac{4\pi\delta^3}{3}\overline{G}_{klmn}(\mathbf{x}',\mathbf{x}^0)+\frac{\delta^5}{10(1-\nu)(|\mathbf{x}'-\mathbf{x}^0|)^5}[\delta_{kl}\delta_{mn}+\delta_{km}\delta_{ln}\\ &+\delta_{kn}\delta_{lm}-5\big(\delta_{kl}n_m n_n+\delta_{mn}n_k n_l+\delta_{km}n_n n_l+\delta_{lm}n_n n_k\\ &+\delta_{kn}n_m n_l+\delta_{ln}n_m n_k\big)+35 n_k n_l n_m n_n]\end{aligned} \tag{4.62}$$

Here the Eshelby's tensor can evaluate the strain caused by a unit uniform eigenstrain over a source domain, which can be generalized to an arbitrary shape. As shown in Chapter 3 with four lemmas, which will lead to two hypotheses, Eshelby's tensor for uniform eigenstrain in an arbitrary inclusion exhibits the following properties:

1) Eshelby's tensor **S** is dimensionless, and its distribution depends on the shape of the source domain and relative location. The integral of Eshelby's tensor of an inclusion over the inclusion itself is proportional to the area or volume, referred to as the similar shape of inclusion with unit area or volume for 2D and 3D cases, respectively.
2) Although Eshelby's tensor **S** may exhibit a $\frac{1}{r}$ singularity at the neighborhood of a vertex, its integral over the inclusion is positive definite and bounded with $\mathbf{0}<\int_{\Omega}\mathbf{S}(\mathbf{x}')d\mathbf{x}'<V\mathbf{I}$, where V is the volume of the inclusion and the unit tensor $I_{klmn}=\frac{1}{2}(\delta_{km}\delta_{ln}+\delta_{kn}\delta_{lm})$.
3) Eshelby's tensor of an area formed by the complement of a large inclusion including a small one is the difference of Eshelby's tensor of the larger triangle from the smaller one.

Now consider the inhomogeneity problem in Fig. 4.6(a). Eshelby [2] proposed the equivalent inclusion method (EIM) to solve the inhomogeneity problem by an equivalent inclusion problem in Fig. 4.6(b), where the eigenstrain ε^*_{ij} is to be determined by the stress equivalent condition. In the following, we will predict the eigenstrain distribution in the triangle in Fig. 4.6(b). First, let $\Omega^{\delta}\in\Omega$, so that its center $\mathbf{x}^0$ cannot be either on the boundary of or outside Ω. The integral in Eq. (4.56) can be written in two parts from Ω^δ and $\overline{\Omega}$ with their areas indicated by A^δ and $\overline{A}$. Because Eq. (4.55) is supposed to be satisfied at every point on the triangle Ω, by Eq. (4.56) we can write the integral of Eq. (4.55) over Ω^δ as

$$\Delta C_{ijkl}\left[A^\delta \varepsilon^0_{kl} + \int_{\Omega^\delta} S^\delta_{klmn}(\mathbf{x}' - \mathbf{x}^0)\varepsilon^*_{mn}(\mathbf{x}')d\mathbf{x}' + \int_{\overline{\Omega}} S^\delta_{klmn}(\mathbf{x}' - \mathbf{x}^0)\varepsilon^*_{mn}(\mathbf{x}')d\mathbf{x}'\right] + A^\delta C^0_{ijkl} : \varepsilon^{*\delta}_{kl} = 0 \tag{4.63}$$

where $A^\delta = \pi\delta^2$ is the area of Ω^δ, $\Delta C_{ijkl} = C^1_{ijkl} - C^0_{ijkl}$, and $\varepsilon_{mn}{}^{*\delta}$ is the average eigenstrain of the circle. Eshelby's tensor in the second term of the above equation is a constant, namely S^c_{klmn}, as the eigenstrain is within the circle. The third term is well defined because $|\mathbf{x}' - \mathbf{x}^0| > \delta$ in the domain $\overline{\Omega}$. Due to the continuity of eigenstrain, using the mean value theorem (MVT), we can rewrite Eq. (4.63) as

$$\varepsilon_{ij}{}^{*\delta} = -\left(S^c_{ijkl} + \Delta C^{-1}_{ijpq} C^0_{pqkl}\right)^{-1}\left[\varepsilon^0_{kl} + \frac{1}{A^\delta}\int_{\overline{\Omega}} S^\delta_{klmn}(\mathbf{x}' - \mathbf{x}^0)d\mathbf{x}'\overline{\varepsilon}^*_{mn}\right] \tag{4.64}$$

where $\overline{\varepsilon}^*_{mn}$ is the eigenstrain at a certain point in $\overline{\Omega}$. Using Eq. (4.62), as $\delta \to 0$, the higher-order term of δ^4 rapidly decays to zero, and we can rewrite the last term in the above equation as

$$\frac{1}{A^\delta}\int_{\overline{\Omega}} S^\delta_{klmn}(\mathbf{x}' - \mathbf{x}^0)d\mathbf{x}' = \int_{\overline{\Omega}} \overline{G}_{klmn}(\mathbf{x}', \mathbf{x}^0)d\mathbf{x}' = S^\Omega_{klmn}(\mathbf{x}^0) - S^c_{klmn} \tag{4.65}$$

where $S^\Omega_{klmn}(\mathbf{x}^0)$ is Eshelby's tensor of the large triangle Ω at $\mathbf{x}^0$. Here ewe used the 3rd property of Eshelby's tensor. As $\delta \to 0$, Eq. (4.64) defines the eigenstrain at $\mathbf{x}^0$ as

$$\varepsilon_{ij}{}^{*\delta} = -\left(S^c_{ijkl} + \Delta C^{-1}_{ijpq} C^0_{pqkl}\right)^{-1}\left[\varepsilon^0_{kl} + \left(S^\Omega_{klmn}(\mathbf{x}^0) - S^c_{klmn}\right)\overline{\varepsilon}^*_{mn}\right] \tag{4.66}$$

If $\mathbf{x}^0$ is far from the vertices, then $S^\Omega_{klmn}(\mathbf{x}^0)$ is well defined and bounded. Based on the 2nd property of Eshelby's tensor, $S^c_{ijkl} + \Delta C^{-1}_{ijpq} C^0_{pqkl} \neq \mathbf{0}$. Therefore the above equation can be solved. As $\mathbf{x}^0 \to \mathbf{x}^0$, $S^\Omega_{klmn}(\mathbf{x}^0) \propto \ln|\mathbf{x}^0|$. Therefore $\varepsilon_{ij}{}^{*\delta}$ may exhibit a singularity of $\ln r$ when $\overline{\varepsilon}^*_{mn}$ is nonzero.

Notice that $\overline{\varepsilon}^*_{mn}$ is located within $\overline{\Omega}$, but it can change with $\mathbf{x}^0$ and approach zero. Therefore it can be related to $\varepsilon_{ij}{}^{*\delta}$ as

$$\varepsilon_{ij}{}^{*\delta} = R_{ijmn}\overline{\varepsilon}^*_{mn} \tag{4.67}$$

where R_{ijmn} defines their relationship. Therefore we can rewrite Eq. (4.66) as

$$\varepsilon_{ij}^{*\delta} = -\left[S_{ijkl}^{c} + \Delta C_{ijpq}^{-1} C_{pqkl}^{0} + (S_{ijmn}^{\Omega}(\mathbf{x}^0) - S_{ijmn}^{c}) R_{mnkl}^{-1}\right]^{-1} \varepsilon_{kl}^{0} \quad (4.68)$$

If $\mathbf{R}$ is a definite tensor between $\mathbf{0}$ and ∞, then Eq. (4.68) will provide $\varepsilon_{ij}^{*\delta} \to \mathbf{0}$ because $S_{ijmn}^{\Omega}(\mathbf{x}^0) \to \infty$, which still makes $\mathbf{R} \to 0$. Moreover, as $\mathbf{R} \to 0$, in Eq. (4.66), we can obtain $\varepsilon_{kl}^{0} + \left(S_{klmn}^{\Omega}(\mathbf{x}^0) - S_{klmn}^{c}\right) \overline{\varepsilon}_{mn}^{*} \to \mathbf{0}$, and thus $\overline{\varepsilon}_{mn}^{*} \to \frac{1}{\ln r}$. Therefore $\varepsilon_{ij}^{*\delta}$ can only be either $\mathbf{0}$ or ∞.

For the eigenstrain with singularity level at $\ln r$, the stress or strain field at the vertex exhibits a singularity level at $\ln^2 |r|$, which can be shown as follows. In Eq. (3.3), the Green's function is defined with ϕ and ψ for the convenience of the domain integral. Considering a uniform eigenstrain, to obtain the strain field, the domain integral is partially differentiated twice, and both the harmonic and biharmonic potentials exhibit a $\ln(r)$ singularity. Similarly, let the eigenstrain distribution around the vertex become $\ln |\mathbf{x} - \mathbf{x}'|$. Then the order of singularity can be observed with derivatives of ϕ.

The closed-form domain integral of $\phi \ln |\mathbf{x} - \mathbf{x}'|$ is derived as follows. Because the strain field is obtained by a second-order partial differentiation, we can apply the Green's theorem in Eq. (4.69):

$$\begin{aligned}
\Phi_f^{\log} &= \frac{\partial}{\partial x_i} \int_{\Omega} \phi \ln |\mathbf{x} - \mathbf{x}'| d\mathbf{x}' = \sum_{f=1}^{N_F} (\lambda_f^0)_i \int_{l_f^-}^{l_f^+} \frac{1}{2} (\ln |b_f^2 + \eta^2|)^2 d\eta \\
&= \frac{b_f}{16} \Big[14(l_f^+ - l_f^-) + 3 l_f^- \ln[b_f^2 + (l_f^-)^2] \\
&\quad - l_f^- \ln[b_f^2 + (l_f^-)^2](-3 + \ln[b_f^2 + (l_f^-)^2]) \\
&\quad - 3 l_f^+ \ln[b_f^2 + (l_f^+)^2] + l_f^+ \ln[b_f^2 + (l_f^+)^2](-3 + \ln[b_f^2 + (l_f^+)^2]) \\
&\quad + 2 b_f (\tan^{-1}[\frac{l_f^+}{b_f}] - \tan^{-1}[\frac{l_f^-}{b_f}])\big(-6 + 4\ln[2 b_f]\big) \\
&\quad + 4 i b_f ((\tan^{-1}[\frac{l_f^+}{b_f}])^2 - (\tan^{-1}[\frac{l_f^-}{b_f}])^2 \\
&\quad + Li_2[-e^{i(2\tan^{-1}[\frac{l_f^-}{b_f}])}] - Li_2[-e^{i(2\tan^{-1}[\frac{l_f^+}{b_f}])}] \Big]
\end{aligned} \quad (4.69)$$

where b_f, l_f^-, and l_f^+ are the perpendicular distance and the distances to two neighbor nodes of the edge. Notice that in Eq. (4.69), Li_2 is Spence's function or the polylogarithm function with base 2, and the imaginary part will be canceled out. To obtain the strain field, the partial derivative is conducted following the chain rule $-(\lambda_f^0)_j \partial b_f - (\eta_f^0)_j (\partial l_f^+ - \partial l_f^-)$ [42]. Therefore we need to observe the $\frac{\partial \Phi_f^{\log}}{\partial b_f}$ and $\frac{\partial \Phi_f^{\log}}{\partial l_f^{\pm}}$ for their limits as $b_f \to 0$ and l_f^+ or $l_f^- \to 0$:

(1) the limit of $\frac{\partial \Phi_f^{\log}}{\partial b_f}$ is $\frac{-i\pi^2}{6}$, where the imaginary part will be canceled out when combining with the other part $l_f^- \neq 0$ or $l_f^+ \neq 0$.

(2) the limit of $\frac{\partial \Phi_f^{\log}}{\partial l_f^{\pm}}$ is $2\ln^2[b_f]$, and $b_f \to \mathbf{0}$.

Based on the limits of the derivatives, given a $\ln[r]$ eigenstrain distribution, the stress or strain in the neighborhood of the vertex shall exhibit a singularity level of $\ln^2[r]$, which is only caused by the derivatives with respect to the edge parameter instead of b_f.

Similarly, for 3D cases, Eshelby's tensor exhibits a singularity level of $\frac{1}{r}$. We can obtain a similar conclusion by replacing $\ln r$ in 2D cases with $\frac{1}{r}$ in 3D cases, and obtain the following hypothesis:

Hypothesis 1. *When an angular elastic inhomogeneity embedded in another elastic solid is subject to an external load in 2D cases, the mismatch eigenstrain approaches either zero or infinity at rate* $\ln r$*, which leads to a stress singularity up to* $\ln^2 r$*. Similarly for 3D cases, the mismatch eigenstrain approaches either zero or infinity at rate* $\frac{1}{r}$*, which leads to the stress singularity up to* $\frac{1}{r^2}$*.*

This is different from the traditional view of the singularity of wedges in solids. More investigation will be done for clarification and validation.

Now consider a continuously distributed eigenstrain field with a singular eigenstrain in the neighborhood of the vertex, namely Ω^δ, in Fig. 4.6(b) by moving the circle center $\mathbf{x}^0$ to $\mathbf{0}$. Based on Lemma 4, the contribution of eigenstrain in $\overline{\Omega}$ to the strain at the vertex decays in a logarithmic scale, which is not singular, so the singularity of elastic stress or strain will increase proportionally only with the eigenstrain ε_{mn}^{*0} in Ω^δ. Because Eshelby's tensor exhibits a $\ln r$ singularity, the stress at the vertex will exhibit a $\ln^2 r$ singularity in case that the eigenstrain approaches infinity at rate $\ln r$. It is still a weaker singularity than r^{-n} with $n > 0$, but it is higher than the inclusion problem with a constant eigenstrain.

Similarly, the contribution of the eigenstrain in Ω^δ to the strain at a point $\mathbf{x}$ far from Ω^δ can be written as

$$\varepsilon_{kl}'(\mathbf{x}) = \int_{\Omega^\delta} \overline{G}_{klmn}(\mathbf{x}, \mathbf{x}')\varepsilon_{mn}^*(\mathbf{x}')d\mathbf{x}' = \overline{G}_{klmn}(\mathbf{x}, \mathbf{x}^a)\int_{\Omega^\delta} \varepsilon_{mn}^*(\mathbf{x}')d\mathbf{x}' \quad (4.70)$$

where $\mathbf{x}^a \in \Omega^\delta$. Here we used the MVT, because $\overline{G}_{klmn}(\mathbf{x}, \mathbf{x}')$ is well defined for $\mathbf{x}$ out of Ω^δ and $\varepsilon_{mn}^*(\mathbf{x}')d\mathbf{x}' = \varepsilon_{mn}^*(\mathbf{x}')r dr d\theta$ with θ varying between 0 to 2π is well defined too [1]. In other words, a singularity of the eigenstrain at rate $\ln r$ in the expression of $\ln r d\mathbf{x}'$ is transferred to $r \ln r dr d\theta$, which exhibits no singularity, so that the MVT is applicable.

For a weak singularity of the eigenstrain at rate $\ln r$, the integral over Ω^δ is definite. As $\delta \to 0$, we have $\mathbf{x}^a \to \mathbf{0}$, so that the contribution of the singular eigenstrain at the vertex decays at rate $\frac{1}{r^2}$ based on the definition of $\overline{G}_{klmn}$ in Eq. (4.57). Therefore, although the value of eigenstrain on the vertex is not defined, its effect on the overall solution is minimal. Similarly, for 3D cases, the influence of the singular eigenstrain decays at rate $\frac{1}{r^3}$ based on the definition of $\overline{G}_{klmn}$ in Eq. (4.60). Therefore we have the following hypothesis:

Hypothesis 2. *When an angular elastic inhomogeneity embedded in another elastic solid is subject to an external load, although the mismatch eigenstrain at the vertex exhibits a singularity at* $\ln r$ *and* $\frac{1}{r}$ *for 2D and 3D cases, respectively, its influence is limited to the neighborhood of the vertex and decays at* $\frac{1}{r^2}$ *and* $\frac{1}{r^3}$ *in the far-field for 2D and 3D cases, respectively.*

Moreover, the term with ΔC^{-1}_{klij} in Eq. (4.68) plays a unique role in the solution for eigenstrain in the neighborhood of the vertex. When the stiffness of the inhomogeneity becomes the same as the matrix, $\Delta C^{-1}_{klij} \to \infty$, so the mismatch eigenstrain is always zero because of no material mismatch. When the material approaches perfectly rigid, this term becomes zero. When the inhomogeneity is very compliant, it approaches $-\mathbf{I}$.

Notice that for 2D cases, the stress singularity level of $\ln^2 r$ is lower than the classic crack at the order of $1/\sqrt{r}$ measured by the stress intensity factor, which can, however, be obtained by the present method taking the limit of $\theta = 0$ and $C^1_{ijkl} = 0$. Therefore eigenstrain and its distribution at the neighborhood of the vertex can serve as the measurement of stress singularity, which is similar to the stress intensity factor for cracks. As many researchers found it difficult to use $1/\sqrt{\mathbf{x}'}$ to explain the size effect with fracture strength [98], the stress singularity of $\ln^2|\mathbf{x}'|$ may shed some light on the size effect of materials.

Comparing the 2D and 3D cases, their main differences are as follows:

1. Eshelby's tensor exhibits a $1/r$ singularity in 3D vs. $\ln r$ in 2D;
2. The eigenstrain in the neighborhood of the vertex exhibits a $1/r$ singularity in 3D vs. $\ln r$ in 2D;
3. The stress in the neighborhood of the vertex exhibits a $1/r^2$ singularity in 3D vs. $\ln^2 r$ in 2D;
4. The influence of the eigenstrain singularity decays at rate $1/r^3$ in 3D vs. $1/r^2$ in 2D.

*4.6 Extension to an semi-infinite domain

EIM is first used to solve for the elastic field of one inhomogeneity embedded in a semi-infinite domain and is then generalized to the multiple-inhomogeneity case.

*4.6.1 One inhomogeneity

In the last section, the semi-infinite domain D with stiffness $\mathbf{C}$ contains an inclusion Ω with an eigenstrain field $\boldsymbol{\varepsilon}^*(\mathbf{x})$. The elastic field has been solved in Eq. (3.79). In this section, the subdomain Ω in Fig. 3.7 is replaced by an inhomogeneity with a different stiffness $\mathbf{C}^1$. A uniform stress $\boldsymbol{\sigma}^0$ is applied in the far-field. Due to the material mismatch, the local field in the neighborhood of Ω will be disturbed. To solve this problem, in this section, we will use the concept of Eshelby's EIM [2,3]. The particle Ω is firstly filled with the same matrix material of $\mathbf{C}$, but an eigenstrain $\boldsymbol{\varepsilon}^*(\mathbf{x})$ is applied in Ω to make the stress Ω in the two cases equivalent. Then the BVPs can be equivalently solved.

For one inhomogeneity in an infinite solid subjected to a uniform far-field stress, the eigenstrain is uniform [1–3]. However, due to the uniform boundary on the semi-infinite solid, the eigenstrain is no longer constant. The variation of $\boldsymbol{\varepsilon}^*(\mathbf{x})$ should be continuous over Ω, which can be written in terms of a polynomial of $\mathbf{x} - \mathbf{x}^c$ with the origin at the center of the ellipsoid as

$$\varepsilon_{ij}^*(\mathbf{x}) = E_{ij}^0 + E_{ijk}^1\left(x_k - x_k^c\right) + E_{ijkl}^2\left(x_k - x_k^c\right)\left(x_l - x_l^c\right) + \cdots, \quad \mathbf{x} \in \Omega \tag{4.71}$$

where the eigenstrain shares the same physical meaning of polarization strain as Eshelby's problem.

Substituting the constitutive law into the equilibrium equation yields

$$C_{ijkl}\left(u_{k,li} - \varepsilon_{kl,i}^*\right) = C_{ijkl}^1 u_{k,li} \tag{4.72}$$

Therefore, if the eigenstrain can be derived from the above equation, then we can obtain the elastic field. Using Taylor's expansion, we can write the disturbed strain inside the particle as

$$\begin{aligned}
\varepsilon_{ij}'(\mathbf{x}) = &-\frac{1}{4\pi\mu_0}\Big[D_{ijkl}^0\left(\mathbf{x}^c\right)C_{klmn}E_{mn}^0 + D_{pijkl}^1\left(\mathbf{x}^c\right)C_{klmn}E_{mnp}^1 \\
&+ D_{stijkl}^2\left(\mathbf{x}^c\right)C_{klmn}E_{mnst}^2\Big] \\
&-\frac{1}{4\pi\mu_0}\begin{bmatrix} D_{ijkl,r}^0\left(\mathbf{x}^c\right)C_{klmn}E_{mn}^0 \\ +D_{pijkl,r}^1\left(\mathbf{x}^c\right)C_{klmn}E_{mnp}^1 \\ +D_{stijkl,r}^2\left(\mathbf{x}^c\right)C_{klmn}E_{mnst}^2 \end{bmatrix}\left(x_r - x_r^c\right) \\
&-\frac{1}{8\pi\mu_0}\begin{bmatrix} D_{ijkl,yz}^0\left(\mathbf{x}^c\right)C_{klmn}E_{mn}^0 \\ +D_{pijkl,yz}^1\left(\mathbf{x}^c\right)C_{klmn}E_{mnp}^1 \\ +D_{stijkl,yz}^2\left(\mathbf{x}^c\right)C_{klmn}E_{mnst}^2 \end{bmatrix}\left(x_y - x_y^c\right)\left(x_z - x_z^c\right)
\end{aligned} \tag{4.73}$$

When a uniform force is applied on the boundary of the semi-infinite domain, a uniform average stress σ^0 will be caused. For the equivalent inclusion problem, a uniform strain $\varepsilon_{ij}^0 = C_{ijkl}^{-1}\sigma_{kl}^0$ will be induced, and the equivalent inclusion condition is written as

$$C_{ijkl}^1\left(\varepsilon_{kl}^0 + \varepsilon_{kl}'\right) = C_{ijkl}\left(\varepsilon_{kl}^0 + \varepsilon_{kl}' - \varepsilon_{kl}^*\right) \tag{4.74}$$

which can be rewritten as

$$\Delta\mathbf{C} : \left(\boldsymbol{\varepsilon}^0 + \boldsymbol{\varepsilon}'\right) = -\mathbf{C} : \boldsymbol{\varepsilon}^* \tag{4.75}$$

where $\Delta\mathbf{C} = \mathbf{C}^1 - \mathbf{C}$. Substituting Eq. (4.73) into Eq. (4.75) and comparing the coefficients of the constant, linear, and quadratic terms at left and right sides, we obtain the following equations system:

$$\begin{cases} \begin{bmatrix} \left[\Delta C_{ijkl}D_{klst}(\mathbf{x}^c)C_{stmn} - 4\pi\mu_0 C_{ijmn}\right]E_{mn}^0 \\ +\Delta C_{ijkl}D_{pklst}(\mathbf{x}^c)C_{stmn}E_{mnp}^1 + \Delta C_{ijkl}D_{pqklst}(\mathbf{x}^c)C_{stmn}E_{mnpq}^2 \end{bmatrix} \\ = 4\pi\mu_0\Delta C_{ijkl}\varepsilon_{kl}^0 \\ \begin{bmatrix} \left[\Delta C_{ijkl}D_{pklst,r}(\mathbf{x}^c)C_{stmn} - 4\pi\mu_0 C_{ijmn}\delta_{pr}\right]E_{mnp}^1 \\ +\Delta C_{ijkl}D_{klst,r}(\mathbf{x}^c)C_{stmn}E_{mn}^0 + \Delta C_{ijkl}D_{pqklst,r}(\mathbf{x}^c)C_{stmn}E_{mnpq}^2 \end{bmatrix} = 0 \\ \begin{bmatrix} \Delta C_{ijkl}D_{klst,yz}(\mathbf{x}^c)C_{stmn}E_{mn}^0 + \Delta C_{ijkl}D_{pklst,yz}(\mathbf{x}^c)C_{stmn}E_{mnp}^1 \\ +\left[\Delta C_{ijkl}D_{pqklst,yz}(\mathbf{x}^c)C_{stmn} - 8\pi\mu_0 C_{ijmn}\delta_{py}\delta_{qz}\right]E_{mnpq}^2 \end{bmatrix} = 0 \end{cases} \tag{4.76}$$

Solving this linear equation system, we can obtain the eigenstrain terms of $\mathbf{E}^0$, $\mathbf{E}^1$, and $\mathbf{E}^2$, with which the displacement is written as

$$\begin{aligned} u_i(\mathbf{x}) &= -\int_\Omega \Gamma_{imn}C_{mnkl}\left(E_{kl}^0 + E_{klp}^1(x_p' - x_p^c) + E_{klpq}^2(x_p' - x_p^c)(x_q' - x_q^c)\right)d\mathbf{x}' \\ &= -\frac{1}{4\pi\mu_0}\left(M_{imn}^0 C_{mnkl}E_{kl}^0 + N_{pimn}^1 C_{mnkl}E_{klp}^1 + T_{pqimn}^2 C_{mnkl}E_{klpq}^2\right) \end{aligned} \tag{4.77}$$

where the explicit forms of M_{imn}^0, N_{pimn}^1, and T_{pqimn}^2 are available in Section 3.5.1. The strain is calculated by Eq. (3.79), and the stress can be obtained by the constitutive law.

*4.6.2 Multiple inhomogeneities

The above formulation can be extended from a single particle to multiple particles in a very straightforward way. When many particles are inserted into

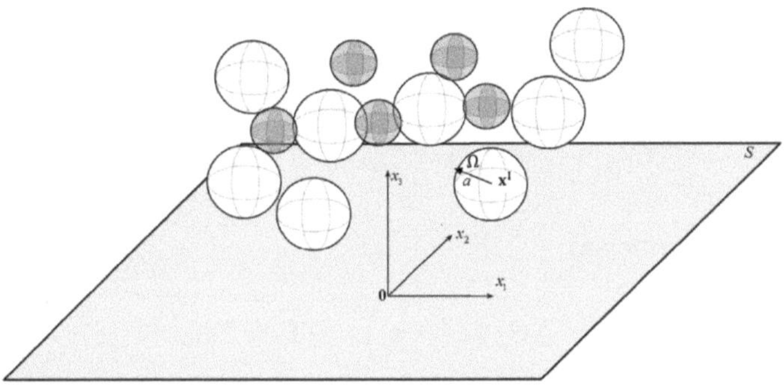

FIGURE 4.7 A semi-infinite domain D ($x_3 \geq 0$) containing many particles.

the semi-infinite domain, their interactions significantly affect the elastic fields. Fig. 4.7 illustrates the whole system with many particles. To formulate this problem, a global coordinate system has been set up in Fig. 4.7, and each particle, say Ω^I, is centered at $\mathbf{x}^I = (x_1^I, x_2^I, x_3^I)$. All particles treated as the same material properties with the matrix, the eigenstrain can be introduced with a polynomial form as

$$\varepsilon_{ij}^{*,I}(\mathbf{x}) = E_{ij}^{0,I} + E_{ijk}^{1,I}\left(x_k - x_k^I\right) + E_{ijkl}^{2,I}\left(x_k - x_k^I\right)\left(x_l - x_l^I\right) + \cdots, \quad \mathbf{x} \in \Omega^I \tag{4.78}$$

where no summation is applied to the superscript, whereas repeated subscripts still mean summation. Then the equivalent condition (4.74) for one of the particles, say Ω^J, reads

$$C_{ijkl}^1\left(\varepsilon_{kl}^0 + \varepsilon_{kl}'\right) = C_{ijkl}\left(\varepsilon_{kl}^0 + \varepsilon_{kl}' - \varepsilon_{kl}^{*,J}\right), \quad \mathbf{x} \in \Omega^J \tag{4.79}$$

The displacement of any point in the domain is caused by the eigenstrain of all particles, which can be obtained by the superposition as follows:

$$u_i(\mathbf{x}) = -\frac{1}{4\pi\mu_0}\sum_{I=1}^{N}\left(M_{imn}^{0,I}C_{mnkl}E_{kl}^{0,I} + N_{pimn}^{1,I}C_{mnkl}E_{klp}^{1,I}\right.$$
$$\left. + T_{pqimn}^{2,I}C_{mnkl}E_{klpq}^{2,I}\right) \tag{4.80}$$

Expanding the disturbed strain on particle Ω^J and substituting it into Eq. (4.79), we obtain the following equation system by comparing the coefficients for different orders of $\left(x_r - x_r^J\right)$ terms:

$$
\left\{
\begin{array}{l}
\left[
\begin{array}{l}
\sum_{I=1}^{N} \Delta C_{ijkl} D^{0}_{klst}\left(\mathbf{x}^{c}\right) C_{stmn} E^{0,I}_{mn} - 4\pi \mu_0 C_{ijmn} E^{0,J}_{mn} \\
\qquad + \sum_{I=1}^{N} \Delta C_{ijkl} D^{1,I}_{pklst}\left(\mathbf{x}^{c}\right) C_{stmn} E^{1,I}_{mnp} \\
\qquad + \sum_{I=1}^{N} \Delta C_{ijkl} D^{2,I}_{pqklst}\left(\mathbf{x}^{c}\right) C_{stmn} E^{2,I}_{mnpq}
\end{array}
\right] = 4\pi \mu_0 \Delta C_{ijkl} \varepsilon^{0}_{kl} \\
\left[
\begin{array}{l}
\sum_{I=1}^{N} \Delta C_{ijkl} D^{1,I}_{pklst,r}\left(\mathbf{x}^{c}\right) C_{stmn} E^{1,I}_{mnp} - 4\pi \mu_0 C_{ijmn} \delta_{pr} E^{1,J}_{mnp} \\
\qquad + \sum_{I=1}^{N} \Delta C_{ijkl} D^{0,I}_{klst,r}\left(\mathbf{x}^{c}\right) C_{stmn} E^{0,I}_{mn} \\
\qquad + \sum_{I=1}^{N} \Delta C_{ijkl} D^{2,I}_{pqklst,r}\left(\mathbf{x}^{c}\right) C_{stmn} E^{2,I}_{mnpq}
\end{array}
\right] = 0 \\
\left[
\begin{array}{l}
\sum_{I=1}^{N} \Delta C_{ijkl} D^{0,I}_{klst,yz}\left(\mathbf{x}^{c}\right) C_{stmn} E^{0,I}_{mn} \\
+ \sum_{I=1}^{N} \Delta C_{ijkl} D^{1,I}_{pklst,yz}\left(\mathbf{x}^{c}\right) C_{stmn} E^{1,I}_{mnp} \\
+ \sum_{I=1}^{N} \Delta C_{ijkl} D^{2,I}_{pqklst,yz}\left(\mathbf{x}^{c}\right) C_{stmn} E^{2,I}_{mnpq} \\
\qquad - 8\pi \mu_0 C_{ijmn} \delta_{py} \delta_{qz} E^{2,J}_{mnpq}
\end{array}
\right] = 0
\end{array}
\right.
\tag{4.81}
$$

Due to symmetry, there are in total 60 independent components of the eigenstrain for each particle when quadratic terms are considered. Therefore Eq. (4.81) can be rewritten in the matrix notation as

$$
[A]^{J}_{60\times 60n}\,[E]_{60n} = [F]^{J}_{60} \tag{4.82}
$$

where the eigenstrain terms of all the particles are represented by a vector $[E]$ of length 60 n, $[A]^J$ is the local coefficient matrix of dimensions 60×60 n, whose elements are the coefficients in Eq. (4.81), and $[F]^J$ is the local force vector of length 60, whose elements are the sums of the body forces multiplied by the corresponding coefficients.

For each particle, we can write one set of equations as before. Rewriting all the equations in the matrix notation and assembling all $[A]^J$ into a global coefficient matrix and all $[F]^J$ into a global force vector, we obtain

$$[A]_{60\text{n}\times 60\text{n}}[E]_{60\text{n}} = [F]_{60\text{n}} \tag{4.83}$$

Solving this global linear equation system, we obtain the eigenstrain terms of all particles. Substituting them into Eq. (4.80), we can also obtain the displacement field and thus the strain and stress fields.

In the numerical implementation of this algorithm, for one particle with uniform, linear, and quadratic eigenstrain distribution, only 6, 18, and 60 degrees of freedom are respectively needed. To investigate the elastic fields of a sample with N particles, we only need $60N$ degrees of freedom, which can provide quadratic accuracy. In comparison, FEM needs to mesh the entire domain, and the mesh near the particles needs to be very fine, so as to obtain accurate results. Generally, hundreds of thousands of degrees of freedom are needed for 3D FEM simulation of a large domain with several particles. Therefore the computational cost of our approach is fairly low compared with the FEM modeling.

Note that the boundary value problem is solved by a continuum mechanics-based approach, and the particle interactions have been implicitly included. The pairwise interaction exists between any pair of particles, and each particle is subjected to the effects of all other particles, which generally decay with interparticle distance rapidly. Therefore we can set a cut-off distance when $[A]^J_{60\times 60\text{n}}$ in Eq. (4.82) is calculated, which will make the global coefficient matrix $[A]_{60\text{n}\times 60\text{n}}$ a sparse matrix for a large-scale particle system. Below we demonstrate the method with a few examples.

*4.6.3 Case study – one particle

Consider that an aluminum particle with radius $a = 10$ μm is embedded in a semi-infinite HDPE at $(0, 0, h)$, where $h = 11, 20$, and 30 μm to the boundary. Young's moduli and Poisson's ratios for aluminum and HDPE are $E_1 = 69$ GPa, $E_0 = 0.8$ GPa, $\nu_1 = 0.334$, and $\nu_0 = 0.38$, respectively [13,99,100]. The strain components corresponding to the uniform load, with $\varepsilon^0_{33} = 0.001$ and $\varepsilon^0_{11} = \varepsilon^0_{22} = -\nu_0\varepsilon^0_{33}$, are applied to the system to represent a far-field tension $\sigma^0_{33} = 0.8$ MPa. Under this far-field tension, the elastic field is investigated.

Fig. 4.8 illustrates the distribution of the eigenstrain component ε^*_{33} inside the inhomogeneity at the x_1–x_3 plane, when the aluminum particle is located at $h = 11$ μm, which means 1-μm spacing to the boundary of a particle with a diameter of 20 μm. A significant difference is observed when the eigenstrain is assumed to be uniformly (cf. Fig. 4.8a), linearly (cf. Fig. 4.8b), and quadratically distributed (cf. Fig. 4.8c), respectively. As Eshelby [2,3] pointed out, the eigenstrain inside an inhomogeneity should be uniform when the inhomogeneity is embedded into an infinite domain subjected to a uniform far-field stress. However, due to the boundary effect, obviously, the eigenstrain is not uniform anymore. A polynomial form of eigenstrain may approach the exact solution

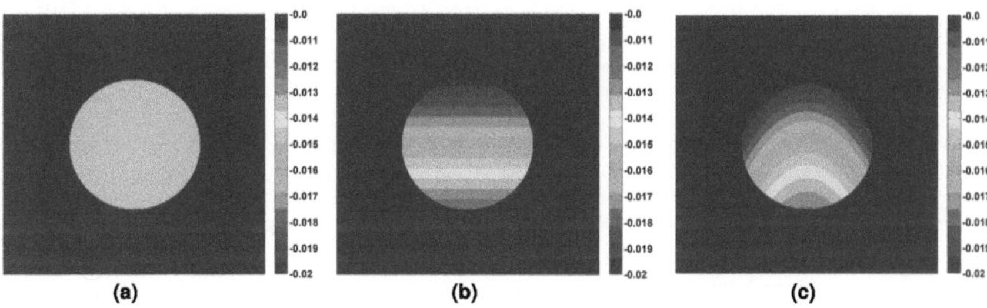

FIGURE 4.8 Distribution of ε_{33}^{*} in the inclusion: (a) uniform distribution; (b) linear distribution; (c) quadratic distribution.

when higher-order terms are used. In the following, we will compare the above three solutions with the FEM result to verify their accuracy. FEM models built with commercial software ABAQUS are used to verify our analytical solution. We use an axisymmetric domain with 0.5 mm × 1 mm to represent the semi-infinite domain with free boundary at $x_3 = 0$ and a total of 92,000 unstructured 6-node triangle elements to approximate the solutions.

Fig. 4.9a illustrates the variation of stress σ_{33} along the x_3 axis ($x_1 = x_2 = 0$) when $h = 11$ μm. We can observe a considerable difference in the results predicted by EIM with a constant eigenstrain from the FEM result. When linear terms are introduced, much better results are obtained. When quadratic terms are considered, the results almost overlap with the FEM results, which indicates that the quadratic form of eigenstrain is sufficient for the present elastic analysis when the spacing to the boundary is higher than 5% of the particle size. When the particle is embedded at $h = 20$ μm (Fig. 4.9b), the results obtained by uniformly and linearly distributed eigenstrains are much closer to the FEM results than those in the previous case because the boundary effect decays with h. With h increasing, the difference in the results from uniform, linear, and quadratic distributed eigenstrain decreases. When h is large enough, the model reduces to one particle embedded into an infinite domain [2,3], in which the eigenstrain is uniform inside the inhomogeneity, and higher-order terms disappear.

In Fig. 4.9a, σ_{33} is equal to the far-field tension at the boundary $x_3 = 0$ and decreases with x_3 to the bottom of the particle. Inside the particle, it increases with x_3 to the top of the particle. Then a sudden change of the slope is observed, and σ_{33} increases quickly to its maximum value, which is about $2\sigma_{33}^{0}$. Then it decreases with x_3 and finally converges to the far-field tension σ_{33}^{0}. Fig. 4.9b illustrates the variation of σ_{33} with x_3. When the aluminum particle is embedded at $h = 20$ μm, whose trend is quite different from that for $h = 11$ μm. The stress σ_{33} is equal to the far-field tension σ_{33}^{0} at the boundary and then increases with x_3 and reaches the peak value near the bottom of the particle. Then it decreases with x_3 to the bottom of the particle. Inside the particle, σ_{33} increases with x_3. We can observe here that the change of σ_{33} inside the particle is much smaller than that for $h = 11$ μm. Then it goes outside of the particle and increases to the maximum value quickly near the top of the particle. Finally, it

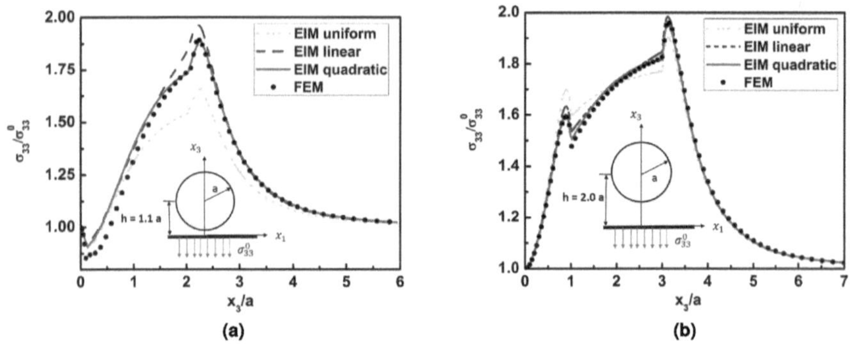

FIGURE 4.9 Variation of stress σ_{33} with x_3: (a) $h = 1.1a$ and (b) $h = 2a$.

decreases with x_3 and converges to the far-field load. In Fig. 4.9b the stress distribution on the particle is not symmetric either, and the peak value near the bottom of the particle is smaller than those near the top. Actually, when the distance between the particle and boundary is large enough, the boundary effect is negligible, so that the result will reduce to Eshelby's original problem for one particle embedded in an infinite domain, with which the curve becomes absolutely symmetric, and the stress inside the particle is uniform.

To investigate the shear deformation, we apply a uniform shear load $\sigma_{31}^0 = 0.58\,\text{MPa}$ at the above system, in which $h = 11$ μm. Fig. 4.10a illustrates the variation of the shear stress with x_3. The shear stress σ_{31} equals the far-field shear stress σ_{31}^0 at the surface and then increases rapidly with x_3 and reaches the inflection point at the bottom of the aluminum particle. It continuously increases inside the particle and arrives at the maximum value at the top of the particle. It then decreases with x_3 fast and reaches the minimum value $\sim 0.75\sigma_{31}^0$. It increases again and then converges to the far-field shear load.

We compare the results from EIM with different eigenstrain forms. Observations show that the stress obtained from uniformly distributed eigenstrain is much lower than those calculated from linear and quadratic distributed eigenstrain. Because the shear stress distribution in the particle is fairly close to a linear trend, both the linear and quadratic forms of eigenstrains provide a similar prediction. Therefore the quadratic distributed eigenstrain can be considered as the convergent solution from the asymptotic analysis point of view.

To investigate the boundary effect, σ_{31} vs. x_3 curves for three cases of $h = 11$, 20, and 30 μm are illustrated in Fig. 4.10b. The shear stresses at the surface are equal to the far field load. When $h = 11$ μm, the stress increases with x_3 to the inflection point at the bottom of the particle, whereas for the cases of $h = 20$ and 30 μm, the stress first decreases along the x_3 axis and then increases to the inflection point at the bottom of the particle. Inside the particle, σ_{31} increases with x_3. The variation of stress in the particle decreases for a higher distance value of h/a. We can predict that the stress distribution in the particle will approach a constant when the distance between the particle and boundary

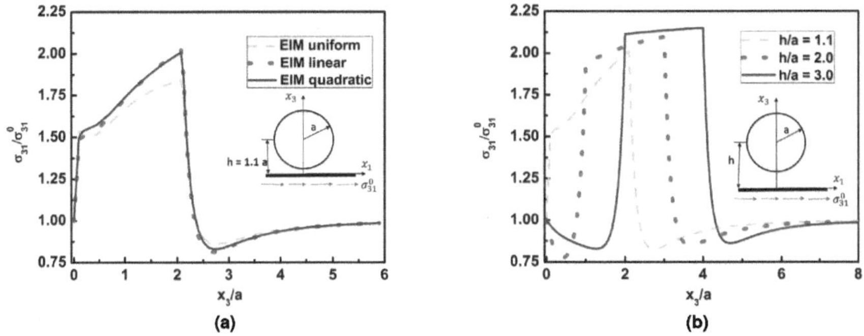

FIGURE 4.10 Variation of shear stress σ_{31} with x_3: (a) $h = 11$ μm for different forms of eigenstrain distribution; (b) quadratic form of eigenstrain for different h values.

is high enough, which makes the case reduce to the classical Eshelby problem of one particle embedded into an infinite domain. Notice that the simple shear problem is not axisymmetric, and thus FEM modeling has not been conducted for the high computational cost because it is not the focus of this chapter.

*4.6.3.1 Case study – two top-down particles

When two aluminum particles are embedded into the same semi-infinite HDPE domain, the problem becomes more complicated due to the coupling of the particle–boundary interaction and the particle–particle interaction [4]. For simplicity, we consider two relative positions of particles: top-down and side-by-side. For the top-down case, the two particles are located at $(0, 0, h)$ and $(0, 0, h + s)$, where s is the center–center distance of two particles. Obviously, s should be higher than $2a$ to avoid the overlap of particles. The same uniaxial load of $\sigma_{33}^0 = 0.8MPa$ is applied to the system corresponding to $\varepsilon_{33}^0 = 0.001$ and $\varepsilon_{11}^0 = \varepsilon_{22}^0 = -\nu_0\varepsilon_{33}^0$.

First, an extreme case is considered first with $h = 2.0a$ and $s = 2.5a$. The results from EIM using uniform, linear, and quadratic forms of eigenstrains are compared with the FEM results in Fig. 4.11a. FEM results are obtained by using an axisymmetric model of commercial software ABAQUS. The solution is approximated by about 135,000 unstructured 6-node triangle elements. We can find that σ_{33} increases with x_3 from the boundary to a specific peak value and then decreases a little bit when reaches the bottom of the first particle. Then it increases again until it reaches the midpoint of two particles, $(0, 0, h + s/2)$. Then it decreases with x_3 until the top of the second particle. After a slight increment, it decreases again and finally converges to the far-field tension load. The four curves are fairly different. EIM with uniform and linearly distributed eigenstrain cannot accurately address the stress distribution. The maximum stress from uniform distributed eigenstrain has a relative error of 30%. However, the EIM with the quadratic form of eigenstrain agrees with the FEM results very well, in which the difference of the maximum stress is less than 3%.

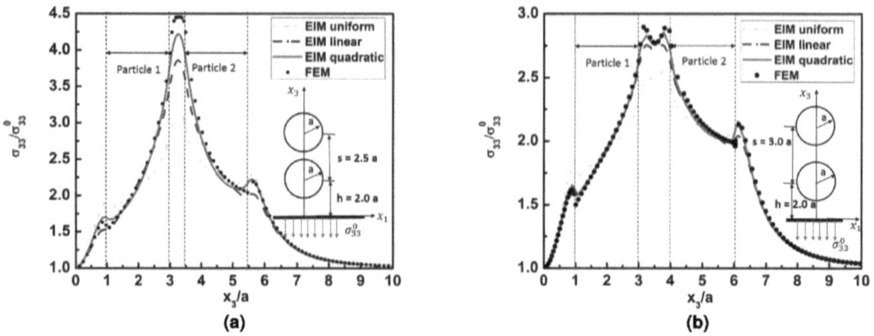

FIGURE 4.11 Variation of stress σ_{33} with x_3 when two top-down particles are embedded into a semi-infinite domain: (a) $s = 2.5a$ and (b) $s = 3.0a$.

When the distance between two particles is larger, the eigenstrain is simpler due to the decay of particle–particle interaction. Fig. 4.11b illustrates the variation of σ_{33} with x_3 when $s = 3.0a$. We can see that the difference between the four curves is smaller although the EIM results for uniform eigenstrain are still fairly distinct from the other three. When we introduce quadratic terms in the eigenstrain, we obtain much better agreement with the FEM results, which highlights the advantage of using the quadratic form of eigenstrain.

Notice that when two particles are closer (cf. Fig. 4.11a), the maximum stress locates in the middle of the two particles, whereas when their distance is longer (cf. Fig. 4.11b), the maximum stress locates close to the boundary of the two particles.

*4.6.3.2 *Case study – two side-by-side particles*

When the relative positions of two particles are side-by-side, the problem is no longer axisymmetric. Therefore FEM modeling will not be conducted. The two aluminum particles are centered at $(-s/2, 0, h)$ and $(s/2, 0, h)$. Under the same far-field load of $\sigma_{33}^0 = 0.8\,\text{MPa}$, the stress profile is investigated. Fig. 4.12a illustrates the variation of σ_{11} with x_1 for $x_3 = h$. In this case, $s = 2.5a$, whereas $h = 1.1a$. We can observe that σ_{11} is negative, which indicates that the particle interaction produces compression between particles in the x_1 direction. The curve is symmetric. For $x_1 \leq 0$, $|\sigma_{11}|$ is zero at far-field and then increases with x_1 and reaches the local peak value near the left particle at about $0.45\sigma_{33}^0$. Then $|\sigma_{11}|$ decreases to a local minimum value, say $0.34\sigma_{33}^0$, in the particle. After that, $|\sigma_{11}|$ increases fast and reaches the maximum value $1.47\sigma_{33}^0$ at the middle point between two particles. We can conclude that the center of the two particles is under high compression due to the particle interaction. We compare the EIM results with uniform, linear, and quadratic forms of eigenstrain. The maximum values from a uniform and linear distributed eigenstrain are much smaller than those from quadratic ones. Because h and s are fairly small, the effects of boundary and particle–particle interactions are very significant, so that the eigenstrain

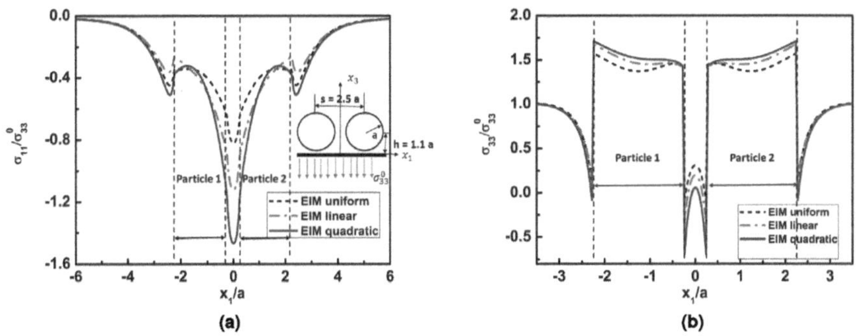

FIGURE 4.12 Variation of stress (a) σ_{11} and (b) σ_{33} vs. x_1/a ($a = 10\,\mu\text{m}$).

distribution is very complex. Therefore the EIM with constant and linear terms cannot provide accurate results for this case.

Fig. 4.12b illustrates the symmetric distribution of σ_{33} with x_1. For $x_1 \leq 0$, the stress σ_{33} decreases from σ_{33}^0 at the far-field to a small negative value at the boundary of the left particle. Then a large jump to $1.71\sigma_{33}^0$ is observed at the particle boundary. Then it decreases with x_1 inside the particle. A large drop is also observed at the other boundary of the particle, where σ_{33} jumps from about $1.43\sigma_{33}^0$ to about $-0.69\sigma_{33}^0$. Finally, it increases to a small positive value at the center of the two particles. The three forms of eigenstrain distribution exhibit a noticeable difference on the particle domain, but the difference between the linear and quadratic forms is smaller than that between the uniform and linear forms, which indicates that the quadratic form solution should be close to the convergent solution.

To investigate the particle interactions and boundary we consider different configurations of h/a and s/a. Fig. 4.13a illustrates the stress σ_{33} distribution along x_1 in the case of $x_3 = h$ and $s/a = 2.5$. Only the right half of configuration is illustrated considering the symmetry. For the three cases of $h = 1.1a$, $h = 2.0a$, and $h = 3.0a$, the stress σ_{33} inside the particle caused by σ_{33}^0 increases with h. The difference in σ_{33} between $h = 2.0a$ and $3.0a$ is much less than that between $h = 1.1a$ and $h = 2.0a$, which means that the boundary effect to the local field in the neighborhood of particles is small for $h \geq 2.0a$. Moreover, σ_{33} outside the particle is almost the same for all three cases. Fig. 4.13b illustrates the stress σ_{33} distribution for different center–center distances, $s = 2.5a$, $s = 3.0a$, and $s = 4.0a$, whereas $h = 1.1a$. The stress component σ_{33} is greater than zero at the midpoint of two particles, i.e., $x_1 = 0$, and the value increases with s. When $s = 3.0a$, σ_{33} is close to the far-field applied stress. The peak value of σ_{33} appears at the left and right boundaries of the particle. When the particle–particle distance is larger, the peak value of stress is higher. Moreover, the stress near the boundary of the particle is negative. The absolute value is higher when s is smaller.

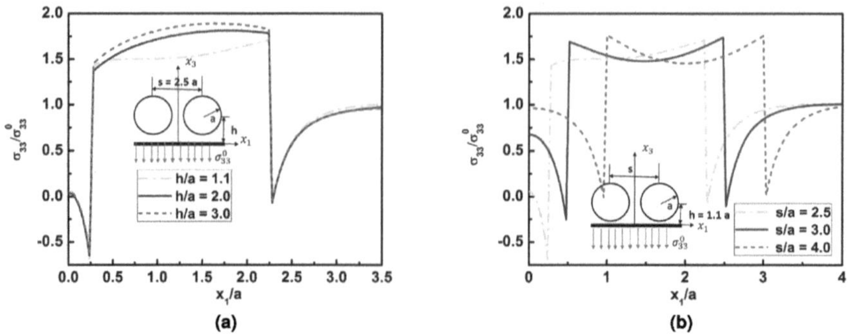

FIGURE 4.13 Variation of σ_{33} with x_1: (a) comparison of different h and (b) comparison of different s.

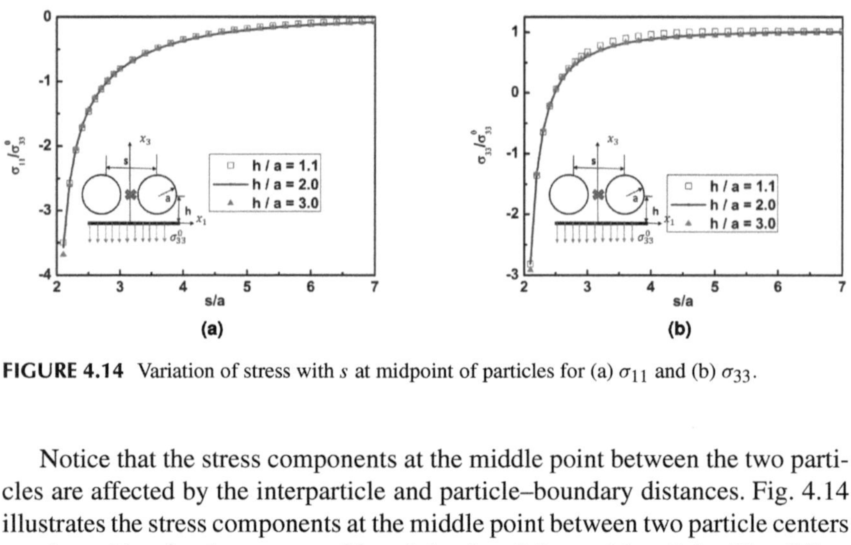

FIGURE 4.14 Variation of stress with s at midpoint of particles for (a) σ_{11} and (b) σ_{33}.

Notice that the stress components at the middle point between the two particles are affected by the interparticle and particle–boundary distances. Fig. 4.14 illustrates the stress components at the middle point between two particle centers varying with s for three cases of $h = 1.1a$, $h = 2.0a$, and $h = 3.0a$. The difference between the three cases is fairly small, which indicates that the boundary effect on the stress at the middle point of two particles is limited. However, the particle interaction produces a significant effect. When the particle–particle distance is very close, say $s = 2a$, although only a tension stress σ_{33}^0 is applied, the induced stress σ_{11} is near $-4.0\sigma_{33}^0$, whereas σ_{33} is near $-3.0\sigma_{33}^0$, which indicates that the center of two particles is under compression. With the increase of s, σ_{11} and σ_{33} increase and finally approach 0 and σ_{33}^0, respectively. It is interesting that σ_{33} is negative for $s = 2a - 2.4a$ and then turn to positive for $s > 2.4a$, which indicates that the stress direction can change with the particle–particle distance.

In summary, the well-known Mindlin solution, reorganized in an elegant concise form, is used to study the elastic field for one particle in a semi-infinite domain. Although previous work has considered a force concentrated on a point or a prescribed strain in an inclusion, it is the first work to use it to study the inhomogeneous materials. The solution has been extended to multiparticle systems for actual composite material samples. Because each particle is represented by an eigenstrain function, this analytical treatment of particles enables us to simulate hundreds to thousands of particles without the necessity of meshing the particles [37]. To our knowledge, this can be among or beyond the highest number of particles that FEM has simulated in the literature [101].

Appendix 4.A Domain integral with quadratic shape function

In this subsection, we discuss its extension to the quadratic shape function for 2D and 3D problems, respectively. As illustration, let us consider the following types of elements: 1) 6-node quadratic triangular element for 2D and 2) 10-node quadratic tetrahedral element for 3D.

Without the loss of generality, consider one tetrahedral element with 10 nodes, say (x_1^i, x_2^i, x_3^i), $i = 1, 2, 3, 4$, are four corner nodes, and $i = 5, \ldots, 10$ are six mid-nodes. Then the eigenstrain distribution in the element can be written in terms of eigenstrain on the nodes as

$$\varepsilon_{ij}^*(\mathbf{x}') = \sum_{n=1}^{4}(2L^n - 1)L^n\,\varepsilon_{ij}^{*n} + \sum_{n=5}^{10} 4L^I L^J\,\varepsilon_{ij}^{*n} \tag{4.84}$$

where the superscript n represents the quantity associated with node n, and again no index summation rule is applied with it; α^I, β^I, γ^I, and χ^I are the components of the volume coordinates (linear shape function) of the Ith local corner node, and the complete form can be written as $L^I = \alpha^I + \beta^I x_1 + \gamma^I x_2 + \chi^I x_3$ (let the other three local corner nodes be J, K, L). The detailed expression of the shape function is shown as Eq. (4.47). Differently from linear shape function, obviously, the product of two volume coordinates, say $L^I L^J$, contains linear and quadratic source terms, which can be treated in the same fashion as we derived the linear and quadratic potentials by separating the source terms into the field and distance vector components. Since the shape functions include the products of two corner nodes, we list the domain integrals of L^I and $L^I L^J$ with $|\mathbf{x} - \mathbf{x}'|$ as an example to rearrange the terms of same order:

$$\Lambda^I = \int_\Omega L^I(\mathbf{x}')|\mathbf{x} - \mathbf{x}'|d\mathbf{x}' = \alpha^I\Psi + \beta^I\Psi_1 + \gamma^I\Psi_2 + \chi^I\Psi_3 \tag{4.85a}$$

$$\begin{aligned}\Lambda^{IJ} &= \int_\Omega L^I L^J(\mathbf{x}')|\mathbf{x} - \mathbf{x}'|d\mathbf{x}' \\ &= \alpha^I\alpha^J\Psi + (\alpha^I\beta^J + \alpha^J\beta^I)\Psi_1 + (\alpha^I\gamma^J + \alpha^J\gamma^I)\Psi_2 \\ &\quad + (\alpha^I\chi^J + \alpha^J\chi^I)\Psi_3 + (\beta^I\gamma^J + \beta^J\gamma^I)\Psi_{12}\end{aligned}$$

$$+(\beta^I\chi^J+\beta^J\chi^I)\Psi_{13}+(\beta^I\chi^J+\beta^J\chi^I)\Psi_{23}$$
$$+\beta^I\beta^J\Psi_{11}+\gamma^I\gamma^J\Psi_{22}+\chi^I\chi^J\Psi_{33} \tag{4.85b}$$

Similarly, the domain integrals of L^I and $L^I L^J$ with $|\mathbf{x}-\mathbf{x}'|^{-1}$ can be expressed by Γ^I and Γ^{IJ}, respectively. Then, using the technique of Green's function, the disturbed displacement field $u_i(\mathbf{x})$ is analytically formulated as Eq. (4.48), but the integrals should be changed as follows:

$$\mathcal{A}_{ij}=\sum_{w=1}^{E^I}\sum_{n=1}^{4}(2\Lambda^{II}-\Lambda^I)(\varepsilon_{ij}^*)^n-\sum_{n=5}^{10}4\Lambda^{IJ}(\varepsilon_{ij}^*)^n \tag{4.86a}$$

$$\mathcal{B}_{ij}=\sum_{w=1}^{E^I}\sum_{n=1}^{4}(2\Gamma^{II}-\Gamma^I)(\varepsilon_{ij}^*)^n-\sum_{n=5}^{10}4\Gamma^{IJ}(\varepsilon_{ij}^*)^n \tag{4.86b}$$

Similarly, following the same fashion in the 3D problem, consider a 6-node quadratic triangular element in which (x_1^1, x_2^1), (x_1^2, x_2^2), and (x_1^3, x_2^3) are the corner nodes and (x_1^4, x_2^4), (x_1^5, x_2^5), and (x_1^6, x_2^6) are the midpoints on each side of the triangle, which can be written in term of the first three nodes. The eigenstrain distribution in the element can be written in terms of eigenstrain on the nodes through their shape functions as

$$\varepsilon_{ij}^*(\mathbf{x}')=\sum_{n=1}^{6}(\varepsilon_{ij}^*)^n N^n \tag{4.87}$$

where the shape function for 6-node quadratic triangular element is

$$N^n=\begin{cases}2(L^n)^2-L^n & n=1,2,3,\ \text{corner node}\\ 4L^I L^J & n=4,5,6,\ \text{midnode of } I\text{th \& } J\text{th nodes}\end{cases} \tag{4.88}$$

in which the superscript n represents the quantity associated with node n, and no index summation rule is applied with it; $L^n=\alpha^n+\beta^n x_1+\gamma x_2$ is the area coordinates of the triangular element; α^I, β^I, and γ^I are the components of the shape functions of the Ith local node, and the superscripts J and K represent the other two corner nodes. The detailed expression of the shape function is shown as Eq. (4.51), and the domain integrals $\mathcal{A}_{ij}$ and $\mathcal{B}_{ij}$ in Eq. (4.48) are rewritten as

$$\mathcal{A}_{ij}=\sum_{n=1}^{3}(2\Lambda^{II}-\Lambda^I)(\varepsilon_{ij}^*)^n-\sum_{n=4}^{6}4\Lambda^{IJ}(\varepsilon_{ij}^*)^n \tag{4.89a}$$

$$\mathcal{B}_{ij}=\sum_{n=1}^{3}(2\Gamma^{II}-\Gamma^I)(\varepsilon_{ij}^*)^n-\sum_{n=4}^{6}4\Gamma^{IJ}(\varepsilon_{ij}^*)^n \tag{4.89b}$$

where

$$\Lambda^I = \int_\Omega L^I(\mathbf{x}')|\mathbf{x}-\mathbf{x}'|^2 \frac{ln|\mathbf{x}-\mathbf{x}'|^2-1}{2} d\mathbf{x}' = \alpha^I \Psi + \beta^I \Psi_1 + \gamma^I \Psi_2^I$$

$$\begin{aligned}\Lambda^{IJ} &= \int_\Omega L^I L^J(\mathbf{x}')|\mathbf{x}-\mathbf{x}'|^2 \frac{ln|\mathbf{x}-\mathbf{x}'|^2-1}{2} d\mathbf{x}' \\ &= \alpha^I\alpha^J\Psi + (\alpha^I\beta^J + \alpha^J\beta^I)\Psi_1 + (\alpha^I\gamma^J + \alpha^J\gamma^I)\Psi_2 \\ &+ (\beta^I\gamma^J + \beta^J\gamma^I)\Psi_{12} + \beta^I\beta^J\Psi_{11} + \gamma^I\gamma^J\Psi_{22}\end{aligned} \tag{4.90}$$

Substituting ϕ into Eq. (4.90), we could write Γ^I and Γ^{IJ} as well. Although the implementation of the quadratic element takes extra efforts in implementation and consumes more computational resources in constructing the stress equivalence equations in the global matrix, however, it exhibits the advantages of good convergence in comparison with FEM.

Appendix 4.B Domain integral with bilinear/quadratic shape function

In the previous sections, we have demonstrated the application with linear and quadratic shape functions. However, we could follow the same procedure to derive domain integrals combined with other elements, such as 4-node quadrilateral element and 8-node hexahedral element, whose advantage is simple implementation for the numerical integral (isoparametric natural coordinates). It is natural to use its numerical features instead of transforming it back to the area/volume coordinates, although partial merits of the method are lost by dropping the closed-form domain integral. Using the Gauss quadrature as in [102,103], the domain integrals (including the interior points) can be handled without singularity because the stress equivalent equations are set up at the Gaussian integral point of the elements.

Appendix 4.C Combination of several types of elements

In the previous derivations, we introduced the implementation of linear, bilinear, and quadratic elements. The most straightforward way, of course, is using one of those elements instead of combining them. However, the use of quadratic elements consumes relatively more computational resources to construct the stress equivalent conditions because of the long formulae of domain integrals. The mesh strategy and its influence on the accuracy of the local elastic fields have been discussed in detail. The improvement (or refinement) of the mesh only exhibits better accuracy in the neighborhood of the vertices. In such cases, because the eigenfields may not change rapidly in the interior parts, for the interior part, linear elements can provide good results. Although higher-order elements, say cubic order of eigenstrain, can provide sound improvement in both mesh quality and convergence, the quadratic elements perform well with a few more elements as shown in the following case studies, so that this chapter has implemented the algorithm up to quadratic shape functions so far.

Chapter 5

The iBEM formulation and implementation

Ellipsoidal inhomogeneities in a bounded domain

In this chapter, we introduce the inclusion-based boundary element method, which focuses on the elastic field with both boundary and volume integrals. Firstly, using a polynomial form of eigenstrains in inclusions with boundary integral, we can calculate the elastic field by inclusions embedded in a finite domain with ellipsoidal shapes. Then the EIM is used to consider the material mismatch. Virtual experiments can be conducted for a finite composite containing ellipsoidal particles under arbitrary testing loads.

5.1 Introduction to BIE and iBEM

Boundary integral equations (BIEs) are a wonderfully efficient way to solve PDE boundary value problems (BVPs). Since nearly all the phenomena occurring in physics are governed by partial differential equations with prescribed boundary conditions, the boundary element method or boundary integral equations, which are numerical tools for solving BVPs, have attracted a great deal of attentions. Its essential idea is to consider two correlated states; one is the true state, and the other is referring to the fundamental solution state for the infinite domain. With the help of reciprocal energy theorem, the boundary integral equations can be derived as follows: Take the elastic problem as an example and consider a domain D with unknown field quantity $\boldsymbol{u}$ subjected to body force $\boldsymbol{b}$ and surface traction $\boldsymbol{t}$. In elasticity the virtual work principle reads

$$\int_{\Omega} \sigma_{ij} u'_{i,j} \mathrm{d}\mathbf{x} = \int_{\partial\Omega} t_i u'_i \mathrm{d}\mathbf{x} + \int_{\Omega} b_i u'_i \mathrm{d}\mathbf{x} \tag{5.1}$$

where u'_i is an admissible virtual displacement.

Use the true displacement u_i as a virtual displacement. Let σ'_{ij} be a new stress state associated with another equilibrium state caused by $\boldsymbol{t}'$ and $\boldsymbol{b}'$. Applying the virtual work principle in the above equation, we can obtain

$$\int_{\partial D} t_i u'_i \mathrm{d}\mathbf{x} + \int_{D} b_i u'_i \mathrm{d}\mathbf{x} = \int_{\partial D} t'_i u_i \mathrm{d}\mathbf{x} + \int_{D} b'_i u_i \mathrm{d}\mathbf{x} \tag{5.2}$$

The Inclusion-Based Boundary Element Method (iBEM). https://doi.org/10.1016/B978-0-12-819384-6.00013-9

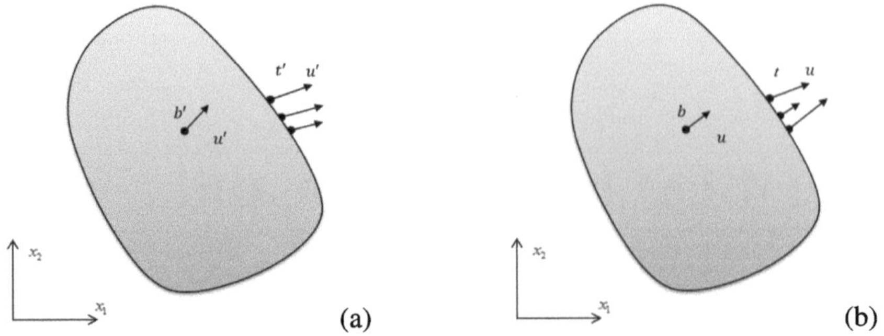

FIGURE 5.1 Application of the Betti reciprocal theorem for two loading cases.

which is the well-known Maxwell–Betti reciprocal theorem. The two loading cases are shown in Fig. 5.1.

Using the equilibrium $b_i' = -\sigma'_{ji,j}$, we obtain the following equation:

$$\int_D \sigma'_{ji,j} u_i \mathrm{d}\mathbf{x} + \int_D b_i u_i' \mathrm{d}\mathbf{x} = \int_{\partial D} t_i' u_i \mathrm{d}\mathbf{x} - \int_{\partial D} t_i u_i' \mathrm{d}\mathbf{x} \tag{5.3}$$

Use the fundamental solution to the following equilibrium equation for the new stress state under a unit point force at $\mathbf{x}'$:

$$\sigma'_{ji,j}(\mathbf{x}) + \delta_i(\mathbf{x} - \mathbf{x}') = 0 \tag{5.4}$$

where $\delta_i(\mathbf{x} - \mathbf{x}')$ is the Dirac delta function indicating a unit point force in a certain direction acting at position $\mathbf{x}'$ with component d_i. Therefore d_i exhibits the following properties:

$$\begin{cases} \int_D \delta_i(\mathbf{x} - \mathbf{x}')\mathrm{d}\mathbf{x} = d_i \\ |\mathbf{d}| = \sqrt{d_i d_i} = 1 \end{cases}$$

This fictitious stress state can be obtained by the classic solution for an infinite domain containing a point force associated with the corresponding stress vector on the boundary. Using the Dirac delta function, we can write

$$\int_D \sigma'_{ji,j}(\mathbf{x}) u_i(\mathbf{x}) \mathrm{d}\mathbf{x} = -\int_D \delta_i(\mathbf{x} - \mathbf{x}') u_i(\mathbf{x}) \mathrm{d}\mathbf{x} = -d_i u_i(\mathbf{x}') \tag{5.5}$$

Now, to solve the displacement at any point $\mathbf{x}'$, we can use the fundamental solution for a fictitious unit point force at that point. The fictitious elastic field will depend on both variables of the fictitious loading point $\mathbf{x}'$ and the field point $\mathbf{x}$.

Combining Eqs. (5.3) and (5.5), we obtain the displacement at any point $\mathbf{x}'$ as follows:

$$u_i(\mathbf{x}')d_i = \int_{\partial D} u_i'(\mathbf{x}, \mathbf{x}')t_i(\mathbf{x})\mathrm{d}\mathbf{x} - \int_{\partial D} t_i'(\mathbf{x}, \mathbf{x}')u_i(\mathbf{x})\mathrm{d}\mathbf{x} + \int_D u_i'(\mathbf{x}, \mathbf{x}')b_i(\mathbf{x})\mathrm{d}\mathbf{x} \tag{5.6}$$

Using the Green's function technique, we can obtain the fictitious elastic field as follows:

$$u_j'(\mathbf{x}, \mathbf{x}') = G_{ij}(\mathbf{x}, \mathbf{x}')d_i \tag{5.7}$$

$$t_j'(\mathbf{x}, \mathbf{x}') = T_{ij}(\mathbf{x}, \mathbf{x}')d_i \tag{5.8}$$

We can easily see the reciprocal property of the above tensors,

$$G_{ij}(\mathbf{x}, \mathbf{x}') = G_{ij}(\mathbf{x}', \mathbf{x}) \text{ and } T_{ij}(\mathbf{x}, \mathbf{x}') = T_{ij}(\mathbf{x}', \mathbf{x}) \tag{5.9}$$

Finally, we can obtain the following boundary integral equation:

$$\begin{aligned} d_i u_i(\mathbf{x}') = {} & d_i \int_{\partial D} G_{ij}(\mathbf{x}, \mathbf{x}')t_j(\mathbf{x})\mathrm{d}\mathbf{x} - d_i \int_{\partial D} T_{ij}(\mathbf{x}, \mathbf{x}')u_j(\mathbf{x})\mathrm{d}\mathbf{x} \\ & + d_i \int_D G_{ij}(\mathbf{x}, \mathbf{x}')b_j(\mathbf{x})\mathrm{d}\mathbf{x} \end{aligned} \tag{5.10}$$

Therefore, to obtain the displacement at any point $\mathbf{x}'$ in a specific direction, we can apply a unit point force along that direction represented by d_i and derive the displacement through the above integrals of $\mathbf{x}$. For notational simplicity, we can switch $\mathbf{x}$ and $\mathbf{x}'$ in the above equation, so that $\mathbf{x}'$ stands for the integral variable and $\mathbf{x}$ is the point to solve for displacement. In addition, the above equation is applicable to an arbitrary fictitious unit force, so d_i at both sides of the above equation can be eliminated. Therefore we can obtain

$$\begin{aligned} u_i(\mathbf{x}) = {} & \int_{\partial D} G_{ij}(\mathbf{x}, \mathbf{x}')t_j(\mathbf{x}')\mathrm{d}\mathbf{x}' - \int_{\partial D} T_{ij}(\mathbf{x}, \mathbf{x}')u_j(\mathbf{x}')\mathrm{d}\mathbf{x}' \\ & + \int_D G_{ij}(\mathbf{x}, \mathbf{x}')b_j(\mathbf{x}')\mathrm{d}\mathbf{x}' \end{aligned} \tag{5.11}$$

where we used the reciprocal property in Eq. (5.9). Note that for an equilibrium state in the absence of body force, the last term does not exist, so we can obtain the displacement by the first two boundary integral terms, which are the classic BIEs.

5.2 Inclusion problems with both boundary and volume integrals

Now we still take an elastic problem as an example for demonstration of the method. Consider an unbound homogeneous elastic domain D containing an inclusion in a subdomain Ω, where an eigenstrain $\varepsilon^*_{ij}(\mathbf{x})$ is applied on Ω but is zero on $D-\Omega$. The elastic modulus is the same in both the inclusion and the rest matrix. The term "inclusion" is to differentiate another counterpart of "inhomogeneity". An inhomogeneity is a subdomain in a homogeneous elastic body with different material properties from the rest.

Consider an ellipsoidal inclusion Ω embedded in a homogeneous elastic domain, where the ellipsoidal inclusion is defined by the function $\frac{x_1^2}{a_1^2}+\frac{x_2^2}{a_2^2}+\frac{x_3^2}{a_3^2}\leq 1$ with a_i $(i=1,2,3)$ being the three principal half-axes of the ellipsoid. The elastic modulus of the inclusion is the same as the matrix written as C. An eigenstrain $\varepsilon^*_{kl}(\mathbf{x})$ is distributed in the domain Ω but is zero in the matrix.

Based on the homogeneous constitutive law, the stress can be expressed as

$$\sigma_{ij}(\mathbf{x})=C_{ijkl}\left[\varepsilon_{kl}(\mathbf{x})-\varepsilon^*_{kl}(\mathbf{x})\right],\quad \mathbf{x}\in\Omega \tag{5.12}$$

Using Eq. (5.12), we can write the equilibrium equation in the absence of body force as

$$C_{ijkl}\varepsilon_{kl,i}(\mathbf{x})=C_{ijkl}\varepsilon^*_{kl,i}(\mathbf{x}),\quad \mathbf{x}\in\Omega \tag{5.13}$$

As discussed in Chapter 3, the right-hand side of the equation can be mathematically treated as a body force, i.e.,

$$b_j(\mathbf{x})=-C_{ijkl}\varepsilon^*_{kl,i}(\mathbf{x})$$

Using the Green's function technique or the fundamental solution [10,95], we can write the displacement fields in the domain D in terms of the integrals of the eigenstrain ε^*_{kl} with the tensorial Green's function $\mathbf{G}$ as follows:

$$u_m(\mathbf{x})=-\int_\Omega C_{ijkl}\frac{\partial\varepsilon^*_{kl}}{\partial x'_i}(\mathbf{x}')G_{jm}(\mathbf{x},\mathbf{x}')\mathrm{d}\mathbf{x}' \tag{5.14}$$

Using the Gauss theorem, we can rewrite Eq. (5.14) as

$$u_m(\mathbf{x})=-\int_\Omega C_{ijkl}\varepsilon^*_{kl}(\mathbf{x}')G_{jm,i}(\mathbf{x},\mathbf{x}')\mathrm{d}\mathbf{x}' \tag{5.15}$$

where we used the eigenstrain in the matrix $\varepsilon^*_{kl}=0$, and $G_{jm,i}$ is also called the modified Green's function [104].

Consider the eigenstrain expanded in a polynomial form as

$$\varepsilon^*_{kl}=\varepsilon^0_{kl}+\varepsilon^1_{klp}x_p+\cdots,\quad \mathbf{x}\in\Omega \tag{5.16}$$

We can obtain the displacement caused by the eigenstrain as

$$u_m(\mathbf{x})=D_{mkl}(\mathbf{x})\varepsilon^0_{kl}+D_{mklp}(\mathbf{x})\varepsilon^1_{klp}+\cdots \tag{5.17}$$

where

$$D_{imn} = -C_{ljmn} \int G_{ij,l}(\mathbf{x}, \mathbf{x}')d\mathbf{x}'$$

$$D_{imnp} = -C_{ljmn} \int G_{ij,l}(\mathbf{x}, \mathbf{x}')x'_p d\mathbf{x}'$$

Now the combination of the inclusion and boundary value problems together represents an inclusion problem in a bounded domain. The total displacements of the inclusion problem in a bounded domain are the sums of the boundary integrals (5.11) and the displacement induced by eigenstrain (5.17):

$$u_i(\mathbf{x}') = \int_{\partial D} G_{ji}(\mathbf{x}, \mathbf{x}')t_j(\mathbf{x})d\mathbf{x} - \int_{\partial D} T_{ji}(\mathbf{x}, \mathbf{x}')u_j(\mathbf{x})d\mathbf{x} + D_{imn}(\mathbf{x}')\varepsilon^0_{mn} + D_{imnp}(\mathbf{x}')\varepsilon^1_{mnp} \tag{5.18}$$

which can include higher-order terms of eigenstrains, although in the above equation shown up to linear terms.

5.3 Equivalent inclusion method for inhomogeneity problems

For the linear elastic problem with isotropic stiffness tensor $C_{ijkl} = \lambda\delta_{ij}\delta_{kl} + \mu\left(\delta_{ik}\delta_{jl} + \delta_{il}\delta_{jk}\right)$, the governing equations are the following equilibrium equations in the absence of body force in terms of the displacements:

$$(\lambda + \mu)\nabla(\nabla \cdot \mathbf{u}) + \mu\nabla^2\mathbf{u} = 0 \tag{5.19}$$

and the Green's functions (discussed in the previous chapter) G and T are

$$G_{ij}(\mathbf{x}, \mathbf{x}') = \frac{1}{4\pi\mu}\frac{\delta_{ij}}{|\mathbf{x} - \mathbf{x}'|} - \frac{1}{16\pi\mu(1-\nu)}\frac{\partial|\mathbf{x} - \mathbf{x}'|}{\partial x_i \partial x_j} \tag{5.20}$$

$$T_{km}(\mathbf{x}, \mathbf{x}') = \frac{1}{2}C_{ijkl}\left(G_{im,j}(\mathbf{x}, \mathbf{x}') + G_{jm,i}(\mathbf{x}, \mathbf{x}')\right)n_l \tag{5.21}$$

In the ellipsoidal domain, the eigenstrain is continuously distributed, so that we can expand them in the polynomial form using the Taylor expansion. Then the integral can be explicitly written with the aid of the identities of the integrals of ϕ and ψ using the elliptic integrals [1,10,95,105], where ϕ and ψ are defined as

$$\phi(\mathbf{x}, \mathbf{x}') = \frac{1}{|\mathbf{x} - \mathbf{x}'|}, \quad \psi(\mathbf{x}, \mathbf{x}') = |\mathbf{x} - \mathbf{x}'| \tag{5.22}$$

As discussed in Chapter 3 about the definition of Φ and Ψ, the tensor D can be expressed by Φ and Ψ as follows:

$$\begin{aligned} D_{imn} &= -C_{ljmn} \int G_{ij,l}(\mathbf{x}, \mathbf{x}') d\mathbf{x}' \\ &= -C_{ljmn} \left(\frac{\delta_{ij}}{4\pi\mu} \Phi_{,l} - \frac{1}{16\pi\mu(1-\nu)} \Psi_{,ijl} \right) \end{aligned} \tag{5.23}$$

$$\begin{aligned} D_{imnp} &= -C_{ljmn} \int G_{ij,l}(\mathbf{x}, \mathbf{x}') x'_p d\mathbf{x}' \\ &= -C_{ljmn} \left(\frac{\delta_{ij}}{4\pi\mu} \Phi_{p,l} - \frac{1}{16\pi\mu(1-\nu)} \Psi_{p,ijl} \right) \end{aligned} \tag{5.24}$$

Now consider multiple-particle cases $I = 1, \ldots, n$. The stain field can be derived based on the derivative of the total displacement:

$$\begin{aligned} \varepsilon_{ij}(\mathbf{x}') &= \frac{1}{2} \left(u_{i,j} + u_{j,i} \right) \\ &= -\int_{\partial D} \frac{1}{2} \left(G_{mi,j}(\mathbf{x}, \mathbf{x}') + G_{mj,i}(\mathbf{x}, \mathbf{x}') \right) t_m(\mathbf{x}) \, d\mathbf{x} \\ &+ \int_{\partial D} \frac{1}{2} \left(T_{mi,j}(\mathbf{x}, \mathbf{x}') + T_{mj,i}(\mathbf{x}, \mathbf{x}') \right) u_m(\mathbf{x}) \, d\mathbf{x} \\ &+ \sum_{I=1}^{n} \frac{1}{2} \left(D^{I0}_{ijmn}(\mathbf{x}') + D^{I0}_{jimn}(\mathbf{x}') \right) \varepsilon^{I0}_{mn} \\ &+ \sum_{I=1}^{n} \frac{1}{2} \left(D^{I1}_{ijmnp}(\mathbf{x}') + D^{I1}_{jimnp}(\mathbf{x}') \right) \varepsilon^{I1}_{mnp} \end{aligned} \tag{5.25}$$

where the properties $G_{,i} = -G_{,i'}$ and $T_{,i} = -T_{,i'}$ are applied.

Eshelby's original equivalent inclusion method (EIM) is to set up the stress equivalent conditions to find the equivalent inclusion problem for the inhomogeneity problem. It has been successfully applied to other physics problems [10,95,106]. Generally speaking, the equivalence can be built on the stress as follows for each inhomogeneity:

$$C^0_{ijkl}(\varepsilon_{kl} - \varepsilon^*_{kl}) = C^1_{ijkl}\varepsilon_{kl} \tag{5.26}$$

By applying Taylor's expansion technique at the center of particle I to the Ith equation on both sides of Eq. (5.26) and matching the coefficients at the same orders of x, we write the equivalence in terms of coefficients for each order of the polynomial evaluated at the center of each particle as follows:

$$C^{0}_{ijmn}\left(\varepsilon_{mn}(\mathbf{0})-\varepsilon^{I0}_{mn}\right)=C^{I}_{ijmn}\varepsilon_{mn}(\mathbf{0})\quad \text{for } I=1:n \tag{5.27}$$

$$C^{0}_{ijmn}\left(\varepsilon_{mn,r}(\mathbf{0})-\varepsilon^{I1}_{mnr}\right)=C^{I}_{ijmn}\varepsilon_{mn,r}(\mathbf{0})\quad \text{for } I=1:n \tag{5.28}$$

When higher-order terms of the polynomial are used in Eq. (5.16), a higher accuracy can be achieved, and more complex eigenstrain field can be described [35]. For simplicity, here we present only linear Taylor's expansion.

5.4 The architecture of iBEM software development

The above formulation has been implemented in a C++ package iBEM, which stands for a BEM code addressing inhomogeneity or inclusion problems. The programming structure is presented in Fig. 5.2, in which each step is elaborated as follows:

1) Given a solid containing a certain number of inhomogeneities in the matrix with perfectly bonded interfaces, the external geometry (∂D) is discretized into boundary elements. This step can be treated as a preprocess. Since only the surface is meshed, the number of elements will be reduced significantly compared with the finite element method (FEM) or the general BEM.
2) Go through all the collection points on the boundary to calculate the individual element matrix for each collection point based on traditional boundary element integration. This step is to complete the boundary element formulation. Therefore the prescribed boundary conditions are taken into account.
3) Go through each inhomogeneity to build up stress equivalence into the global matrix. This is the key step of the proposed iBEM algorithm. Not only the interaction between particles but also the boundary effects will be taken into account through Eqs. (5.27) and (5.28), respectively. Notice that if more higher-order terms of eigenstrain are considered, then more equations for each inhomogeneity are required. The completeness of the equations can always be achieved. The numbers of unknowns for each inhomogeneity with polynomial eigenstrains are 6, 24, and 60 for uniform, linear, and quadratic terms with tailorable accuracy,

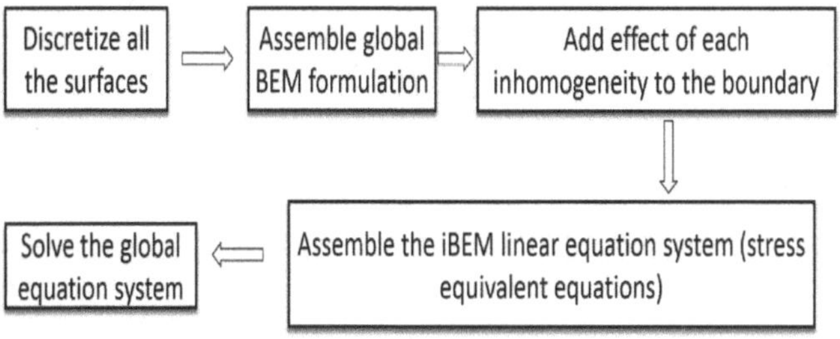

FIGURE 5.2 Implementation flow of the iBEM, a software package introducing inhomogeneity/inclusion problems into the BEM code.

and the number of unknowns for each boundary node is 3. Therefore in the linear eigenstrain case the total number of unknowns is 24 × the number of inhomogeneity and 3 × the number of boundary nodes. The traditional BEM will provide three equations for each node, which makes 3 × the number of boundary nodes equations as well. In addition, the equivalence of stress conditions will also provide 24 × the number of inhomogeneity equations for the whole system. In the present method the number of equations and unknowns is the same as indicated in Eq. (5.29), and therefore a closed linear equation system can be formed for the numerical method. If higher-order terms of eigenstrain are used, then more unknowns in each particle can be used, but the same number of equations will also be obtained.

$$\begin{bmatrix} Q+H & \cdots & -g^{0I} & -g^{1I} & -g^{2I} & \cdots \\ \vdots & \vdots & \vdots & \vdots & \vdots & \vdots \\ \Delta C^I H^{1I} & \cdots & \Delta C^I S^{0I} & \Delta C^I S^{1I} & \Delta C^I S^{2I} & \cdots \\ \Delta C^I H^{2I} & \cdots & \Delta C^I S^{0I'} & \Delta C^I S^{1I'} & \Delta C^I S^{2I'} & \cdots \\ \Delta C^I H^{3I} & \cdots & \Delta C^I S^{0I''} & \Delta C^I S^{1I''} & \Delta C^I S^{2I''} & \cdots \\ \vdots & \vdots & \vdots & \vdots & \vdots & \vdots \end{bmatrix}_{(NN+NI)\times(NN+NI)} \begin{bmatrix} u \\ \vdots \\ \varepsilon^{*0I} \\ \varepsilon^{*1I} \\ \varepsilon^{*2I} \\ \vdots \end{bmatrix}_{(NN+NI)}$$

$$= \begin{bmatrix} U \\ \vdots \\ -\Delta C^I U^{1I} \\ -\Delta C^I U^{2I} \\ -\Delta C^I U^{3I} \\ \vdots \end{bmatrix}_{(NN+NI)\times(NN)} \underset{(NN)}{[t]} \tag{5.29}$$

4) Solve the global linear equation system to obtain the unknown boundary response and unknown eigenstrain. The full LU decomposition [107] with the pivot method has been implemented to calculate the numerical results from the linear equation system.

5) Use the eigenstrain and boundary traction/displacement to calculate the elastic fields, including the stress, strain, and displacement fields for elastic analysis. This step can be treated as a postprocess. As the unknown boundary response and eigenstrains are all obtained, from Eq. (5.18) we can get the displacement at any field point, and therefore from Eq. (5.25) we can also calculate the internal strain field.

In Eq. (5.29), the displacements of boundary nodes are related to $Q + H$, written in two parts of a diagonal matrix Q and a regular matrix from the boundary integral with zero diagonal element. Because the fundamental solution for the traction exhibits strong singularity when the field point is on the boundary, there exists numerical difficulty to calculate the diagonal elements of the matrix containing Cauchy principal value integrals. Instead of direct mathematical integral, Beer et al. [21] proposed an indirect method using the rigid-body motion scheme to calculate the diagonal element of the matrix. Consider a unit displacement field for all nodes (like a rigid-body motion) without any traction on the boundary, which should be an allowable solution and satisfy the same linear equation system. Because the displacement and traction are given as 1 and 0, respectively, we can use the linear equation system to calculate the diagonal element of H instead. Therefore the Cauchy principal value integrals are avoided, and the diagonal components of matrix are calculated as the summation of other elements in the same row of H, written as Q. We use $Q + H$ to indicate this process.

5.5 Periodic boundary conditions for periodic microstructure

To consider the effective material properties of a composite, PBC is a common approach using a unit cell to represent the microstructure features. Take a 2D problem as an example with the constraint equations of displacements and surface tractions (stress vector) be written as

$$\begin{aligned} u_i^T = u_i^B + \bar{\varepsilon_{yy}} l_y, \quad u_i^R = u_i^L + \bar{\varepsilon_{xx}} l_x \\ \sigma_{ij}^T = \sigma_{ij}^B, \quad \sigma_{ij}^R = \sigma_{ij}^L \end{aligned} \tag{5.30}$$

where T, B, R, and L represent the top, bottom, right, and left lines of the square, $\bar{\epsilon_{ij}}$ is the macroaveraged strain, and l_x and l_y are the lengths of the square in horizontal and vertical directions. However, the boundary conditions in Eq. (5.30) cannot be directly integrated into the iBEM global matrix, since they are constraint equations instead of prescribed displacements or surface traction [108]. The boundary constraint equations increase the size of the linear equation system (5.29), which leads to higher demand for computation resources. To implement the aforementioned PBC in iBEM without increasing the size of the global matrix, we employ the following constraint equations to reduce unknowns, which combine the coefficients of constrained quantities, such as quantities at left/right or top/bottom edges:

$$\begin{bmatrix} H_T + H_B & H_R + H_L & \dots & -g^{fI} & \dots \\ \vdots & \vdots & \vdots & \vdots & \vdots \\ \Delta C^I (H_T^{fI} + H_B^{fI}) & \Delta C^I (H_R^{fI} + H_L^{fI}) & \dots & \Delta C^I S^{fI} & \dots \\ \vdots & \vdots & \vdots & \vdots & \vdots \end{bmatrix} \begin{bmatrix} u_T \\ u_R \\ \vdots \\ \varepsilon^{*fI} \\ \vdots \end{bmatrix}$$

$$
-\begin{bmatrix} U_T+U_B & U_R+U_L \\ \vdots & \vdots \\ -\Delta C^I(U_T^{fI}-U_B^{fI}) & -\Delta C^I(U_R^{fI}-U_L^{fI}) \\ \vdots & \vdots \end{bmatrix}\begin{bmatrix} t_T \\ t_R \end{bmatrix}
$$

$$
=\begin{bmatrix} H_B & H_L \\ \vdots & \vdots \\ -\Delta C^I H_B^{fI} & -\Delta C H_L^{fI} \\ \vdots & \vdots \end{bmatrix}\begin{bmatrix} \bar{\varepsilon_{yy}} l_y \\ \bar{\varepsilon_{xx}} l_x \end{bmatrix} \tag{5.31}
$$

where the superscripts f and I represent the expansion order of eigenstrain and the Ith inhomogeneity, respectively. Equation (5.31) is derived from Eq. (5.29) by applying all the constraint equations to combine the coefficients of (right/left, top/bottom) unknown quantities.

However, close attention should be paid to address the corner nodes or edge nodes in 3D problems on the boundary as they are subjected to more than one constraint equations because they exhibit periodic feature in different directions. For example, for a 2D problem with four corner points, assigning the displacements for one corner node, the displacements at the other three corner nodes are known, which makes the number of unknowns less than the original iBEM. A least-square solver could address the issue [109] while it requires to assign displacement at least to one node. The treatment in Eq. (5.31) avoids the problem by keeping the number of the equations and unknowns.

*5.6 Numerical verification and comparison of iBEM with FEM

The above-mentioned algorithm has been implemented in the iBEM software package as shown in Eq. (5.29) to solve elasticity problems with various prescribed boundary conditions, including the PBC. To verify the algorithm and accuracy of eigenstrain expansions of different orders $(0, 1, 2)$, a numerical study with Neumann and Dirichlet boundary conditions was set up. In general, higher-order expansion of eigenstrain would improve the coverage of the field when the inhomogeneities are close to the boundary and there exist interactions between inhomogeneities. In the following subsections, we compare the stress fields solved by the present iBEM with the FEM results. Since the ordinary EIM proposed by Eshelby was proved to be accurate in the infinite space, we focused the comparison on close inhomogeneity interactions and inhomogeneity–boundary interactions. The boundary conditions in the entire section are as follows: (1) the top surface(3D)/line(2D) is subjected to a uniform downward pressure 10^4 Pa, (2) the side surfaces(3D)/lines(2D) are free, and (3) the displacements of bottom surface(3D)/line (2D) are constrained at zero. The elastic properties for the matrix material are $E^0 = 10^6$ Pa for Young's modulus and $\nu^0 = 0.25$ for Poisson's ratio, whereas the elastic properties for the stiffer particle are $E^0 = 2 \times 10^6$ Pa for Young's modulus and $\nu^0 = 0.25$ for Poisson's ratio. In all the virtual experiments, the boundary mesh (line, surface) keeps the same: (i) For 2D line mesh, the 3-node quadratic element is selected, and the numbers of nodes and elements are 400 and 200, respectively; (ii) for 3D surface mesh, the 4-node quadrilateral element is selected, and the numbers of nodes and elements are 1178 and 1176, respectively. Subsequently, in the following, three case studies were conducted and compared with FEM: 1) one particle boundary interactions, 2) equal-sized particle interactions, and 3) large–small particle interactions. More examples of particles, such as softer particles and 7-particle interactions, are available in the Appendix at the end of the chapter.

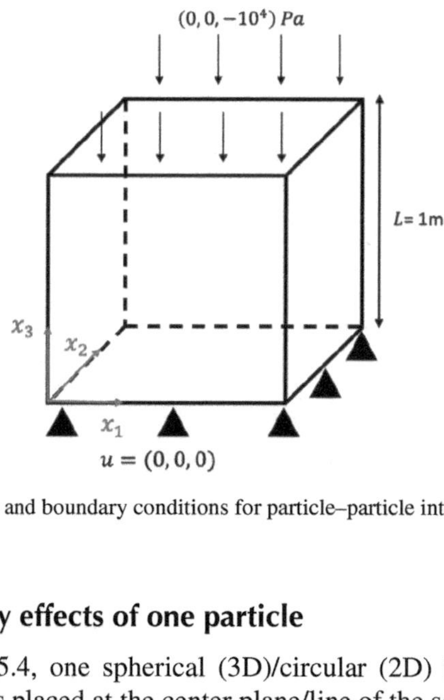

FIGURE 5.3 Geometry and boundary conditions for particle–particle interaction.

*5.6.1 Boundary effects of one particle

As shown in Fig. 5.4, one spherical (3D)/circular (2D) inhomogeneity with radius $a = 0.05$ m is placed at the center plane/line of the solid body with a distance of $0.5a$ to the bottom boundary for 3D and 2D cases, respectively. The results for both 2D and 3D cases are illustrated in Figs. 5.4a and 5.4b. (i) For 2D surface mesh, the 4-node quadrilateral element is selected, and the numbers of nodes and elements are 132142 and 43789, respectively. (ii) For 3D volume mesh, the 4-node tetrahedral element is selected, and the numbers of nodes and elements are 1607282 and 1179599, respectively. For all cases, the curves for iBEM-quadratic case agree well with the FEM results in the normal stresses of σ_{11}, σ_{22}, or σ_{33} along the vertical centerline. Relatively, the linear expansion is acceptable if the computational cost is considered. The 2D and 3D results show similar comparisons.

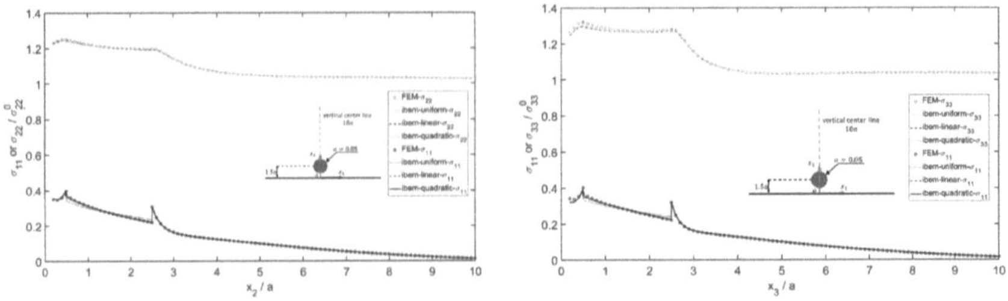

(a) Stiffer inhomogeneity–boundary interaction **(b)** Softer inhomogeneity–boundary interaction

FIGURE 5.4 Comparison of the stress distribution along x_2 or x_3 between FEM and iBEM (3 levels of accuracy) on inhomogeneity–boundary interaction for 2D/3D cases.

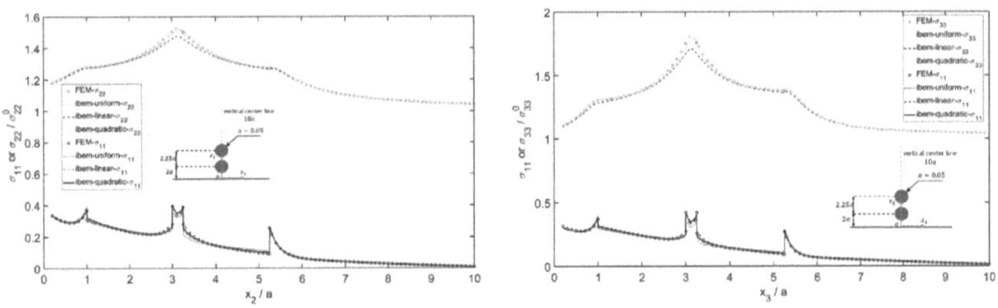

(a) Equal-sized inhomogeneities (2D) **(b)** Equal-sized inhomogeneities (3D)

FIGURE 5.5 Comparison of the stress distribution along x_2 or x_3 between FEM and iBEM (3 levels of accuracy) on top-down inhomogeneity interaction case for 2D/3D cases.

*5.6.2 Equal-sized particle interaction

Besides the inhomogeneity–boundary interaction, another important issue is the interactions between inhomogeneities, especially when they are close, which is very common for particulate composites. Firstly, the stress field of two stiffer inhomogeneities with radius $a = 0.05$ m under the same boundary conditions shown in Fig. 5.3. Along with the numerical comparison (Fig. 5.5), the two inhomogeneities were placed as top-down positions. Instead of specifying the distance, the ratio of radius to the distance to the boundary and the ratio of radius to the distance between two inhomogeneities are selected as 1.5 and 2 for the top-down case and 1 and 1.5 for the side-by-side case. (i) For 2D surface mesh, the 4-node quadrilateral element is selected, and the numbers of nodes and elements are 151,295 and 50,172, respectively. (ii) For 3D volume mesh, the 4-node tetrahedral element is selected, and the numbers of nodes and elements are 1,719,570 and 1,263,575, respectively.

*5.6.3 Large–small particle interaction

In this subsection, we consider the particles with large size difference. In FEM, such a large difference of particle dimensions requires much more efforts in

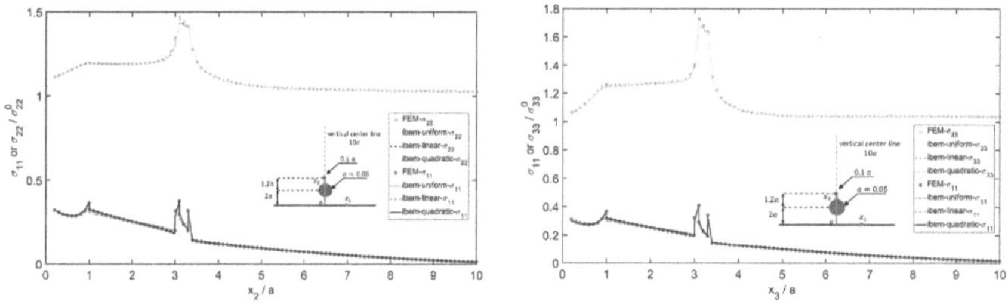

(a) Large–small inhomogeneities (2D) (b) Large–small inhomogeneities (3D)

FIGURE 5.6 Comparison of the stress distribution along x_2 or x_3 between FEM and iBEM (3 levels of accuracy) on top-down inhomogeneity interaction case for 2D/3D cases.

meshing the particle interfaces because a smooth and reliable solution requires the growth ratio of element size to be strictly constrained (1.1–1.5) for numerical stability. As a minimum number of elements are required to describe the geometry of small particles, a much large number of elements is needed for large particles. Under these circumstances, the degree of freedoms increases inevitably and significantly, which consumes much higher computer memory and storage, even if the FEM only needs to solve the sparse matrix instead of a dense one. In the following, two FEM-verified case studies of large–small particle interactions will be used to illustrate the efficiency of iBEM, and, subsequently, examples of many particles will be introduced for virtual experiments to obtain the effective material behavior including the periodic boundary conditions. Adjust the radius of one inhomogeneity to be $0.1a$. In that case, to generate a good-conditioned volume, dramatic increase of degree of freedom (DOF) in FEM is required, since the dimension of the inhomogeneities is relatively small, which requires much more refined elements to catch the geometry for accuracy. (i) For 2D surface mesh, the 4-node quadrilateral element is selected, and the numbers of nodes and elements are 364,319 and 121,008, respectively. (ii) For 3D volume mesh, the 4-node tetrahedral element is selected, and the numbers of nodes and elements are 1,737,524 and 1,276,973, respectively.

The normal stress σ_{xx}, σ_{yy}, or σ_{zz} were compared, and the difference between "quadratic term" and FEM was very small. The other two cases with uniform and linear terms approached the FEM results in general, but they could not capture some extreme points in the curve. Since the stress field was observed along the centerline, the middle point in the curve had the shortest distance, which exhibited the largest difference between those two curves. Hence, if the inhomogeneity interaction does not dominate, then the "uniform" and "linear" eigenstrain distributions could still provide good results in comparison with the FEM results. The 3D cases exhibit higher variation for the stress distribution compared with the 2D alternatives. The iBEM can capture particle–particle and particle–boundary interactions accurately for different sizes of particles with the same computational cost.

*5.7 Virtual experiments of particulate composite with spherical particles

In Section 5.6, the algorithm of iBEM for the elasticity package has been verified with FEM in three cases of boundary effects and particle interactions. In the following, we will use the advantages of iBEM to investigate the effective material behavior of the composites containing spherical particles. Before considering its applications, as demonstrated in Section 5.5, a unit cell with PBC or a representative volume element (RVE) under load test (i.e., uniaxial loading) containing a large number of particles can be used to obtain the effective moduli. Hence, a "cut-of" distance ratio strategy based on the features of Green's function is proposed to reduce consumption in computational resource while compromising very limited accuracy of 0.2%. Subsequently, one case study with PBC at different volume fractions and the other case study with size gradation are provided below. More examples, such as a random distribution and the same volume fraction but with different numbers of particles, are available in the Appendix at the end of this chapter.

*5.7.1 A "cut-off" strategy in the simulation of a large number of particles

To simulate a system (virtual experiment) of a large number of particles, the most accurate way, of course, is involving interactions of every single particle and their effects on the boundary (when using the uniaxial loading for effective moduli). However, although iBEM exhibits a prevailing advantage dealing with the increase of the number of particles, where the degrees of freedom only increase at constant rate. When the number of particles reaches, say 3000, the computational load is still too large for a personal computer. Therefore, in this section, we introduce a "cut-of"" distance ratio to minimize the resource requirements but provide reliable test results.

In Chapter 4, two case studies of particle interactions were used to demonstrate the tailorable accuracy of a single polynomial form eigenstrain for ellipsoidal particles. An initial conclusion was obtained that the larger the distance between two particles, the less the interactions, and thus lower-order polynomial terms are required to describe it. Moreover, because the uniform Eshelby tensor (which is dominant) is dimensionless, the distance ratio $\frac{d}{a}$ has been used to quantify the intensity of interactions. Note that a is the characteristic length of the particle, for example, the average length of an elliptical particle. For angular particles, the characteristic length will be discussed in Chapter 6. Without loss of generality, let us consider the 3D problem, in which the second derivative of Green's function has order $1/d^3$. When the distance d increases, the contribution or effect from the other particle drops rapidly, which enables us to set up a "cut-off" distance ratio for large-scale simulations, and a similar conclusion can be applied to the displacement as well. For a system of large number particles:

(1) Consider dropping the higher-order polynomial terms, such as linear and quadratic terms when the distance ratio ρ is larger than 2.
(2) Set up the "cut-of" ratio by calculating the distance between two particle centers, say d_c. Then the distance ratio $\rho = d_c - a_1 - a_2$, where a_1 and a_2 are the characteristic lengths of particles 1 and 2 in the pairs. For the readers' need on the accuracy, $\rho = 8$–12 provides reliable results; for more accurate results, $\rho = 5$–8 can be used.

*5.7.2 Effective modulus under periodic boundary condition

Virtual experiments are conducted for periodic particulate composites. The matrix material properties remain the same as in Section 5.6, and the corresponding stiffer inhomogeneity is applied, i.e., $E^0 = 1 \times 10^6$ Pa and $\nu^0 = 0.25$, or $E^I = 2 \times 10^6$ Pa and $\nu^I = 0.25$. For the convenience of virtual experiments, the macroaveraged strain is $\bar{\varepsilon_{xx}} = -0.01$ for both 2D and 3D cases with all other strain components being 0. In the PBC geometry the inhomogeneity was placed at the center of the composite body and through changing the volume fraction from 5% toward 40%. According to the surface tractions on the boundary, the effective Young modulus and Poisson ratio are calculated and shown in Fig. 5.7. The effective Poisson ratio is evaluated by the average surface traction through applying the Green's theorem, $\nu = \frac{\sigma_{yy}}{\sigma_{yy}+\sigma_{xx}}$. With the increase of the volume fraction, the effective Young modulus increases, and the Poisson ratio decreases. Notice that although both the matrix and inhomogeneity exhibit the same Poisson ratio, the effective Poisson ratio reduces with the volume fraction of stiffer particles.

Both 2D and 3D cases provide a similar trend, but the 3D results of Young's moduli are slightly higher than that of the 2D case. The 3D virtual experimental results are compared for the Mori–Tanaka model and the dilute model. Because no particle interaction is considered, it provides the lowest prediction of the effective Young modulus; the Mori–Tanaka model addresses the effect of particle interaction, and it gives a better prediction, whereas the current method with PBC provides the highest prediction, which is consistent with the comparison in the literature [1,60].

*5.7.3 Case study of size gradation of random distribution on effective modulus

The radius of inhomogeneities obeys the uniform distribution in the range of $[0.03, 0.075]$. Then 1154 nonoverlapping inhomogeneities were randomly distributed in the composites. As shown in Fig. 5.9, with the increase of the volume fraction, it is obvious that the effective Young modulus increases and the Poisson ratio decreases, which agrees well with Fig. 5.7b. Compared with the Mori–Tanaka model in Fig. 5.7b, the differences between the virtual experiment and

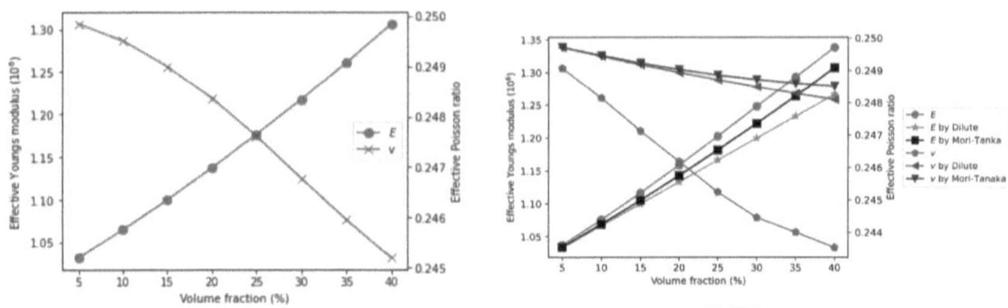

(a) 2D case

(b) 3D case

FIGURE 5.7 Effective modulus with variation of volume fraction of inhomogeneity located at the centroid.

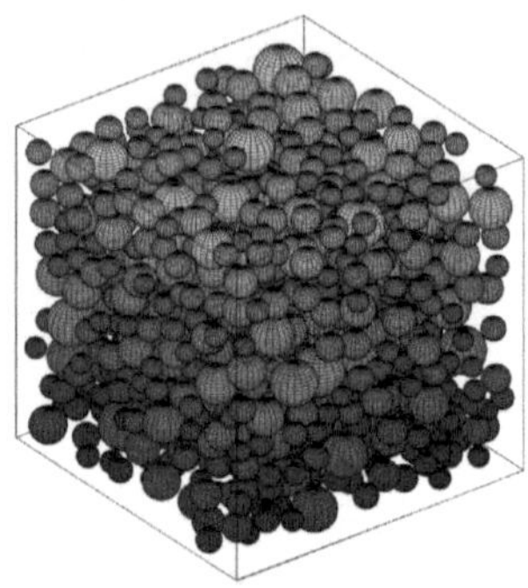

FIGURE 5.8 Microstructure of randomly distributed inhomogeneities with size gradation.

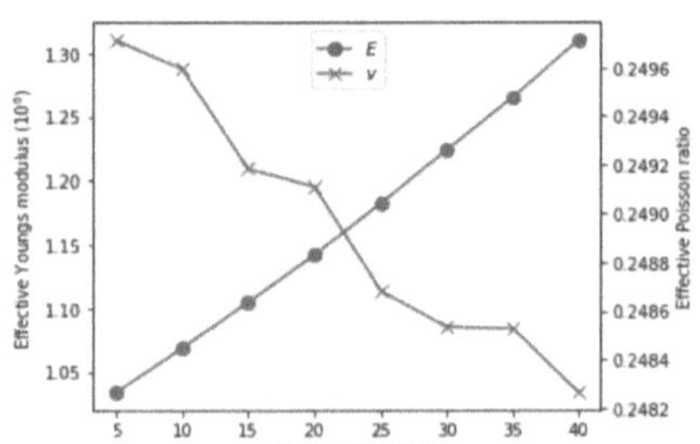

FIGURE 5.9 Effective modulus of various and randomly distributed inhomogeneities.

micromechanical model are 0.17% and 0.1% for effective Young's modulus and Poisson's ratio. In summary, the iBEM is a powerful tool for virtual experiments of particulate composites for both 2D and 3D cases. Because no mesh is needed for particles, it is ideal for composites with particle size gradation. It can be used for numerical experiments for actual composite samples. (See Fig. 5.8.)

Appendix 5.A Examples of particle interactions

In Section 5.6, some numerical comparisons with FEM are illustrated, i.e., boundary effects and equal-sized particle interactions. In the following, we show some interesting case studies through changing the elastic moduli, alignments, dimensions, and distribution of the inhomogeneities.

5.A.1 Softer inhomogeneity–boundary interaction

In Section 5.6.1, the boundary effect of a particle with stiffer elastic moduli is investigated and compared with FEM. As a extension, we substitute the particle with softer elastic moduli ($E^{II} = 1 \times 10^5$ Pa and $\nu^{II} = 0.25$), whereas the alignment of the particle and boundary conditions remain the same as in Fig. 5.3. As indicated in Chapter 4, the intensity of interactions is determined by the distance ratio $\frac{d}{a}$, and hence the same alignment ensures the same level of intensity for case study in Section 5.6.1 and this section. (See Fig. 5.10.)

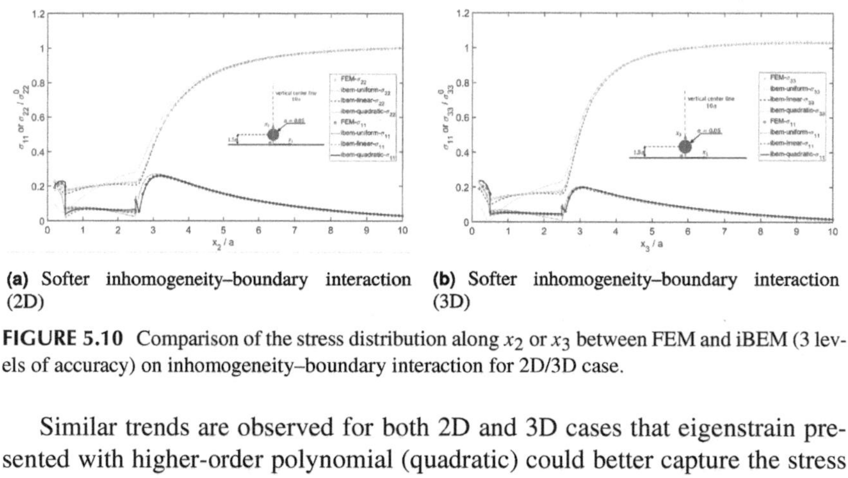

(a) Softer inhomogeneity–boundary interaction (2D) **(b)** Softer inhomogeneity–boundary interaction (3D)

FIGURE 5.10 Comparison of the stress distribution along x_2 or x_3 between FEM and iBEM (3 levels of accuracy) on inhomogeneity–boundary interaction for 2D/3D case.

Similar trends are observed for both 2D and 3D cases that eigenstrain presented with higher-order polynomial (quadratic) could better capture the stress variations, and the same conclusion can be made in examples in Figs. 4.2–4.5. However, for softer inhomogeneities, eigenstrains with uniform linear order polynomial expressions seldom describe the stress/strain change in the neighborhood of the inhomogeneities. To interpret such a phenomenon, the magnitude of stress variations should be considered that, in general cases, the eigenstrain field for a stiffer inhomogeneity varies less than a softer one, which is why it can be expressed by a lower-order polynomial.

5.A.2 One softer center inhomogeneity

In Section 5.6.1 and the previous section, case studies of boundary effects are investigated and compared with the stiffer inhomogeneities; the softer one usually requires a higher-order polynomial-form eigenstrain to describe the stress variation. In the following, we will use the softer inhomogeneity far from boundary to illustrate how boundary effects influence the polynomial order of the eigenstrain field.

When the inhomogeneity moves to the center, Fig. 5.11 shows the stress comparisons again. Since the distance to the boundary is comparatively large to the radius, the boundary effect of the loading produces little influence on the stress field around the inhomogeneity. Thus all of the three orders of expansion (uniform, linear, and quadratic) exhibit similar results. If we assume the ratio of inhomogeneity radius and its distance to the boundary to be large enough, then the loading could be regarded as far loading, and a uniform eigenstrain will be induced according to Eshelby's conclusion. In general, the variation of the stress field in a circular or spherical inhomogeneity is small, so that a uniform or linear eigenstrain may produce acceptable accuracy. However, when the inhomogeneity is close to the boundary or another inhomogeneity, the variation can be considerable, particularly, when the inhomogeneity is softer than the matrix. Therefore the quadratic eigenstrain can provide higher accuracy. In the following, we consider the particle interaction.

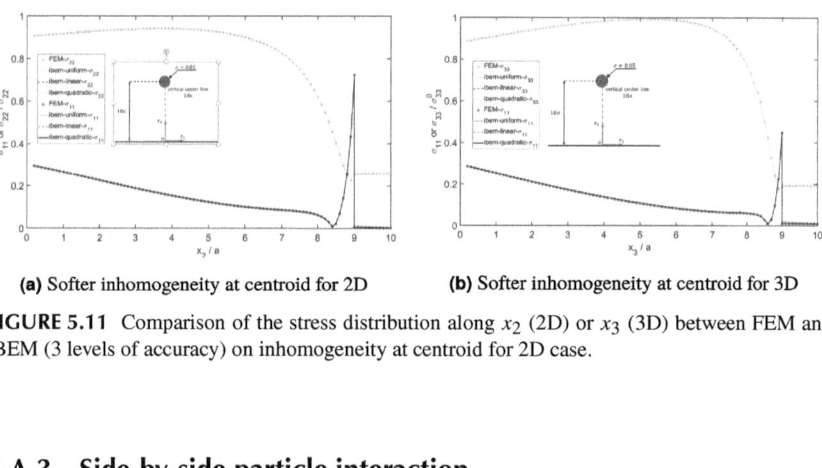

(a) Softer inhomogeneity at centroid for 2D **(b)** Softer inhomogeneity at centroid for 3D

FIGURE 5.11 Comparison of the stress distribution along x_2 (2D) or x_3 (3D) between FEM and iBEM (3 levels of accuracy) on inhomogeneity at centroid for 2D case.

5.A.3 Side-by-side particle interaction

In Sections 5.6.2 and 5.6.3, we investigated the interactions of inhomogeneities when two (equal-sized & large–small) inhomogeneities are aligned vertically. In the following, the elastic moduli, dimension of the inhomogeneities remain the same, but they are aligned horizontally as shown in Figs. 5.12–5.15. Because the geometric properties are similar to case studies in Sections 5.6.2 and 5.6.3, the mesh information (number of elements and nodes) for FEM is close. As shown, the quadratic term always provides better description than other order terms, which agrees with the previous conclusion made. Similarly to Figs. 5.5–5.6, stress concentration is observed at the interface of inhomogeneities and matrix.

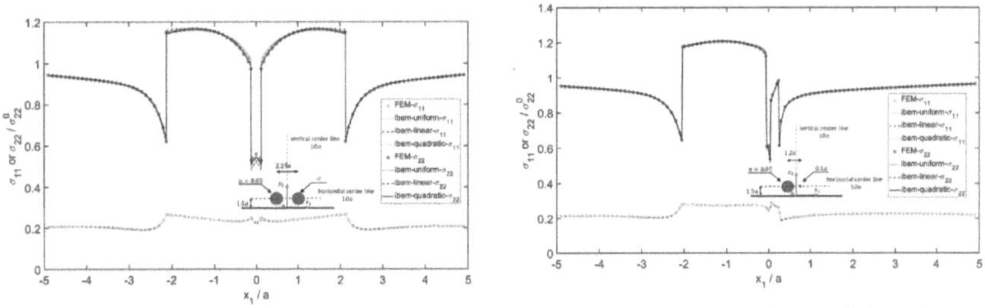

(a) Equal-sized inhomogeneities

(b) Large ($r = a$) and small ($r = 0.1a$) inhomogeneities

FIGURE 5.12 Comparison of the stress distribution along x_1 between FEM and iBEM (3 levels of accuracy) on side-by-side inhomogeneity interaction case for 2D.

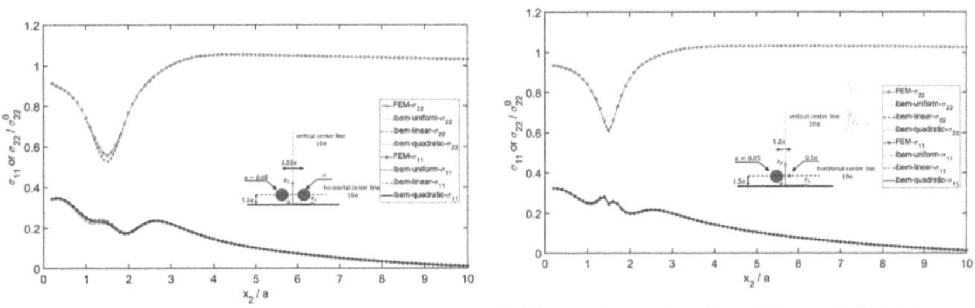

(a) Equal-sized inhomogeneities

(b) Large ($r = a$) and small ($r = 0.1a$) inhomogeneities

FIGURE 5.13 Comparison of the stress distribution along x_2 between FEM and iBEM (3 levels of accuracy) on side-by-side inhomogeneity interaction case for 2D.

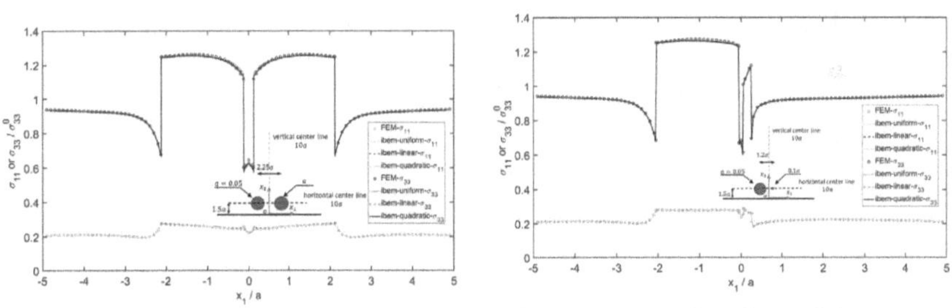

(a) Equal-sized inhomogeneities

(b) Large ($r = a$) and small ($r = 0.1a$) inhomogeneities

FIGURE 5.14 Comparison of the stress distribution along x_1 between FEM and iBEM (3 levels of accuracy) on side-by-side inhomogeneity interaction case for 3D.

5.A.4 7-particle interaction

As shown in Fig. 5.16, 7 inhomogeneities with radius $a = 0.05$ were placed around the center of the block domain, and the ratio of radius to the distances between neighbor inhomogeneities is selected as 1.5. The domain mesh in FEM:

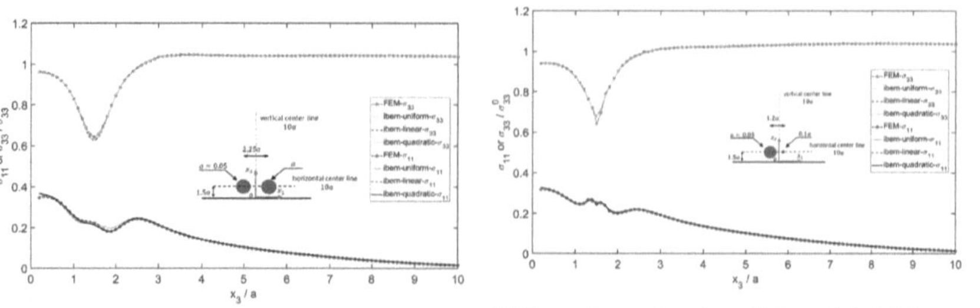

(a) Equal-sized inhomogeneities

(b) Large ($r = a$) and small ($r = 0.1a$) inhomogeneities

FIGURE 5.15 Comparison of the stress distribution along x_3 between FEM and iBEM (3 levels of accuracy) on side-by-side inhomogeneity interaction case for 3D.

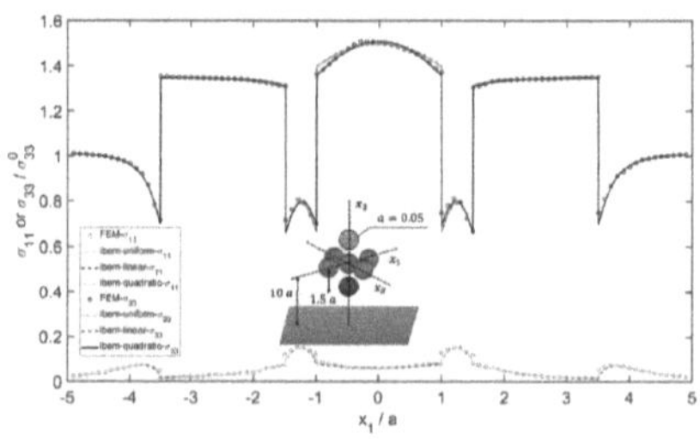

FIGURE 5.16 Comparison of the stress distribution along x_1 between FEM and iBEM (3 levels of accuracy) on 7 inhomogeneity interaction case for 3D.

the 4-node tetrahedral element is selected, and the numbers of nodes and elements are 1,467,304 and 1,074,979, respectively.

5.A.5 Examples of virtual experiments with large number of spherical particles

5.A.5.1 Effect of number of inhomogeneities on effective modulus

In Section 5.7.2, the effective modulus has been obtained under PBC with one inhomogeneity embedded at the centroid. In this section, we investigate the effect of number of inhomogeneities through keeping the same volume fraction 40% with uniform spatial distribution. Considering boundary effect, in contrast to the PBC, the boundary conditions were set to be a standard uniaxial loading test with $\sigma_{xx} = -10^4$ Pa. For 2D case, 9, 49, 100, 400, and 900 equal-sized inhomogeneities will be uniformly distributed in the composites in the simple square lattice, whereas for 3D case, 8, 64, 125, 512, 1000, and 3375 in the simple cubic lattice will be added. For both 3D and 2D cases, with the increase of the inhomogeneities, the effective Young modulus increases, and the Poisson ratio decreases, which shows the same trend of adding more inhomogeneities with

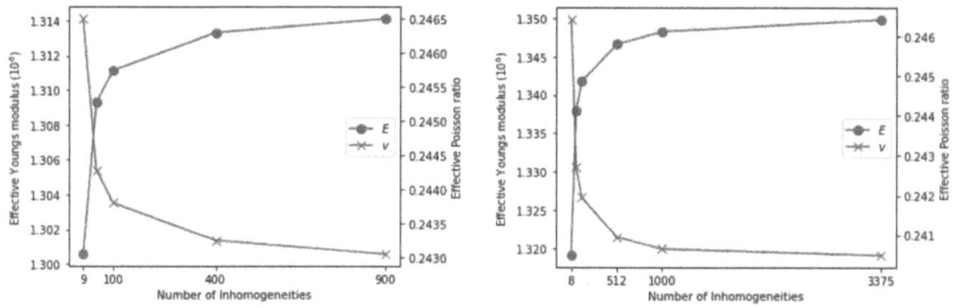

(a) 2D case (b) 3D case

FIGURE 5.17 Effective moduli of the same volume fraction and different numbers of inhomogeneities.

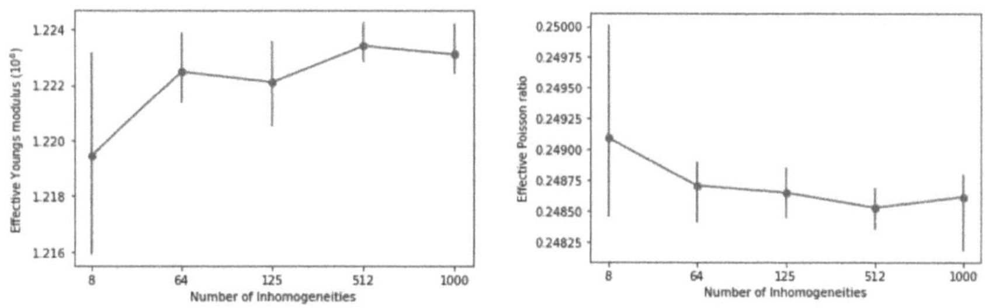

(a) Effective Young's modulus (b) Effective Poisson's ratio

FIGURE 5.18 Effective moduli of randomly distributed equal-sized inhomogeneities at 30% volume fraction.

stiffer material properties into a matrix. As shown in Fig. 5.17a, the differences in 9 and 900 cases are 1.07% and 1.42% for the effective Young modulus and Poisson ratio, respectively. Similarly, for 3D test, the differences are 2.27% and 2.55% for the effective Young modulus and Poisson ratio between 8 and 3375 cases. Compared with the Mori–Tanaka model in Fig. 5.7b, the differences between the virtual experiment (3375 particles) and micromechanical model are 3.17% and 3.33% for the effective Young modulus and Poisson ratio. Due to the particle interaction, the Young modulus increases, whereas the Poisson ratio reduces with the number of particles even if the volume fraction keeps the same. When the number is higher, the effective modulus approaches the constant value.

5.A.5.2 *Effect of random distribution on effective modulus*

In this section, equal-sized stiffer inhomogeneities of 30% volume fraction will be randomly distributed in the composites. The numbers of inhomogeneities in the case study were set as 8, 64, 125, 512, and 1000. Due to the random distribution, there exist fluctuations in the modulus. Hence for each case of the same number of inhomogeneities, five cases were calculated for an error-bar test. In

Fig. 5.18, each marker is placed with the average value. Larger differences in both effective moduli were observed in the case of eight inhomogeneities. As an overall approximate but not exact trend, the higher the number of inhomogeneities, the lower the variation of the effective modulus, the higher the averaged effective Young modulus.

Chapter 6

The iBEM implementation with particle discretization

Polyhedral inhomogeneities

In this chapter, we extend the iBEM implementation to polygonal or polyhedral inhomogeneities. Because a single polynomial function may not provide an accurate description of eigenstrain distribution for irregular shape of particles with singularity effects, the eigenstrain distribution is represented by the piecewise interpolation of the eigenstrain through the particle discretization, which may approach the exact solution by increasing the resolution of the mesh. The domain discretization is general and can be used for arbitrary shape of particles, and therefore iBEM can be applicable for virtual experiments of composite materials with arbitrary particles. Homogenization is introduced to yield effective or nonlocal material behavior from the local fields.

6.1 Introduction to iBEM for composites containing arbitrary inhomogeneities

Recall the closed-form domain integrals on inclusion problems [48–55] and its further combination of shape functions, namely the particle discretization method in Chapters 3 and 4 and [42,43], where the arbitrarily shaped inhomogeneities in the infinite space have been solved. The particle discretization method uses the analytical Eshelby tensor for piecewise continuous eigenstrain at the C^0 continuity, and we have demonstrated its accuracy, numerical stability, and rapid convergence to the exact solution. In Chapter 5, we presented the algorithm of iBEM to solve local fields and effective mechanical modulus under certain boundary conditions, where the discussion was limited for ellipsoids and ellipses. Substituting the polynomial-form eigenstrain assumptions with particle discretization method, the iBEM is particularly suitable for modeling the effective behavior containing a large variety of inhomogeneities in size, shape, and material properties. Particularly, the iBEM is suitable for cross-scale modeling and simulation because Eshelby's tensor is size independent and dimensionless, which avoids several numerical integral issues caused by the size variations of elements. It can predict the effective material performance with given microstructure conveniently because the traction and displacement on the

The Inclusion-Based Boundary Element Method (iBEM). https://doi.org/10.1016/B978-0-12-819384-6.00014-0

boundary are the principal variables to solve [40]. As the effective material behaviors do change with the shapes (i.e., aspect ratio, orientations), size gradation (i.e., dimension), and distribution of the inhomogeneities, the iBEM can capture the differences caused by these microstructural features. Using analytical integrals on inhomogeneities, the iBEM can avoid the numerical integral issues and provide higher accuracy in comparison with other BEM methods, such as fast multipole and dual reciprocal discussed in Section 1.4.

6.2 Implementation for a polynomial eigenstrain on polygonal and polyhedral inhomogeneities

In Chapter 5, is introduced a detailed implementation of ellipsoidal/elliptical inhomogeneities as in Eq. (5.29). Thanks to the geometric shapes of inhomogeneities (which could be described with a quadratic function), the distribution of eigenstrain field is approximated as uniform, linear, and quadratic terms of a polynomial form. As mentioned in Section 4.3, the microstructure singularity effects can significantly disturb the eigenstrain field, which exhibits a large slope around vertices and edges. Hence, in this section, we propose an implementation of particle discretization in iBEM. With the closed-form domain integral of fundamental solution, the above approach can retain the ultra advantage of the boundary element method that the numerical approximation is only applied at the vertices of polyhedrals/polygons, and the whole interior elastic field is computed analytically with high accuracy. Since the boundary–inhomogeneity interaction has been provided in Section 5.3, this section will focus on the stress equivalent conditions.

Consider a finite domain $\mathcal{D}$ containing multiple N^I inclusions Ω^I ($I = 1, 2, 3, \ldots, N^I$) subjected to a certain boundary condition. Some source fields, such as body force or eigenstrains, exist on the inclusions and induce stress or strain in their neighborhood. Since the disturbed fields are evaluated in terms of the shapes of inclusions, let N^{ell} and N^{p} denote the numbers of ellipsoidal/elliptical and polyhedral/polygonal subdomains, respectively. For the elastic stage, the actual elastic field by the loads on the boundary and the source field on the inclusions can be obtained through superposition of the boundary and inclusion integrals. The combination of Eqs. (5.18) and (4.48) provides the displacement and strain at field point $\mathbf{x}$ (within $\mathcal{D}$) in Eqs. (6.1) and (6.2), respectively:

$$u_i(\mathbf{x}) = -\sum_{e=1}^{N^b} H_{ij} u_j^e + \sum_{e=1}^{N^b} U_{ij} t_j^e + \left[\sum_{I=0}^{N^{ell}} g_{ikl} \varepsilon_{kl}^{*0I} + \sum_{I=0}^{N^{ell}} g_{iklp} \varepsilon_{klp}^{*1I} + \sum_{I=0}^{N^{ell}} g_{iklpq} \varepsilon_{klpq}^{*2I} \right]$$

$$+\frac{1}{8\pi(1-\nu)}\left[\sum_{J=0}^{N^p}(\mathcal{A}_{kl,kli}-2\nu\mathcal{B}_{kk,i}-4(1-\nu)\mathcal{B}_{ik,k})\right] \tag{6.1}$$

$$\begin{aligned}\varepsilon_{im}(\mathbf{x}) = &-\sum_{e=1}^{N^b} H_{imj}u_j^e + \sum_{e=1}^{N^b} U_{imj}t_j^e \\ &+\left[\sum_{I=0}^{N^{ell}} S_{imkl}\varepsilon_{kl}^{*0I} + \sum_{I=0}^{N^{ell}} S_{imklp}\varepsilon_{klp}^{*1I} + \sum_{I=0}^{N^{ell}} S_{imklpq}\varepsilon_{klpq}^{*2I}\right] \\ &+\frac{1}{8\pi(1-\nu)}\left\{\sum_{J=0}^{N^p}\left[\mathcal{A}_{kl,klim}-2\nu\mathcal{B}_{kk,im}\right.\right. \\ &\left.\left.-2(1-\nu)(\mathcal{B}_{ik,km}+\mathcal{B}_{mk,ki})\right]\right\}\end{aligned} \tag{6.2}$$

where S_{ijkl}, S_{ijklp}, and S_{ijklpq} are Eshelby's tensors for uniform, linear, and quadratic eigenstrain fields, which can be derived by the compatibility law, i.e., $S_{ijkl} = \frac{1}{2}(g_{ik,lj} + g_{jk,li})$; similarly, H_{imj} and U_{imj} denote the boundary integrals

$$H_{imj} = -\frac{1}{2}\int_{\partial\mathcal{D}}(G_{ni,m}(\mathbf{r}) + G_{nm,i}(\mathbf{r}))N_{nj}(\mathbf{x}')ds(\mathbf{x}') \tag{6.3}$$

$$U_{imj} = -\frac{1}{2}\int_{\partial\mathcal{D}}(T_{ni,m}(\mathbf{r}) + T_{nm,i}(\mathbf{r}))N_{nj}(\mathbf{x}')ds(\mathbf{x}') \tag{6.4}$$

Combining the boundary integral with the BEM and domain integral over the inclusions into Eq. (6.1), the displacement can be solved for the inclusion problem, and thus the strain and stress fields can be derived from the displacements.

6.2.1 Equivalent inclusion method for a bounded domain

In this section, we solve the inhomogeneity problem shown in Fig. 2.1. In the following, we will extend the previous work [35,39,40] to the inhomogeneity problem for an arbitrary shaped polygonal/polyhedral inhomogeneity in a bounded matrix. The key idea of the EIM is introducing an eigenstrain ε_{ij}^* on the particle domain to simulate the material mismatch on the stress field. Therefore we transfer the inhomogeneity problem to an equivalent inclusion problem by using the mismatch eigenstrain at each point on the inhomogeneity to satisfy the stress equivalent condition as follows:

$$C_{ijkl}^0 : (\varepsilon_{kl}^b + \varepsilon_{kl}' - \varepsilon_{kl}^*) = C_{ijkl}^I : (\varepsilon_{kl}^b + \varepsilon_{kl}') \tag{6.5}$$

where ε^b_{kl} is the strain caused by the boundary integral in Eq. (6.6)

$$\varepsilon^b_{kl}(\mathbf{x}) = \frac{1}{2}\int_{\partial D}[G_{kj,l}(\mathbf{x},\mathbf{x}') + G_{lj,k}(\mathbf{x},\mathbf{x}')]t_j(\mathbf{x}') - [T_{kj,l}(\mathbf{x},\mathbf{x}') + T_{lj,k}(\mathbf{x},\mathbf{x}')]u_j(\mathbf{x}')d\mathbf{x}' \tag{6.6}$$

and ε'_{kl} the strain caused by the area integral

$$\varepsilon'_{kl}(\mathbf{x}) = -\frac{1}{2}\int_{\Omega}[G_{ki,jl}(\mathbf{x},\mathbf{x}') + G_{li,jk}(\mathbf{x},\mathbf{x}')]C^0_{ijmn}\varepsilon^*_{mn}(\mathbf{x}')d\mathbf{x}' \tag{6.7}$$

In Section 4.3, to handle the inhomogeneity problem in the infinite space, a subdomain collocation particle domain discretization method is proposed, where the scheme to deal with strong singularities and discontinuities uses the concept of control volume/area. Combining the boundary element part and ellipsoidal/elliptical parts with the particle discretization method, we can rewrite the stress equivalence equations as Eqs. (6.9), (6.10), and (6.11) in the appendix at the end of this chapter.

When the linear equation system for eigenstrains and boundary unknowns is established and solved, we can use them to calculate the displacement field by the domain integral in Eqs. (6.6) and (6.7). Therefore the strain and stress fields can be calculated. Notice that Eshelby's tensor exhibits a $\ln r$ singularity on vertices [42,49] for 2D problems and $1/r$ on vertices and $\ln r$ on edges for 3D problems [43,49]. Therefore the stress singularity in the inhomogeneity problem may end up at $\ln^2 r$ and $1/r^2$ for 2D and 3D problems, respectively. Although the particle discretization using a shape function to provide the eigenstrain distribution exhibits numerical errors depending on the mesh aspects, because the influence zone of the singularity on the vertices is small, and also the domain integral is based on the exact solution, this algorithm can still provide high-fidelity results to the elastic field with low mesh sensitivity. In addition to the physical singularities on the vertices and along the edges, the handling of mathematical singularities at the interior nodes (i.e., post process to solve stress/strain) is also elaborated in Section 4.4.

6.2.2 Implementation of the iBEM algorithm

The iBEM has successfully simulated $>$ 1000 ellipsoidal/elliptical particles [40]. The above stress equivalent equations can be integrated into the global matrix (6.8) for the simulation of polyhedral/polygonal and ellipsoidal/elliptical particles in the bounded domain as follows:

$$
\underset{(dim\times N^b+\frac{13}{2}(dim^2+dim)N^{ell}+\frac{dim^2+dim}{2}N^i)^2}{\begin{bmatrix}
\mathcal{Q}+H & \cdots & -g^{0I} & -g^{1I} & -g^{2I} & \cdots & -\mathcal{K}^J & \cdots \\
\vdots & \vdots & \vdots & \vdots & \vdots & \vdots & \vdots & \vdots \\
\Delta C^I H^{1I} & \cdots & \Delta C^I S^{0I} & \Delta C^I S^{1I} & \Delta C^I S^{2I} & \cdots & \Delta C^I \mathcal{F}^J & \cdots \\
\Delta C^I H^{2I} & \cdots & \Delta C^I S^{0I'} & \Delta C^I S^{1I'} & \Delta C^I S^{2I'} & \cdots & \Delta C^I (\mathcal{F}^J)' & \cdots \\
\Delta C^I H^{3I} & \cdots & \Delta C^I S^{0I''} & \Delta C^I S^{1I''} & \Delta C^I S^{2I''} & \cdots & \Delta C^I (\mathcal{F}^J)'' & \cdots \\
\vdots & \vdots & \vdots & \vdots & \vdots & \vdots & \vdots & \vdots \\
\Delta C^J H^{1J} & \cdots & \Delta C^J S^{0J} & \Delta C^J S^{1J} & \Delta C^J S^{2J} & \cdots & \Delta C^J \mathcal{F}^J & \cdots \\
\vdots & \vdots & \vdots & \vdots & \vdots & \vdots & \vdots & \vdots
\end{bmatrix}}
\underset{(dim\times N^b+\frac{13}{2}(dim^2+dim)N^{ell}+\frac{dim^2+dim}{2}N^i)}{\begin{bmatrix}
u \\ \vdots \\ \varepsilon^{*0I} \\ \varepsilon^{*1I} \\ \varepsilon^{*2I} \\ \vdots \\ \varepsilon^{*J} \\ \vdots
\end{bmatrix}}
=
\underset{(dim\times N^b+\frac{13}{2}(dim^2+dim)N^{ell}+\frac{dim^2+dim}{2}N^i)\times(dim\times N^b)}{\begin{bmatrix}
U \\ \vdots \\ -\Delta C^I U^{1I} \\ -\Delta C^I U^{2I} \\ -\Delta C^I U^{3I} \\ \vdots \\ -\Delta C^I U^{1J} \\ \vdots
\end{bmatrix}}
\underset{(dim\times N^b)}{[t]}
\tag{6.8}
$$

where $dim = 2$ or 3 for the 2D or 3D problem, respectively; $\mathcal{Q}$ is the diagonal matrix introduced by the rigid-body method to avoid the Cauchy principle value integral on the strong singularity for the displacement on the boundary points BEM [21]. For simplicity, in Eqs. (6.1) and (6.2), except the boundary and domain integrals of the ellipsoidal part, let $\mathcal{K}^J$ and $\mathcal{F}^J$ denote the sums of the remaining parts (domain integrals of the Green's function over polyhedral/polygonal domains regarding displacement and strain, respectively). The rank of the matrix is marked under itself. Again, N^{b} is the number of boundary nodes; N^{ell} is the number of elliptical inhomogeneities; N^{i} is the total number of nodes of polygonal/polyhedral inhomogeneities.

Here the interactions between inhomogeneities and boundary are considered as follows:

1) The interactions between inhomogeneities and boundary are introduced by the components g^{0I} and $\mathcal{K}^J$, where the superscripts I and J represent the Ith elliptical inhomogeneity and Jth polygonal internal node, respectively. The superscripts $0, 1, 2$ denote the uniform, linear, and quadratic terms with respect to the eigenstrains, as shown in Appendix 6.B.
2) The interactions between inhomogeneities are considered in stress equivalent equations inserted below the traditional BEM global matrix, where $'$ means the partial derivatives of Eshelby's tensor. In Eq. (6.8), we consider the uniform, linear, and quadratic eigenstrain terms, although lower-order terms can predict the field accurately for some cases, such as dilute particle distributions and stiffer fibers.

Once the eigenstrain distribution is obtained, the displacement and strain/stress fields could be calculated with Eqs. (6.1) and (6.2), respectively. Unlike other methods with numerical integrals or approximation of the internal fields, the iBEM algorithm maintains the advantage of BEM that all the elastic fields can be expressed in an analytical form.

6.3 Continuity and singularity of elastic fields

Similarly to Section 4.5, the continuity and singularity concerns for an angular inhomogeneity embedded in a bounded domain exist along the surfaces, edges, and vertices of the particle. The interior elastic field of stress and strain in inclusions or inhomogeneities should be continuous based on the continuum mechanics assumption. However, for triangular or tetrahedral domain integrals, as the field point coincides with the interior node, there exists a mathematical singularity, which requires proper handling to evaluate the number. In the literature, this problem can be referred in [61,97] as the multiregion nonelastic problem, where some prescribed thermal/inelastic strain is given. The routine method is isolating the singularity as follows:

1) Generate a small circular (2D)/spherical (3D) exclusion zone, where the strain is caused by constant circular/spherical Eshelby's tensor multiplied by eigenstrain at the node;
2) The singularities canceled out themselves at the infinitesimal circumference (2D)/shell (3D), and thus the numerical integrals can be conducted.

Our recent work [43] proposed a simpler alternative scheme, which abandons any numerical handling, and it can be extended to the 2D problem as well.

Due to the assumption of distribution of eigenstrain by shape functions, the domain integral contains uniform, linear, and quadratic terms of the eigenstrain distribution, in which the linear and quadratic Eshelby tensors are written in terms of the uniform Eshelby tensors and the distance vector [42,43], and the singularity is only generated by uniform Eshelby's tensors. When a field point coincides with the interior nodes, all other shape functions at other nodes become 0, which implies the handling on the unit uniform tensors only. Using the divergence theorem, the unit uniform tensor is transferred to contour integrals, where no singularity exists anymore, and the number can be evaluated exactly by avoiding the numerical singularity in the mathematical formulation.

For a field point at a vertex, the singularity of Eshelby's tensor creates the stress singularity in the inclusion problem. Because of the $\ln r$ singularity of Eshelby's tensor at a vertex of a triangular inhomogeneity in 2D problems, the eigenstrain also exhibits a $\ln r$ singularity or approaches zero at the vertex. For the 3D problem, the $\ln r$ singularity of Eshelby's tensor exhibits on the edge of an angular particle similarly to the 2D problem, whereas the $\frac{1}{r}$ singularity can be found at the vertex [43,49]. The singular eigenstrain leads to a higher-order singularity in the stress distribution at $\ln^2 r$ and $\frac{1}{r^2}$ for 2D and 3D cases, respectively, as shown in Section 4.5.

The boundary effect does not change the continuity or singularity order along the interface. However, the boundary confinement does exist to reduce the singularity level of the stress or strain distribution.

6.4 Numerical verification with angular particles

The aforementioned algorithm has been implemented in the software package to conduct virtual experiments with various boundary conditions. To verify its accuracy, this section conducts two numerical case studies: 1) a subdomain with prescribed eigenstrain, i.e., a thermal strain; 2) an inhomogeneity embedded matrix under a vertical uniform load. As shown in Fig. 6.1, the isosceles triangle (2D)/tetrahedral (3D) with edge and height $w = h = 0.1$ m are embedded in the matrix with fixed bottom. The load is applied on the subdomain or the boundary. For the boundary element part, 1) 200 3-node line elements (2D) and 2) 1176 4-node quadrilateral elements (3D) are applied, whereas the domain discretization in the subdomains varies case by case.

6.4.1 Inclusion problem for a subdomain with prescribed eigenstrain

Firstly, we consider the inclusion problem as the subdomain exhibits the same stiffness as the matrix. Structural steel is selected with Young's modulus $E_s = 200\,\text{GPa}$, Poisson's ratio $\nu = 0.3$, and thermal expansion coefficient $\alpha_s = 1.2 \times 10^{-5} \frac{\text{m}}{\text{m}\cdot{}^\circ\text{C}}$. When a triangular/tetrahedral sample is taken out, to repair the hole, a similar larger sample is cooled down ($C^I = C^0$) 20° to tightly refill it without

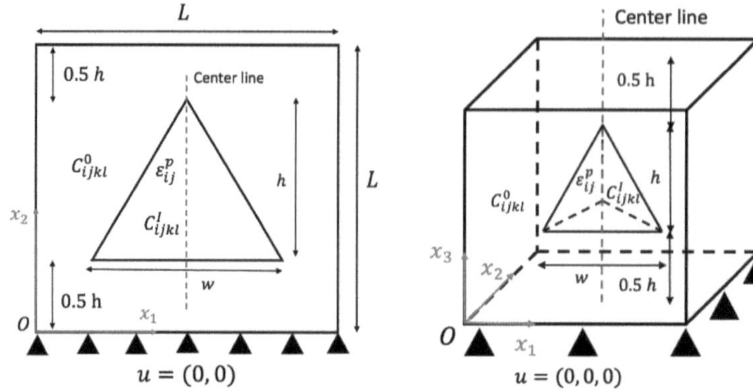

(a) 2D (b) 3D

FIGURE 6.1 Geometry and boundary conditions of the inclusion problem with a prescribed strain ε^p_{ij}.

welding. The misfit strain can be described by the thermal strain as well in the form of a prescribed eigenstrain $\varepsilon^*_{ij} = 2.4 \times 10^{-4}\delta_{ij}$.

Due to the eigenstrain, the displacement/strain field around the neighborhood of the inclusion is disturbed. The elastic field can be analytically expressed by the integral of the inclusion domain and boundary integral as well. Hence the domain discretization in iBEM is unnecessary, and a single element produces the convergent results shown in Fig. 6.2.

However, FEM requires many elements to illustrate the variation of the elastic field due to the singularity at the vertex. A refined mesh is used with the element size at 0.05 cm (178,436 elements) and 0.1 cm (1,418,222 elements) for 2D and 3D cases, respectively. As indicated in Fig. 6.2, a good agreement with the convergent results of FEM is observed with a single element for the inclusion problem of the iBEM, since it is the analytical domain integral of the Green's function. In addition, the cases with coarse FEM meshes at heights 1.25 and 0.3125 cm are illustrated to show the convergence situation of the FEM.

Although overall the FEM can provide convergent results with refined mesh, it is observed that the numerical instability at the vertex, particularly for the 3D case, because it exhibits a higher singularity level of $1/r$ than the 2D case of $\ln r$ [42,43,49]. Because iBEM includes the singularity into the analytical integral of Eshelby's tensor, it is robust and stable once the eigenstrain is prescribed.

6.4.2 Triangular/tetrahedral inhomogeneity under a uniaxial load

In this section, we substitute the subdomain in Section 6.4.1 with an aluminum alloy with Young's modulus $E_a = 70\,\text{GPa}$ and Poisson's ratio $\nu_a = 0.33$. In addition to the boundary conditions prescribed in Fig. 6.1, a downward uniaxial load of 1 MPa is applied at the top line (2D)/surface (3D). Other sides are free

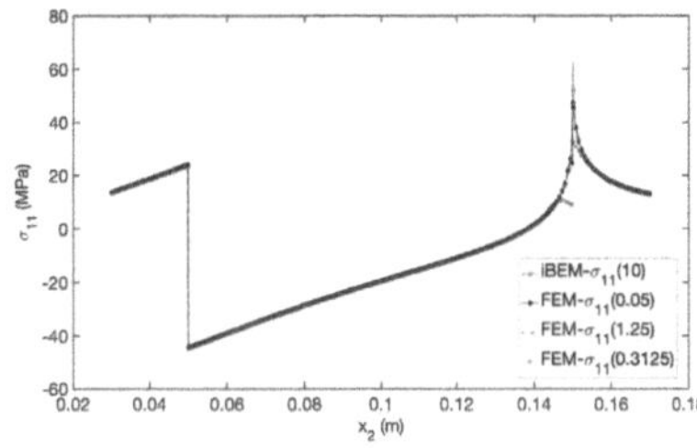

(a) σ_{11} 2D

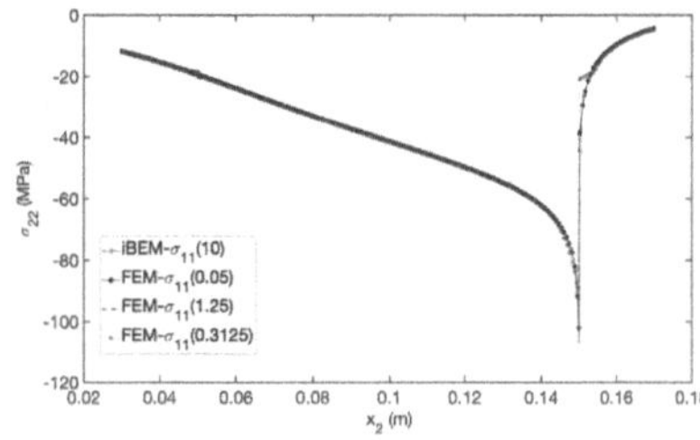

(b) σ_{22} 2D

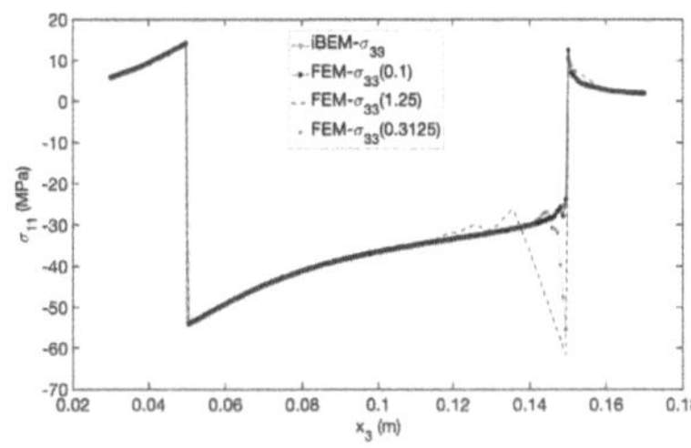

(c) σ_{11} 3D

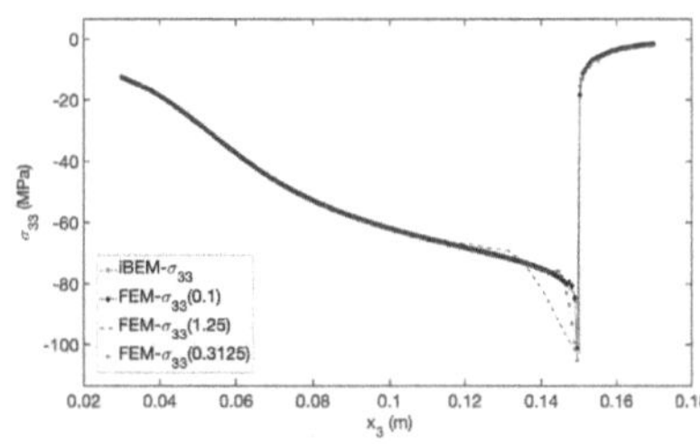

(d) σ_{33} 3D

FIGURE 6.2 Comparison of the stress distribution (a) σ_{11}, (b) σ_{22} (2D), (c) σ_{11}, (d) σ_{33} (3D) along the x_2 and x_3 for 2D and 3D, respectively, in the range of $[3, 17]$ cm between FEM and iBEM on inclusion problem with the prescribed eigenstrain $\varepsilon_{ij}^{p} = 2.4 \times 10^{-4}\delta_{ij}$.

of traction. Therefore it becomes an inhomogeneity problem for the material mismatch between the particle and matrix. Herein the mismatched eigenstrain is used the simulate the material difference in the equivalent inclusion problem, and it varies in an angular particle. We use the domain discretization and compare the results with the FEM results in Fig. 6.3. For the 2D triangular problem, $1, 4$, and 25 elements are applied in the inhomogeneity. It is interesting that when a single element is used on the inclusion, though the solution exhibits a small disparity from the convergent FEM results (178,436 elements), it predicts the trend and singularity of elastic solutions pretty well, because Eshelby's tensor of the particle plays the dominant role and the influence region of the eigenstrain singularity is quite limited. When the number of elements increases, say 4, the difference between FEM and iBEM rapidly reduces, and it well agrees even in a neighborhood of vertex. It is observed that a more refined mesh indeed improves the results as the number of elements increases to 25. However, the disparity between the two cases of iBEM-25 and iBEM-4 is negligible, which implies a good convergence. Similarly, for the 3D problem, $1, 8$, and 64 elements are employed. Due to the higher order of singularities and discontinuity of the surfaces, the 8-element case has little discrepancy around the bottom part of the tetrahedron. The stress curve iBEM-$\sigma_{11}(8)$ is parabolic in $[0.05, 0.10]$ m exactly inside one element, which suggests that a refined mesh will provide better stability. Hence, in the 64 element case, the solution agrees well with the

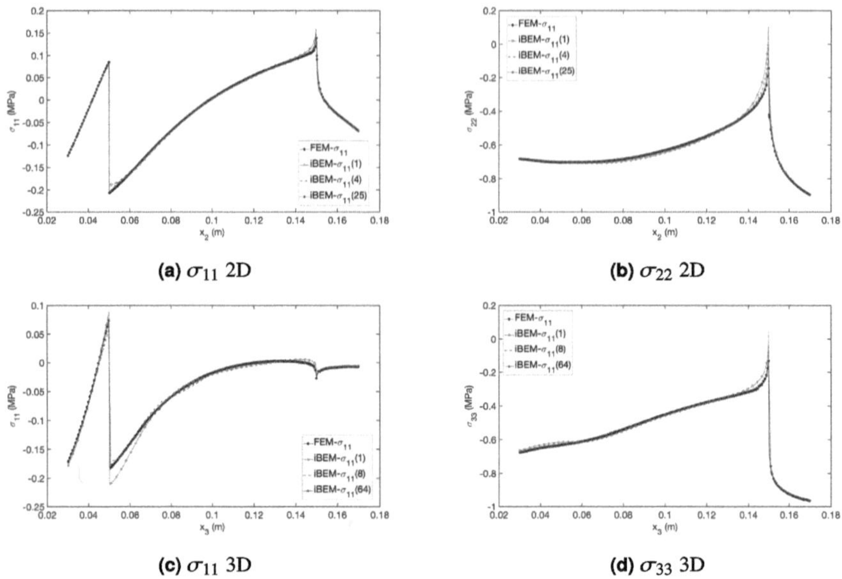

(a) σ_{11} 2D

(b) σ_{22} 2D

(c) σ_{11} 3D

(d) σ_{33} 3D

FIGURE 6.3 Comparison of the normal stress distribution (a) σ_{11}, (b) σ_{22} (2D); (c) σ_{11}, (d) σ_{33} (3D) along the x_2 and x_3 for 2D and 3D, respectively, in the range of $[3, 17]cm$ between FEM and iBEM on inhomogeneity problem with a uniform boundary pressure on the top line / surface of the square / cube.

convergent FEM results (1,418,222 elements). Moreover, for FEM, it is barely possible to describe stress singularities with such a small number of elements, since a high number of nodes is necessary to show the large variation of a curve with displacement on the nodes.

*6.5 Virtual experiments for arbitrary composites

In Section 6.3, we have validated the iBEM algorithm with case studies of inclusion and inhomogeneity problems by comparisons with the FEM results for the local fields. Meanwhile, the convergence test was conducted with various size steps between iBEM and FEM. The iBEM is particularly suitable for the prediction of effective material properties because the displacement and traction on the boundary are explicitly solved. Given a test load on the boundary, say traction, the response of the displacement along the boundary can immediately provide information for the stiffness of the material system. Therefore iBEM can be a powerful tool for virtual experiments. In Fig. 6.8, with a refined global element size, the eigenstrain field exhibits a mesh-related property only in neighborhoods of the vertices, which suggests that the local field shares the same features. To investigate the effective mechanical behavior of the composites, several methods can be applied, such as (1) classical micromechanical analysis, (2) unit cell under the periodic boundary conditions (PBC), and (3) an adequately large RVE with test loading. The limitation of micromechanical models has been discussed in the introduction, and the PBC will be implemented and discussed in this section. In the following, we investigate the effect of refinement of mesh on the effective modulus under uniaxial loading.

*6.5.1 Effective modulus for one triangular/tetrahedral inhomogeneity

In Fig. 6.1 of Section 6.4, one triangular/tetrahedral inhomogeneity is located inside a plate/cube (edge 20 cm). For the effective modulus test, the bottom constrained displacement condition is only applied to the x_2 direction, and the top pressure keeps the same as 1 MPa. As listed in Table 6.1, the effective modulus is collected with different size steps. Since Eshelby's tensor is dimensionless, instead of specifying the global element size, the number of division seeds along the edge is applied.

In Tables 6.1 and 6.2, we can observe a good convergence of iBEM in the modeling of effective modulus. Letting the bottom line data serve as the reference of convergent results, we can see the following aspects of the convergence of elastic moduli:

TABLE 6.1 The effective modulus of the composite containing one 12.5% triangular inhomogeneity with different number of division seed, where soft and stiff represents the aluminum and steel inhomogeneities, respectively.

#*Seed*	# elements	$\overline{E}$ (soft) (GPa)	$\overline{\nu}$ (soft)	$\overline{E}$ (stiff) (GPa)	$\overline{\nu}$ (stiff)
1	1	173.699	0.304714	78.236	0.327353
2	4	173.704	0.304676	78.256	0.327304
3	9	173.706	0.304663	78.259	0.327299
5	25	173.708	0.304656	78.261	0.327294
15	181	173.708	0.304652	78.261	0.327294

TABLE 6.2 The effective modulus of the composite containing one 1.8% tetrahedral inhomogeneity with different number of division seed, where soft and stiff represents the aluminum and steel inhomogeneities, respectively.

#*Seed*	# elements	$\overline{E}$ (soft) (GPa)	$\overline{\nu}$ (soft)	$\overline{E}$ (stiff) (GPa)	$\overline{\nu}$ (stiff)
1	1	196.770	0.300111	71.397	0.329230
2	8	196.771	0.300098	71.411	0.329186
3	27	196.770	0.300094	71.413	0.329174
4	64	196.771	0.300092	71.414	0.329169
8	324	196.771	0.300092	71.414	0.329168

TABLE 6.3 The effective modulus of the composite containing one 30% triangular inhomogeneity with different number of division seed, where soft and stiff represents the aluminum and steel inhomogeneities, respectively.

#Seed	# elements	$\overline{E}$ (soft) (GPa)	$\overline{\nu}$ (soft)	$\overline{E}$ (stiff) (GPa)	$\overline{\nu}$ (stiff)
1	1	143.910	0.313749	91.949	0.324680
2	4	143.947	0.313584	91.999	0.324628
3	9	143.960	0.313544	92.003	0.324624
5	25	143.966	0.313522	92.008	0.324619
15	181	143.970	0.313511	92.009	0.324618
FEM	20, 228	143.974	0.313353	92.004	0.3247831

1) For softer triangular inhomogeneity, $5.18 \times 10^{-3}\%, 2.30 \times 10^{-3}\%, 1.15 \times 10^{-4}\%$, and 0;
2) For stiffer triangular inhomogeneity, $3.19 \times 10^{-2}\%, 6.39 \times 10^{-3}\%, 2.56 \times 10^{-3}\%$, and 0;
3) For softer tetrahedral inhomogeneity, $5.08 \times 10^{-4}\%, 0, 5.08 \times 10^{-4}\%$, and 0;
4) For stiffer tetrahedral inhomogeneity, $2.38 \times 10^{-2}\%, 4.2 \times 10^{-3}\%, 1.4 \times 10^{-3}\%$, and 0.

Notice that the results are round to three decimal places, and thus the difference 0 is actually a nonzero small number and can be neglected in the sense of engineering applications. In the above cases, a single element for one particle can provide an acceptable prediction of the effective elastic constants, which saves efforts in computation and enables virtual experiments of large particle systems. However, the interactions between inhomogeneities and boundary are relatively low due to the low volume fraction. Hence, in the following case study, the volume fractions of triangle and tetrahedral inhomogeneity increase to 30% and 6.6%, and similar trends are exhibited in Tables 6.3 and 6.4, respectively. For a reference, an FEM case study is conducted with 20,228 elements for the 3D case with 30% of particles.

The differences between the 1-element (no discretization) case and the bottom line case do not exceed 0.14%, which implies that even if the boundary interactions are considered, the iBEM still provides good results for very few element meshes. For the 2D case at 30%, comparing the FEM and 1-element iBEM case, the differences are:

1) for soft inhomogeneity, $4.4 \times 10^{-2}\%$ and $1.26 \times 10^{-1}\%$ for Young's modulus and Poisson's ratio, respectively;
2) for stiff inhomogeneity, $5.97 \times 10^{-2}\%$ and $3.17 \times 10^{-2}\%$ for Young's modulus and Poisson's ratio, respectively.

Considering its excellent convergence performance in the prediction of effective elastic constants, even including the boundary and inhomogeneity interactions, only $\#Seed = 1, 2$ will be used in the following section. The case studies of

TABLE 6.4 The effective modulus of the composite containing one 6.6% tetrahedral inhomogeneity with different number of division seed, where soft and stiff represents the aluminum and steel inhomogeneities, respectively.

#*Seed*	# element	$\overline{E}$ (soft) (GPa)	$\overline{\nu}$ (soft)	$\overline{E}$ (stiff) (GPa)	$\overline{\nu}$ (stiff)
1	1	188.627	0.300792	75.328	0.327344
2	8	188.626	0.300730	75.388	0.327203
3	27	188.627	0.300718	75.400	0.327173
4	64	188.700	0.0.300704	75.441	0.327164
8	324	188.711	0.300696	75.450	0.327157

elliptical/ellipsoidal inhomogeneities can be referred in [40], and no repeated discussion will be added below. Here we show the case of multiple triangular/tetrahedral inhomogeneities in the Appendix at the end of this chapter.

*6.5.2 Effective modulus under periodic boundary condition

In 3D problems of a cube containing a regular tetrahedron, the maximum volume fraction of a tetrahedron is $\frac{1}{6}$. To simulate higher-volume fractions, a smaller cube is discretized with 12 tetrahedrons. As shown in Fig. 6.4, the effective moduli are compared with the classic micromechanical models, namely the dilute and Mori–Tanaka models [1] as references. Notice that the two classic micromechanical models are based on Eshelby's tensor of circular or spherical particles, whereas the present iBEM results are based on the angular particles, which are in between the two micromechanical models. Generally, the dilute model does not consider the interactions between inhomogeneities and underestimates the effective elasticity. Therefore the iBEM results are reasonable. Although the local elastic field is significantly affected by the particle shape, the effective elasticity does not change significantly by the particle shape, but closely depends on the particle volume fraction. The iBEM provides a powerful tool to predict the effective stiffness of composites with arbitrarily shaped particles.

*6.5.3 Virtual experiments of size gradation and random distribution of particles on the effective elasticity

In Sections 6.5.1 and 6.5.2, we solved the effective moduli through uniaxial loading on unit cell under periodic boundary conditions and verified them with FEM and micromechanical models. In this section, to exhibit the robustness of the method, instead of considering a RVE or unit cell, we used the virtual experiment with over 1137 tetrahedral and 3470 triangular particles for composites with 50% volume fraction. Shown in Fig. 6.5, for example, 1591 triangular in-

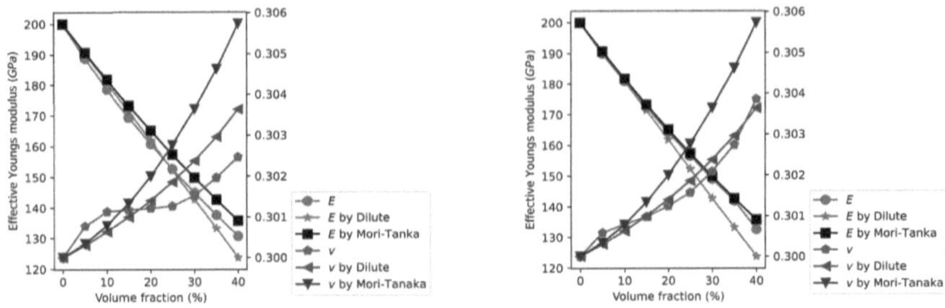

(a) 2D **(b)** 3D

FIGURE 6.4 Comparison of the iBEM results with the dilute and Mori–Tanaka models for the effective elastic moduli of composites under the periodic boundary condition in (a) 2D and (b) 3D cases.

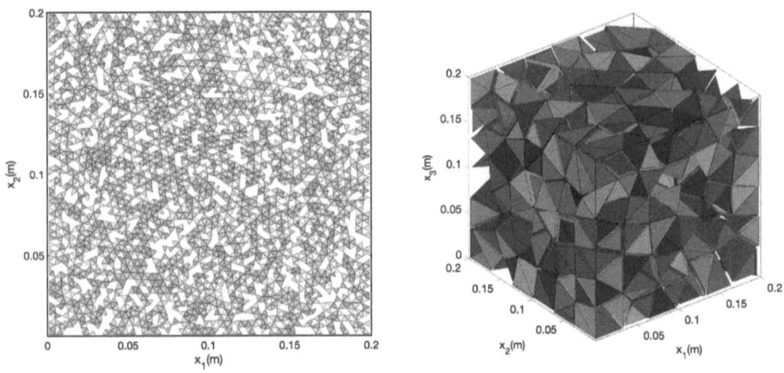

(a) 2D **(b)** 3D

FIGURE 6.5 Schematic plot of 2D and 3D inhomogeneities distribution with 50% volume fraction.

homogeneities are created in a 0.2 m × 0.2 m square plate. The mesh is generated by the following steps:

1) Mesh the square with triangular elements;
2) use a pseudorandom library in C++ to select elements;
3) shrink the elements again using the pseudorandom functions.

In Sections 6.4.1 and 6.4.2, we have demonstrated a good convergence performance with a very few elements to represent an inhomogeneity. Therefore, in this section, triangles and tetrahedrons are represented by one element. Shown in Table 6.5, the effective moduli are compared with different numbers of inhomogeneities: for a 2D problem, 6500 and 3470 particles are included, whereas for a 3D problem, 61,338 and 1137 particles are included, respectively. Although the orientation and size of particles are relatively random, the effective Young modulus and Poisson ratio are stable for the virtual experiments of the samples

TABLE 6.5 The effective modulus of virtual composite samples with 50% of aluminum particles randomly distributed in a steel matrix.

# 2D particles	$\overline{E}$ (GPa) (2D)	$\overline{\nu}$ (2D)	# 3D particles	$\overline{E}$ (GPa) (3D)	$\overline{\nu}$ (3D)
6	115.168	0.311444	61	125.776	0.305335
500	115.509	0.311746	338	124.4481	0.306705
3470	115.334	0.311482	1137	124.698	0.306706

with different numbers of particles under the same volume fraction at 50%. The 3D case shows a higher Young's modulus than the 2D case.

Appendix 6.A Stress equivalent equations with boundary integral equation

In Sections 4.4.1 and 4.4.2, we introduced the particle domain discretization method for polyhedral and polygonal problems, respectively. In Section 5.2 (Eqs. (5.17) and (5.25)), the inclusion problem with boundary integral equations and domain integral of the particles was presented with prescribed eigenstrain. Subsequently, Eqs. (5.27) and (5.28) serve as stress equivalence conditions for uniform and linear eigenstrain, respectively. To extend with arbitrarily shaped particles (polygonal/polyhedral discretization), we just need to replace the domain integral part with summation of small elements. Following the subdomain collocation scheme in Section 4.4.3, the stress equivalence conditions can be set up (adding the boundary integral part) as follows:

$$
\begin{aligned}
&\Delta C^I_{ijkl}\left(-\sum_{e=1}^{N^b} H_{klm}u^e_m + \sum_{e=1}^{N^b} U_{klm}t^e_m + \sum_{I=0}^{N^{ell}} S_{klab}\varepsilon^{*0I}_{ab}\right.\\
&\left.+\sum_{I=0}^{N^{ell}} S_{klabp}\varepsilon^{*1I}_{abp} + \sum_{I=0}^{N^{ell}} S_{klabpq}\varepsilon^{*2I}_{abpq}\right)\\
&+\frac{\Delta C^I_{ijkl}}{8\pi(1-\nu)}\left(\sum_{J=1}^{N^p}\left[\mathcal{A}_{kl,klij} - 2\nu\mathcal{B}_{kk,ij} - 2(1-\nu)(\mathcal{B}_{ik,kj} + \mathcal{B}_{jk,ki})\right]\right)\\
&+C^0_{ijkl}\varepsilon^{*0I}_{kl} = 0
\end{aligned}
\tag{6.9}
$$

$$
\begin{aligned}
&\Delta C^I_{ijkl}\left(-\sum_{e=1}^{N^b} H_{klrm}u^e_m + \sum_{e=1}^{N^b} U_{klrm}t^e_m + \sum_{I=0}^{N^{ell}} S_{klabr}\varepsilon^{*0I}_{ab}\right.\\
&\left.+\sum_{I=0}^{N^{ell}} S_{klabpr}\varepsilon^{*1I}_{abp} + \sum_{I=0}^{N^{ell}} S_{klabpqr}\varepsilon^{*2I}_{abpq}\right)
\end{aligned}
$$

$$+\frac{\Delta C^I_{ijkl}}{8\pi(1-\nu)}\left(\sum_{J=1}^{N^p}\left[\mathcal{A}_{kl,klijr}-2\nu\mathcal{B}_{kk,ijr}-2(1-\nu)(\mathcal{B}_{ik,kjr}+\mathcal{B}_{jk,kir})\right]\right)$$
$$+C^0_{ijkl}\varepsilon^{*1I}_{klr}=0 \tag{6.10}$$

$$\Delta C^I_{ijkl}\left(-\sum_{e=1}^{N^b}H_{klrwm}u^e_m+\sum_{e=1}^{N^b}U_{klwrm}t^e_m+\sum_{I=0}^{N^{ell}}S_{klabrw}\varepsilon^{*0I}_{ab}\right.$$
$$\left.+\sum_{I=0}^{N^{ell}}S_{klabprw}\varepsilon^{*1I}_{abp}+\sum_{I=0}^{N^{ell}}S_{klabpqrw}\varepsilon^{*2I}_{abpq}\right)$$
$$+\frac{\Delta C^I_{ijkl}}{8\pi(1-\nu)}\left(\sum_{J=1}^{N^p}\left[\mathcal{A}_{kl,klijrw}-2\nu\mathcal{B}_{kk,ijrw}\right.\right.$$
$$\left.\left.-2(1-\nu)(\mathcal{B}_{ik,kjrw}+\mathcal{B}_{jk,kirw})\right]\right)+C^0_{ijkl}(\varepsilon^{*2I}_{klrw}+\varepsilon^{*2I}_{klwr})=0 \tag{6.11}$$

Appendix 6.B Stress contour plot of inclusion problem

In this subsection, following the case study in Section 6.4.1 (2D triangular inhomogeneity), the converged stress is exhibited as a contour plot in Fig. 6.6, where the stress concentration and singularity at the neighborhood of the vertices are demonstrated. In the square, 201 × 201 uniformly distributed sample points are applied to provide smooth plot of the stresses close to edges of the triangular domain. It is obvious that there exist stress concentrations (or singularity) at all vertices and stress discontinuity at all edges, except σ_{22} for the bottom edges, which agrees well with the formulations in Chapter 4; σ_{11} and σ_{22} possess different distribution that the variation of σ_{11} exhibits a gradation pattern in the x_2 direction within the subdomain, which is caused by the boundary confinement at the bottom. In addition, though there exists stress singularity at all the vertices, the stress variation around the top vertex is larger than two other bottom vertices.

Appendix 6.C Stress contour plot of inhomogeneity problem

Following the inhomogeneity case study in Section 6.4.2, the convergent stress is plotted in Fig. 6.7. Compared with Fig. 6.6 (inclusion problem with prescribed eigenstrain), though both of them have stress singularities at vertices, the stress varies more intensively for the inhomogeneity problem. To interpret the phenomenon, just consider the distribution of eigenstrain introduced in Section 4.5 that the eigenstrain of inhomogeneity may approach infinity or zero at the vertices. Hence the stress singularity in the inhomogeneity problem can be at higher level than the inclusion problem with uniform eigenstrain. As proved, with a $\ln r$

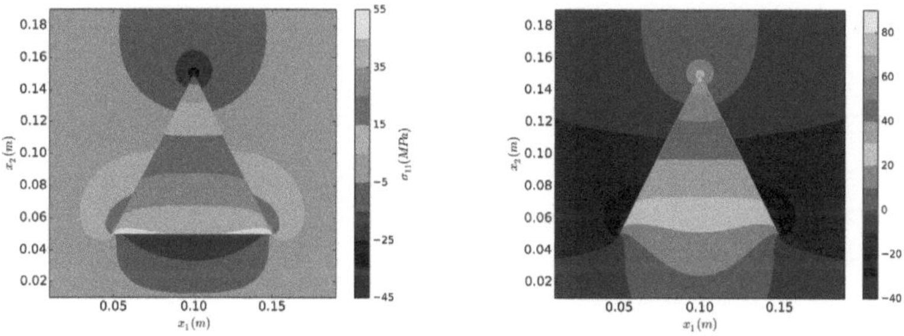

(a) σ_{11} **(b)** σ_{22}

FIGURE 6.6 Contour plot of the normal stresses field in the square of the range $x_1 \in [1, 19]$ cm and $x_2 \in [1, 19]$ cm for the inclusion problem with prescribed eigenstrain for the 2D problem.

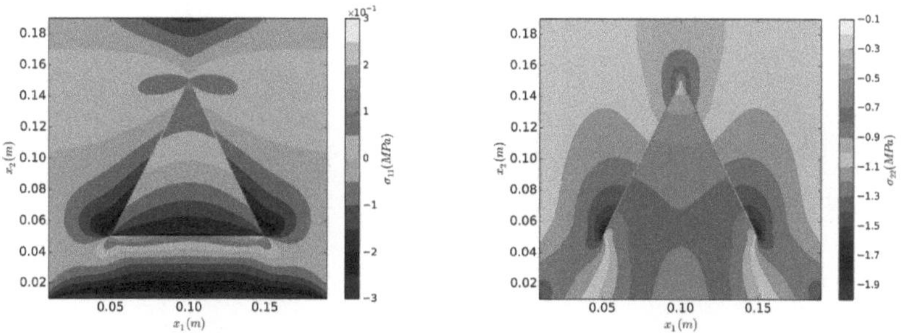

(a) σ_{11} **(b)** σ_{22}

FIGURE 6.7 Contour plot of the normal stresses field in the square of the range $x_1 \in [1, 19]$ cm and $x_2 \in [1, 19]$ cm for the 2D inhomogeneity problem under a uniaxial load.

distribution eigenstrain, a $\ln^2 r$ stress singularity is obtained in the neighborhood of the vertices. In addition, though the distribution of the eigenstrain is not uniform anymore, the gradation pattern still exists due to the boundary confinement.

Appendix 6.D Discussion on mesh strategy of polygonal inhomogeneities

As discussed in Section 6.4.2, there exists singular effect of stresses in the neighborhood of the vertices, where the eigenstrain exhibits a larger variation in comparison to the eigenstrain field of the central part. In this section, we apply the contour plots of eigenstrain field with three different global element sizes: (1) 5 cm (4 elements), (2) 2 cm (25 elements), and (3) 0.5 cm (426 elements). Shown in Fig. 6.8, each contour plot contains 24 levels to exhibit the difference. The ε^*_{22} is concentrated around the middle part, whereas ε^*_{11} concentrates around the two side vertices. Although the overall distribution follows a similar pattern for three element resolutions, smaller elements can obviously capture the eigenstrain singularity better with a larger distribution range. Refiner mesh does not improve much in the middle parts, because the singularity effect

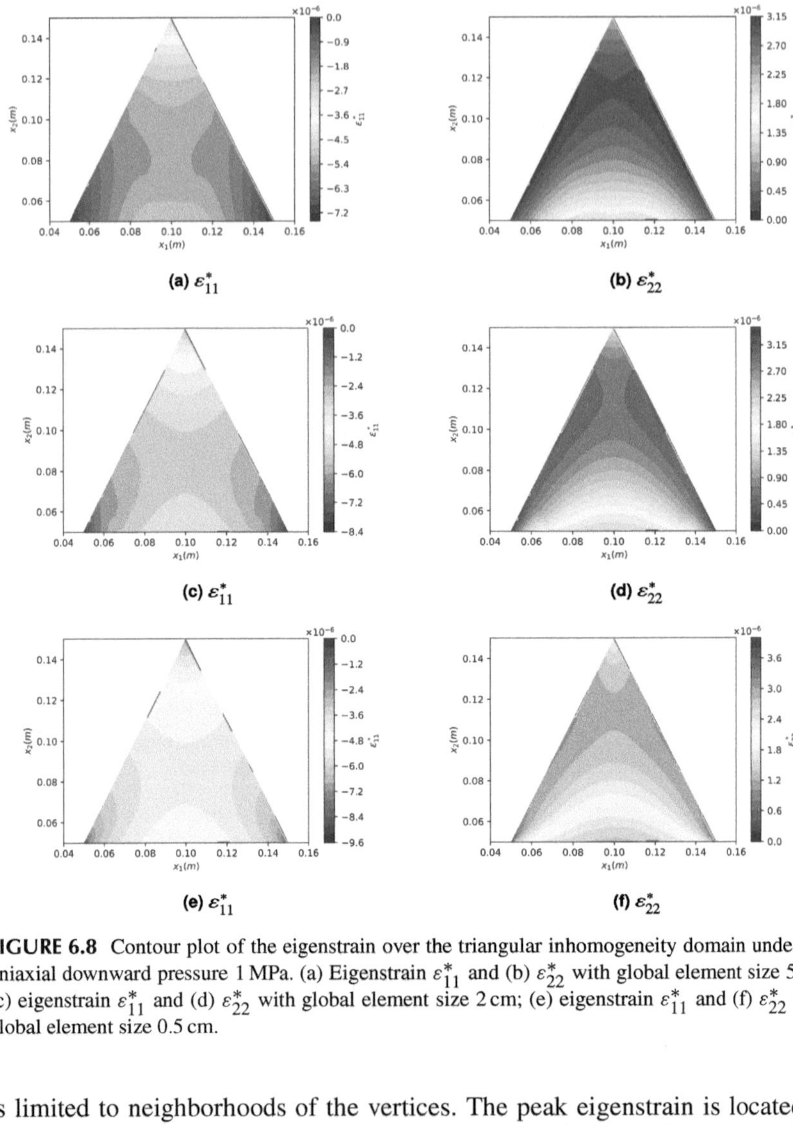

(a) ε_{11}^* (b) ε_{22}^*

(c) ε_{11}^* (d) ε_{22}^*

(e) ε_{11}^* (f) ε_{22}^*

FIGURE 6.8 Contour plot of the eigenstrain over the triangular inhomogeneity domain under the uniaxial downward pressure 1 MPa. (a) Eigenstrain ε_{11}^* and (b) ε_{22}^* with global element size 5 cm; (c) eigenstrain ε_{11}^* and (d) ε_{22}^* with global element size 2 cm; (e) eigenstrain ε_{11}^* and (f) ε_{22}^* with global element size 0.5 cm.

is limited to neighborhoods of the vertices. The peak eigenstrain is located at the top area along the edges, where the dark red area (dark gray in print version) ($\varepsilon_{22} > 2.7 \times 10^{-6}$) reduces as the dimension of global elements decreases.

The above observations provide an alternative mesh strategy to improve both the accuracy and efficiency: (1) for interior middle part, we can simply use larger/coarse mesh, (2) extra attentions can be paid on the neighborhood of the vertex and edges if one pursues the high accuracy. Notice that because Eshelby's tensor is size independent, iBEM can maintain good numerical stability even with very fine elements. However, in comparison with FEM in Fig. 6.3, 4- and 25-element cases have already predicted accurate results, which implies that the change of eigenstrain at some "sensitive" regions (i.e., around vertices or edges) only produces limited influences to the global solutions. In the common sense, no special measurements are required to be applied at those regions during the mesh process; a uniform mesh is acceptable. Because the elastic field is obtained by the integral of eigenstrain over the particle domain and the influence of the eigenstrain at one point decays at $1/r^2$ or $1/r^3$ for 2D or 3D problems, respectively, a minor error at one point may not significantly affect the overall accuracy. Particularly, although the eigenstrains at the vertex and boundary nodes are sensitive to the mesh, their effects on overall results of the stress distribution are localized on their neighborhood. That is the reason why even a single element can provide good prediction of the trend of elastic field variation.

Appendix 6.E Effective modulus with multiple triangular/tetrahedral inhomogeneities

In this section, we introduce the particle interactions by adding more inhomogeneities into the simulation domain. Again, the effective moduli are compared through changing various division seed (#*Seed*) of the edges. Without loss of generality, 25 equal-sized isosceles and 27 tetrahedrons are uniformly distributed in the square/cube with total volume fraction equal to 30% and 6.6%, as illustrated in Section 5.1. Because Eshelby's tensor decreases rapidly with the distance between the field and source point (Appendix 6.D), it is reasonable to set up a global cutoff "distance ratio ρ" to minimize the computation efforts. Taking the 2D domain integral, for instance, define $\rho = \frac{\mathcal{L}}{|\mathbf{x}'-\mathbf{x}|}$, where $\mathcal{L} = \sqrt{2A}$ is the characteristic length of the triangular element, and A is the area. Similarly, define $\rho = (6V)^{1/3}$ for tetrahedral elements in 3D, where V is the volume. When $\rho > 15$, the domain integral plays a small role in the elastic solution, and hence the contribution can be neglected. The cut-off can be redefined with the required accuracy. Notice that the choice of ρ may influence the accuracy of the solution; however, the accuracy can be compromised to save computational efforts if the error is controlled.

Shown in Table 6.6, when $\rho = 15$, the errors increase up to 0.82%, but it is still acceptable with the single element case. Notice that if global cutoff is not considered, then the effective Young moduli are 147.544 GPa and 95.066 GPa

TABLE 6.6 The effective modulus of the composite containing 25 triangular inhomogeneity (30% volume fraction) with different number of division seed, where soft and stiff represents the aluminum and steel inhomogeneities, respectively.

#*Seed*	# elements	$\overline{E}$ (soft) (GPa)	$\overline{\nu}$ (soft)	$\overline{E}$ (stiff) (GPa)	$\overline{\nu}$ (stiff)
1	25	147.568	0.303659	94.958	0.317976
2	100	147.530	0.303955	94.980	0.317757

TABLE 6.7 The effective modulus of the composite containing 27 tetrahedral inhomogeneity (6.7% volume fraction) with different number of division seed, where soft and stiff represents the aluminum and steel inhomogeneities, respectively.

#Seed	# elements	$\overline{E}$ (soft) (GPa)	$\overline{\nu}$ (soft)	$\overline{E}$ (stiff) (GPa)	$\overline{\nu}$ (stiff)
1	27	188.578	0.300192	75.437	0.326844
2	216	188.457	0.300196	75.571	0.326824

for softer and stiffer inhomogeneities, respectively. Moreover, in the case of division seed *#Seed* at 2, the differences are comparatively small, which is due to the larger sizes of the element, so the "cutoff" condition is not frequently triggered. (See Table 6.7.)

Chapter 7

The iBEM for potential problems

Scalar potential flows – heat conduction

Following the application of iBEM for elastic analysis, in this chapter, we introduce the iBEM application and implementation for the scalar potential problems, which is applicable to linear flows such as heat, electric, magnetic, or ground water flows. Using heat conduction as an example, we extend similar development process for elastic problems with displacement potentials to a scalar potential such as temperature. We introduce an eigentemperature gradient to simulate the mismatch of thermal conductivities of inhomogeneities. A case study shows the application of iBEM toward virtual experiments.

7.1 Generalization of the EIM to boundary value problems with inhomogeneities

For ellipsoidal particles embedded in the infinite domain subjected to a uniform far field load, the problem can be similarly solved by the equivalent inclusion method of the elastic analysis. Therefore, the boundary value problem can be formulated and solved by Eshelby's EIM in a similar fashion. Consider a general composite material occupying a domain D. Inside the composite material, there are many subdomains Ω_i whose material property E_i is different from the matrix property E_0 as shown in Fig. 7.1. The field quantity U in Eq. (7.7), which can be a scalar or vector, is governed by a set of partial differential equations. For example, it can be the velocity field in the Stokes equations, the displacement field in the Navier–Cauchy equations, or the temperature field in the potential flow, etc. These equations form a complete system $f(U, E) = 0$ with constitutive law, which characterizes the field response with different material properties. To fully analyze the field in composites, each governing equation $f(U, E) = 0$ in subdomain Ω_i need to be developed, and all the boundary conditions on ∂D and continuity on $\partial\Omega_i$ need to be satisfied as well.

Eshelby [2] proposed EIM to solve the elastic field of one inhomogeneity in an infinite domain by replacing the inhomogeneity problem with an equivalent inclusion problem, in which the material properties are uniform but eigenstrain is introduced to simulate the inhomogeneity. Here we adopt a similar idea but extend to the cases of a finite domain and multiphysical problems. Similarly to the elastic problem in Chapter 5, the overall field can be expressed not only by

The Inclusion-Based Boundary Element Method (iBEM). https://doi.org/10.1016/B978-0-12-819384-6.00015-2

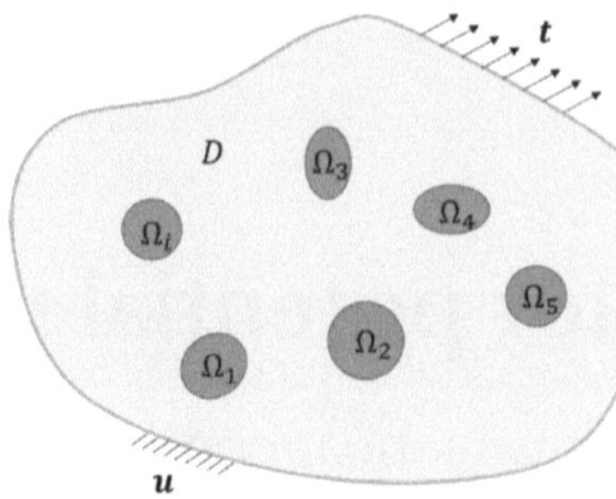

FIGURE 7.1 Problem statement.

the effect from inclusions but also by surface integral of boundary responses [35] as follows:

$$U\left(\mathbf{x}'\right)=\int_{\partial D} G\left(\mathbf{x}',\mathbf{x}\right)t\left(\mathbf{x}'\right)\mathrm{d}\mathbf{x}-\int_{\partial D} T\left(\mathbf{x}',\mathbf{x}\right)u\left(\mathbf{x}\right)\mathrm{d}\mathbf{x}+\sum_{I=1}^{N} D_i^I\left(\mathbf{x}'\right)\varepsilon_i^{I*}, \tag{7.1}$$

where the first two terms are commonly seen in the traditional boundary element method, which is derived from reciprocal work theorem [21], G is Green's function, which is the fundamental solution for a specific physical problem; T describes the first-order derivative of the field, which is derived from G based on fundamental procedure in BEM, and N is the total number of particles. Notice that depending on the physical problems, U, T, G can be in scalar, vector, or tensor forms and the rank of $\mathbf{D}$, $\boldsymbol{\varepsilon}^*$ can be higher as well. Here $\boldsymbol{\varepsilon}^{I*}$ is introduced in the last term to simulate the material mismatch on each particle I; $\mathbf{D}^I$ is a tensor to relate the disturbed field with eigenfield ε^{I*} on the inclusion I. It can be evaluated by an elliptical integral over the particle domain analytically as

$$\mathbf{D}^I(\mathbf{x})=-E^I\int_{\Omega_I}\nabla G\left(\mathbf{x}',\mathbf{x}\right)d\mathbf{x}'. \tag{7.2}$$

Eshelby's original equivalent inclusion method (EIM) is to set up the stress equivalent conditions to find an equivalent inclusion problem for the inhomogeneity problem. It has been successfully applied to other physics problems [10,95,106]. So generally speaking, the equivalence can be built on the flux of the field quantity U as follows:

$$E^0\left(\nabla U-\boldsymbol{\varepsilon}^{I*}\right)=E^I\nabla U \quad \text{for } I=1:n. \tag{7.3}$$

Mura [1] has devoted a lot of effort to improving the accuracy of EIM when dealing with multiple particles by using high-order polynomial form of ε^{I*} in each inclusion. Basically, we assume the following distribution of eigenstrain in each particle:

$$\varepsilon_i^{I*} = \varepsilon_i^{I0} + \varepsilon_{ij}^{I1} x_j + \cdots, \quad \mathbf{x} \in \Omega_I. \tag{7.4}$$

When higher-order terms of the polynomial are used in Eq. (7.4), higher accuracy can be achieved, and more complex eigenstrain field can be described [35]. By applying the Taylor expansion technique at the center of particle I for the Ith equation on both sides of Eq. (7.3) and matching the coefficients at the same power of x, the equivalence on ∇U will be written in terms of coefficients for each order of the polynomial evaluated at the center of each particle as follows:

$$E^0 \left(\nabla U(\mathbf{0}) - \boldsymbol{\varepsilon}^{I0} \right) = E^I \nabla U(\mathbf{0}) \quad \text{for } I = 1:n, \tag{7.5}$$

$$E^0 \left(\nabla\nabla U(\mathbf{0}) - \boldsymbol{\varepsilon}^{I1} \right) = E^I \nabla\nabla U(\mathbf{0}) \quad \text{for } I = 1:n, \tag{7.6}$$

where $\mathbf{0}$ in Eqs. (7.5) and (7.6) indicates the function should be evaluated at the center of particle I as the origin of the local coordinate. Combining with the equations from BEM as Eq. (7.1), the linear equation system can be assembled to determine the unknowns of boundary responses and the eigenfield $\boldsymbol{\varepsilon}^*$ for all inhomogeneity particles at the same time. Therefore the field U will be eventually derived by Eq (7.1) by its corresponding equivalent inclusion problem. Here two facts should be highlighted:

a) There is no discretization on each particle involving in this solving process. Only a few terms of $\boldsymbol{\varepsilon}^*$ need to be introduced to take into account the material mismatch in the particle domain. The BEM formula in Eq. (7.1) will not only have surface integral but also include the effect of inclusions by the volume integral tensor D.
b) The surface integral will appear inside the equivalence conditions in Eqs. (7.5) and (7.6), which is a more general form of Eshelby stress equivalent conditions. If the domain is infinite, then by the property of Green's function $G \to 0$ as $r \to \infty$ the surface integral will vanish, and it will reduce to the classic form.

This general EIM framework will be applied to the heat transfer problem as a scalar potential flows in this chapter and extended to Stokes' flow problem in the next chapter.

7.2 The iBEM for potential flow – heat conduction

For the multiphysical material behavior of an ellipsoidal inclusion in an infinitely extended domain, the governing equation for the steady-state problem can be written as a potential or Poisson equation

$$\nabla^2 U(\mathbf{x}) = f(\mathbf{x}), \tag{7.7}$$

TABLE 7.1 Physical phenomena related to potential flow.

Problems	Potential	Proportionality constant	Physical flow (flux)
Heat transfer	Temperature u	Thermal conductivity k	Heat flux $\mathbf{q}=-k\nabla u$
Ground water flow	Hydradic heat H	Permeability k	Water flow $\mathbf{q}=-k\nabla H$
Electrostatic field	Potential E	Permittivity ε	Electric flow $\mathbf{q}=-\varepsilon\nabla E$
Electric conduction	Electric voltage V	Electric conductivity k	Electric current $\mathbf{I}=-k\nabla V$
Magnetic field	Potential U	Permeability μ	Induction $\mathbf{B}=-\mu\nabla U$
Gravitation	Potential ϕ	Gravitational constant G	Gravitational force $\mathbf{F}=G\nabla\phi$

where U is the potential and f is a source. In the linear range the gradient of the potential is typically proportional to a physical flow (flux). In analogy to the elastic problem, the potential, the gradient, and the flow are mapped into displacement, strain, and stress except that in the potential problem the variables have one rank lower than the elastic problem, i.e., a vector field reduces to a scalar field, and a second-rank tensor reduces to a vector field. The above equation mathematically models many important physical phenomena listed in Table 7.1.

In this section, the steady-state heat conduction problem in an infinite domain is illustrated. The field quantity $U=u$ in Eq. (7.7) is the temperature field inside the domain, which is also a scalar field, and the material property is the thermal conductivity K. The governing equation in absence of a heat source or sink is the Laplace equation

$$-K\nabla^2 u=0, \tag{7.8}$$

and the Green's function G and T in Eq. (7.1) are

$$G(\mathbf{x},\mathbf{x}')=-\frac{1}{4\pi K|\mathbf{x}-\mathbf{x}'|}, \tag{7.9}$$

$$T(\mathbf{x},\mathbf{x}')=-KG_{,i}(\mathbf{x},\mathbf{x}')n_i, \tag{7.10}$$

where $|\mathbf{x}-\mathbf{x}'|=\sqrt{(x_1-x_1')^2+(x_2-x_2')^2+(x_3-x_3')^2}$, and n_i is the normal vector pointing outward from the surface. Similarly to the elastic problem, define the harmonic potential Φ and its derivatives over the domain Ω as follows:

$$\Phi=\int_\Omega \frac{1}{|\mathbf{x}-\mathbf{x}'|}d\mathbf{x}', \tag{7.11}$$

$$\Phi_{pq\cdots,ijk\cdots} = \frac{\partial}{\partial x_i \partial x_j \partial x_k \cdots} \int_\Omega \frac{x'_p x'_q \cdots}{|\mathbf{x} - \mathbf{x}'|} d\mathbf{x}'. \tag{7.12}$$

The explicit expressions for Φ and its derivatives can be found in Chapter 3 for different shapes of inclusions. Then the tensor D in Eq. (7.2) can be expressed by Φ and its derivatives as follows:

$$D_i^0 = K \int_\Omega G_{,i}(\mathbf{x}, \mathbf{x}')d\mathbf{x}' = -\frac{1}{4\pi}\Phi_{,i}, \tag{7.13}$$

$$D_{ip}^1 = K \int_\Omega G_{,i}(\mathbf{x}, \mathbf{x}')x'_p d\mathbf{x}' = -\frac{1}{4\pi}\Phi_{p,i}, \tag{7.14}$$

and the total temperature and the gradient of temperature, which is related to the flux by the Fourier law, can be written as follows based on Eq. (7.1):

$$\begin{aligned} u(\mathbf{x}') = &\int_{\partial D} G(\mathbf{x}, \mathbf{x}')q(\mathbf{x})\mathrm{d}\mathbf{x} - \int_{\partial D} T(\mathbf{x}, \mathbf{x}')u(\mathbf{x})\mathrm{d}\mathbf{x} \\ &+ \sum_{I=1}^{n} D_i^{I0}(\mathbf{x}')\varepsilon_i^{I0} + \sum_{I=1}^{n} D_{ip}^{I1}(\mathbf{x}')\varepsilon_{ip}^{I1}, \end{aligned} \tag{7.15}$$

where

$$\begin{aligned} \varepsilon_i(\mathbf{x}') = u_{,i} = &-\int_{\partial D} G_{,i}(\mathbf{x}, \mathbf{x}')q(\mathbf{x})\mathrm{d}\mathbf{x} + \int_{\partial D} T_{,i}(\mathbf{x}, \mathbf{x}')u(\mathbf{x})\mathrm{d}\mathbf{x} \\ &+ \sum_{I=1}^{n} D_{ij}^{I0}(\mathbf{x}')\varepsilon_j^{N0} + \sum_{I=1}^{n} D_{ijp}^{I1}(\mathbf{x}')\varepsilon_{jp}^{I1}, \end{aligned} \tag{7.16}$$

where the identities of $G_{,i} = -G_{,i'}$ and $T_{,i} = -T_{,i'}$ are applied, and $q = q_i n_i$ is the heat flux cross the interface. The detailed expressions of tensor D can be found in Chapter 3 for different shapes of inclusions. Similarly to elastic problems, Eshelby's equivalent inclusion method (EIM) can be used to solve for heat conduction in inhomogeneity problems for a continuous matrix containing inhomogeneities with different heat conductivities. By introducing a polynomial eigentemperature gradient ε^* the equivalence conditions to determine ε^* are explicitly written as

$$K^0\left(\varepsilon_i(\mathbf{0}) - \varepsilon_i^{I0}\right) = K^I \varepsilon_i(\mathbf{0}) \quad \text{for } I = 1, 2, \ldots, n, \tag{7.17}$$

$$K^0\left(\varepsilon_{i,r}(\mathbf{0}) - \varepsilon_{ir}^{I1}\right) = K^I \varepsilon_{i,r}(\mathbf{0}) \quad \text{for } I = 1, 2, \ldots, n, \tag{7.18}$$

where ε_i^{I0} and ε_{ir}^{I1} are the constant and linear terms of the eigentemperature gradients, and high-order polynomial eigentemperature gradient terms can be added for higher accuracy.

The above formulation can be applicable to different shapes of inhomogeneities, including ellipsoidal, elliptical, polyhedral, and polygonal particles. The domain discretization can be similarly used for an arbitrary shape of particles as well. The implementation of iBEM programming is straightforward following the procedure of the elastic analysis.

7.3 Boundary effect on the heat flow

One advantage of our method is that it is able to capture the boundary effect to the field around particles [110]. To show the boundary effect, here we investigate two cases. One is that a prolate ellipsoidal particle with thermal conductivity $K_1 = 100$ is sitting near the boundary of a cubic domain. The center of the ellipsoidal particle is set as $(x_1, x_2, x_3) = (0, 0, 0.2)$, the two short axes $a_1 = a_3 = 0.1$, and the long axis $a_2 = 0.15$. The other one is that the particle is located at the center that is far from the boundary surfaces. In this case the center of the ellipsoidal particle is set as $(x_1, x_2, x_3) = (0, 0, 1)$. The radius of the cylinder is set as 0.5, and the length is set as 2. We assign the same property $K_0 = 10$ to the matrix in both cases. There is a constant heat flux $t = 1$ flowing out of the cubic at the top. The rest of the surfaces are isolated; in other words, $t = 0$. The bottom temperature T is fixed to be zero.

The iBEM results agree with FEM results very well. By comparing the temperature distributions along the z-direction in these two cases as in Fig. 7.2 it shows that the effect from boundary can be captured by our method. The stress pattern is significantly affected by the existence of the particle. The boundary effect makes the temperature distribution in Fig. 7.2(a) asymmetric, whereas the stress distribution in Fig. 7.2(b) is not significantly affected by the boundary. The distribution from Fig. 7.2(b) is similar to the case where one particle is embedded in the infinite domain.

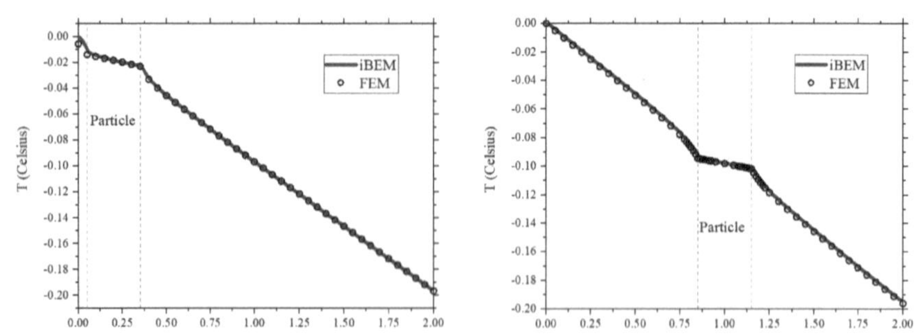

(a) Particle located near the top surface **(b)** Particle located at the center

FIGURE 7.2 Comparison of iBEM and FEM along the z-direction. (a) Particle located near the top surface; (b) particle located at the center.

7.4 Particle interactions in steady-state heat conduction

The disturbance due to one single particle in a composite is easy to solve. Only constant terms in ε^* are enough to represent the effects. However, when two or more particles exist and are close, the disturbed field becomes complex. Asymptotic analysis has shown that linear even quadratic terms are indeed necessary to obtain results of acceptable accuracy [10]. In the following numerical examples, all the quantities follow the SI units. A cube of length 1 is built. Two particles are located at the center of the cubic. The thermal conductivity constants of matrix and particle are $K^0 = 2$ and $K^1 = 20$, respectively. The top surface flux is continuously flowing out at 1, and we fix the temperature at the bottom to be 0. The configuration is shown in Fig. 7.3(a). The temperature along the central axis and around the bottom particle with $h = 0.12$ is recorded. Good agreement on the temperature distribution between iBEM and FEM is obtained in Fig. 7.4. The gradient of the temperature in the particle domain is significantly different from that in the matrix.

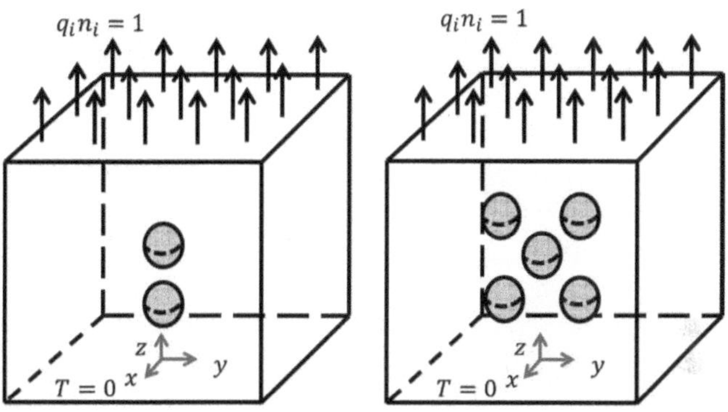

FIGURE 7.3 (a) Configuration about two particles at the center. (b) Configuration about five particles at the center.

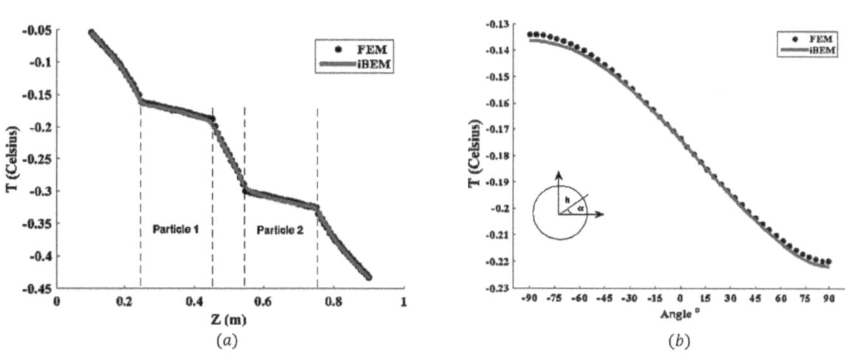

FIGURE 7.4 (a) Comparison between iBEM and FEM along the z direction. (b) Comparison between iBEM and FEM around the bottom particle.

For the next case, five particles with $K^1 = 0.2$ are placed at the center of the matrix with $K^0 = 2$. Here the shape of particles can be a sphere with radius $a = 0.1$ m or prolate with long axis $a_1 = 0.12$ and two short axes $a_2 = a_3 = 0.06$ in Fig. 7.3(b). From Fig. 7.5, the resulting temperature distribution is very complicated near the particles. Due to the symmetry of our problem, the temperature distribution is symmetric to the z-axis. The temperature continuity along the interfaces of particles is indeed preserved in the iBEM results. By arranging the positions and orientations of particles differently we observed distinct patterns in the temperature distributions.

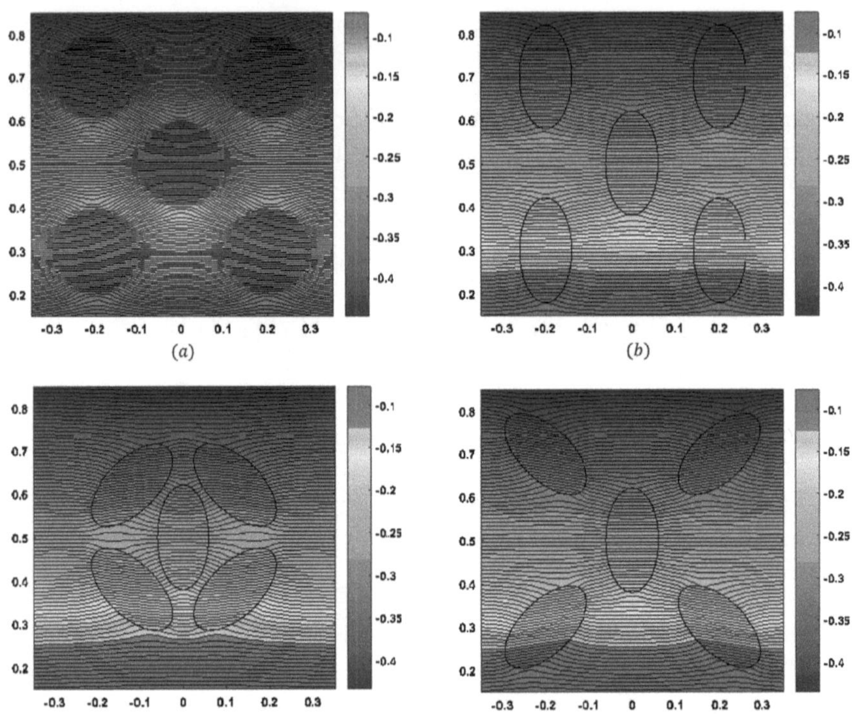

FIGURE 7.5 (a) Temperature distribution around five spherical particles. (b) Temperature distribution around five prolate particles putting inward. (c) Temperature distribution around five prolate particles putting outward. (d) Temperature distribution around five prolate particles putting erect.

*7.5 Homogenization of particle reinforcement composites by iBEM toward virtual experiments

A virtual experiment is conducted to predict the thermal conductivity of particle reinforced composites. In the following derivation, we can show that the homogenized thermal conductivity is only related to the surface integral, which can be directly obtained from our iBEM results. Therefore it is very convenient to calculate the homogenized material property from our iBEM algorithm, which only involves a simple algebraic average in the postprocess procedure. The detailed configuration is shown as follows. A unit cubic RVE containing thousands of elliptical particles is constructed. The bottom temperature T is fixed to be zero. There is a constant heat flux $t = 1$ flowing out of the cubic at the top. The remaining surfaces are isolated; in other words, $t = 0$.

The effective thermal conductivity of the composites is defined by correlating the average gradient of temperature $\langle \nabla u \rangle$ and flux $\langle \mathbf{q} \rangle$ over the volume of RVE by the following equation:

$$\langle \mathbf{q} \rangle \equiv -\langle K \rangle \langle \nabla u \rangle \tag{7.19}$$

where the effective gradient of temperature can be derived by the boundary integral of a continuous temperature field as follows:

$$\langle \nabla u \rangle = \frac{1}{V} \int_D \nabla u d\mathbf{x} = \frac{1}{V} \int_D u_{,i} d\mathbf{x} = \frac{1}{2V} \int_{\partial D} n_i u d\mathbf{x} \tag{7.20}$$

For the thermal loading condition in this specific virtual experiment, the effective gradient of temperature in the loading direction (z-direction) can be written as

$$\langle \nabla u \rangle_3 = \frac{1}{V} \int_{\partial D} n_3 u d\mathbf{x} \tag{7.21}$$

where only the top and bottom surfaces contribute to the integral. By approximation we will average the temperature on the top and bottom surfaces to get $\langle \nabla u \rangle_3$ as

$$\langle \nabla u \rangle_3 = \frac{1}{h} (\overline{u}^{\text{top}} - \overline{u}^{\text{bot}}) \tag{7.22}$$

in which, based on our boundary condition, $\overline{u}^{\text{bot}} = 0$ and $h = 1$. Then the effective gradient of temperature can be calculated straightforwardly.

The average flux can also be written in terms of the boundary response:

$$\begin{aligned} \langle \mathbf{q} \rangle &= \frac{1}{V} \int_D q_i d\mathbf{x} \\ &= \frac{1}{V} \int_D \left[(q_k x_i)_{,k} - q_{k,k} x_i \right] d\mathbf{x} \\ &= \frac{1}{V} \int_{\partial D} (q_k n_k x_i) d\mathbf{x} \\ &= \frac{1}{V} \int_{\partial D} (q x_i) d\mathbf{x} \end{aligned} \tag{7.23}$$

Similarly, only the top surface contributes to the above integral. Therefore the average stress $\langle q \rangle_3$ can be written as the surface integral of a heat flux boundary condition at the top:

$$\langle q \rangle_3 = q^{\text{top}} = 1 \tag{7.24}$$

Finally, the effective thermal conductivity can be expressed as

$$\langle K \rangle = -\langle q \rangle_3 / \langle \nabla u \rangle_3 \tag{7.25}$$

In the first numerical example, we fix the volume fraction of the fibers $\phi = 0.1$. The conductivity constants of matrix and fibers are $K_0 = 2$ and $K_1 = 200$, respectively. Two short axes of the prolate particles are $a_1 = a_2 = a = 0.025$, but the number of particles N varies. So the long axis of the elliptical particles can be determined by $a_3 = 3\phi/(4Na^2)$. The local coordinate of an elliptical particle coincident with the global coordinate, which implies that the long axis is along with the z direction as in Fig. 7.6. In this case the long axis a_3 of the prolate particles shrinks gradually at the increase of the particle number. This process aims to simulate the transformation from the long fiber reinforced composite to the spherical particle reinforced composite. All the particles are generated randomly in all simulations.

In Fig. 7.7, we provide the relation between the effective conductivity and the log scale of the number of particles. The results demonstrate that the fiber composite is more effective than the particulate composite at enhancing the material property in the z direction. A linear relation is observed between the homogenized material property and the log of the number of particles. Because of the number of particles $N = 3\phi(a/a_3)/(4\pi a^3) \propto (a/a_3)$ for a unit volume, we can see that there is an exponential relation between the homogenized material property and the aspect ratio of fibers. In other words, increasing this ratio, which means shrinking the long axis, will decrease the effective property in that direction exponentially under the assumption that the overall volume fraction or total quantity of particle reinforcements in that material is constant.

In the second numerical example, the material properties are the same as the previous ones, and the number and shape of the particles are fixed as $n = 1000$, $a_1 = a_2 = a = 0.02$, and $a_3 = 0.048$. The orientation of particle long axes can vary from along the z direction to perpendicular to the z direction as shown in Fig. 7.8. So the effect of the orientation of particles can be investigated. The microstructure of the composite is a simple lattice. Moreover, a case of randomized orientations of particles is also constructed for comparison.

In Fig. 7.9, we can observe that the orientation of elliptical particles has a great effect on the effective conductivity. The maximum effective conductivity is achieved when the long axis of a particle is along with the enhanced direction. It is also interestingly discovered that the randomized orientation case produces the same results as that around $\theta = 45°$.

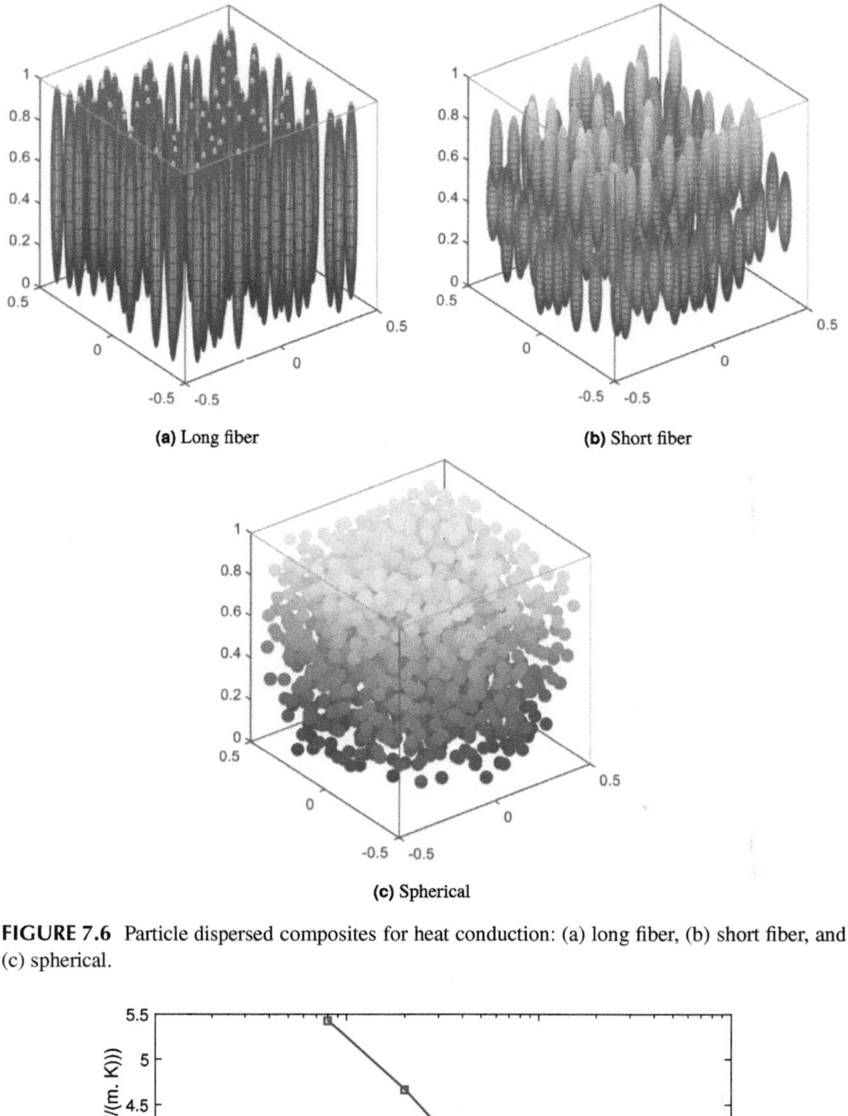

FIGURE 7.6 Particle dispersed composites for heat conduction: (a) long fiber, (b) short fiber, and (c) spherical.

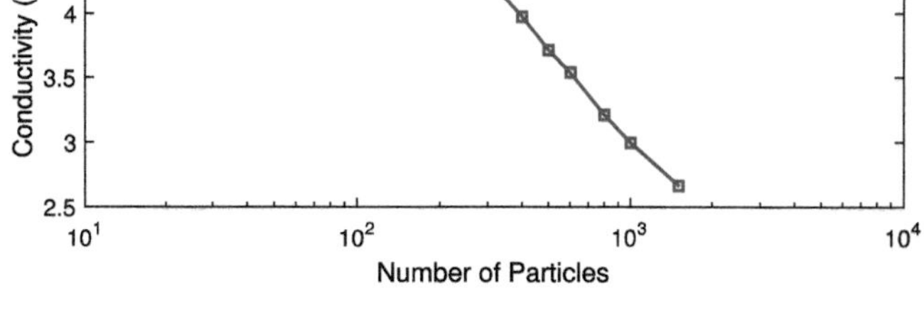

FIGURE 7.7 Variation of the effective heat conductivity with respect to particle numbers.

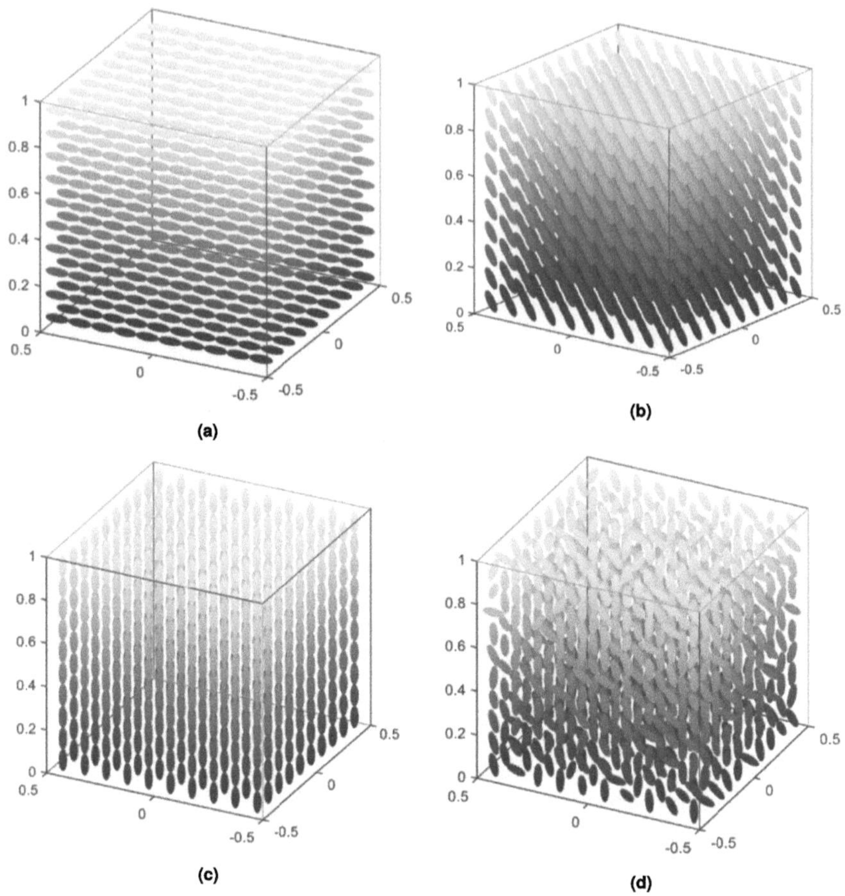

FIGURE 7.8 (a) Short fibers perpendicular to the z direction ($\theta = 0$), (b) $\theta = 60$, (c) short fibers parallel to the z direction ($\theta = 90$), (d) randomized orientation.

*7.6 Heat flow of an infinite bimaterial domain containing inhomogeneities

Similarly to elastic analysis, the heat conduction analysis can be extended to a bimaterial domain. Fig. 7.10 gives an infinite matrix domain D with subdomain Ω centered at $\mathbf{x}^c$ with interface $S(x_3 = 0)$ of zero thermal resistance to separate the bimaterial matrix. The particle thermal conductivity is K^{Ω}. According to Mura [1], an inhomogeneity is a subdomain Ω in a homogeneous solid with different material properties from the rest $(D - \Omega)$, which can be simulated by an eigenstrain. The heat conduction problem can be treated similarly with an eigentemperature-gradient (ETG) on the particle, which is typically continuous over Ω and can be written in terms of the Taylor expansion of $\mathbf{x} - \mathbf{x}^c$

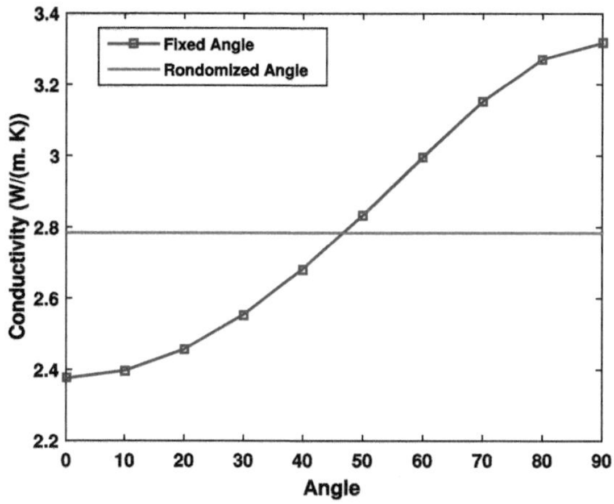

FIGURE 7.9 Variation of the effective heat conductivity with respect to different angles.

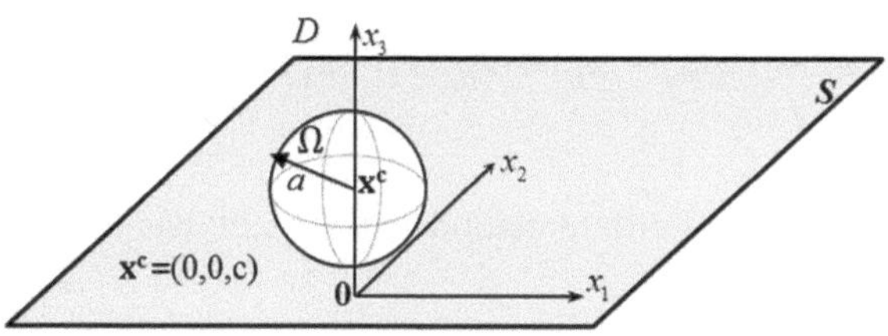

FIGURE 7.10 One particle embedded in the bi-material matrix.

with the reference point at the particle center:

$$T_i^*(\mathbf{x}) = T_i^0 + T_{ki}^1(x_k - x_k^c) + T_{kli}^2(x_k - x_k^c)(x_l - x_l^c) + \cdots, \quad \mathbf{x} \in \Omega \tag{7.26}$$

The constitutive law over the upper semi-infinite matrix with thermal conductivity K' is rewritten as

$$q_i = -K'\delta_{ij}(u_{,j}, -T_j^*) \tag{7.27}$$

where $T_i^* = 0$ for $\mathbf{x} \in (D - \Omega)$. By applying Green's function satisfying the boundary conditions the temperature can be written in terms of the integral of the source of the eigentemperature-gradient as

$$u(\mathbf{x}) = \int_D G(\mathbf{x}, \mathbf{x}')(-K')\delta_{mi}\frac{\partial T_i^*}{\partial x_m'}d\mathbf{x}' \tag{7.28}$$

Applying the Gauss theorem and considering zero eigentemperature-gradient at the boundary of the semi-infinite domain, we can rewrite the above equation as

$$u(\mathbf{x}) = \int_\Omega \frac{\partial G(\mathbf{x}, \mathbf{x}')}{\partial x_i'} K' T_i^* d\mathbf{x}' = \int_\Omega \Gamma_i K' T_i^* d\mathbf{x}' \tag{7.29}$$

where the modified Green's function

$$\Gamma_i = -\frac{1}{4\pi K'} \left((\frac{1}{r})_{,i} + \frac{K' - K''}{K' + K''} M_{ij} (\frac{1}{s})_{,j} \right) \tag{7.30}$$

with $M = \begin{bmatrix} 1 & 0 & 0 \\ 0 & 1 & 0 \\ 0 & 0 & -1 \end{bmatrix}$.

Consider the ETG as a continuous tensor function over the inhomogeneity and write it in a polynomial form in Eq. (7.26). For simplicity, we take into account the constant, linear, and quadratic terms only as

$$T_{,i}^*(\mathbf{x}') = \begin{cases} T_i^0 + T_{ik}^1 (x_k' - x_k^c) + T_{ikl}^2 (x_k' - x_k^c)(x_l' - x_l^c), & \mathbf{x}' \in \Omega \\ 0, & \mathbf{x}' \in (D - \Omega) \end{cases} \tag{7.31}$$

Thus by taking partial differentiation of Eq. (7.29) and using Eq. (7.31), we have

$$\begin{aligned} u_{,i}(\mathbf{x}) &= \int_\Omega \Gamma_{j,i} K' [T_j^0 + T_{jk}^1 (x_k' - x_k^c) + T_{jkl}^2 (x_k' - x_k^c)(x_l' - x_l^c)] d\mathbf{x}' \\ &= -\frac{1}{4\pi} (D_{ij} T_j^0 + D_{ijk} T_{jk}^1 + D_{ijkl} T_{jkl}^2) \end{aligned} \tag{7.32}$$

where D_{ij}, D_{ijk}, D_{ijkl} are derived following the formulation in Yin's book [58]. Take D_{ij} as an example to show the analytical explicit of this formulation:

$$D_{ij} = \int_\Omega [(\frac{1}{r})_{,ij} + \frac{K' - K''}{K' + K''} M_{JJ} (\frac{1}{s})_{,ij}] d\mathbf{x}' = \Phi_{,ij} + \frac{K' - K''}{K' + K''} M_{JJ} \overline{\Phi}_{,ij} \tag{7.33}$$

with $\Phi_{,ij}$ derived as

$$\Phi_{,ij} = \begin{cases} \frac{4\pi a^3}{3r^3} (\delta_{ij} - 3n_i n_j), & x \in (D - \Omega) \\ -\frac{4\pi}{3} \delta_{ij}, & x \in \Omega \end{cases} \tag{7.34}$$

where n_i is defined as $n_i = (x_i - x_i^c)/r$. Again define $\overline{n}_i = (x_i - M_{II} x_i^c)/s$. Similarly, for $\overline{\Phi}$, changing the distance term r to s and n_i to $\overline{n}_i$, we write

$$\overline{\Phi}_{,ij} = \frac{4\pi a^3}{3s^3} (\delta_{ij} - 3\overline{n}_i \overline{n}_j) \tag{7.35}$$

Notice that here $\overline{\Phi}_{ij}$ represents the components of the image part, and thus the integral is always outside the particle. Following a straightforward but lengthy procedure, D_{ijk}, D_{ijkl} and $\Phi_{k,ij}$, $\Phi_{kl,ij}$ and $\overline{\Phi}_{k,ij}$, $\overline{\Phi}_{kl,ij}$ can now be explicitly solved and are listed in the box below. The temperature field is solved in Eq. (7.32) for the case of semi-infinite domain D with thermal conductivity K' containing an inclusion Ω with ETG $T_i^*(\mathbf{x})$.

$$
\begin{aligned}
D_{ijk} &= \int_\Omega [(\frac{1}{r})_{,ij} + \frac{K'-K''}{K'+K''} M_{JJ}(\frac{1}{s})_{,ij}](x'_k - x^c_k)d\mathbf{x}' \\
&= \Phi_{k,ij} + \frac{K'-K''}{K'+K''} M_{JJ} M_{KK} \overline{\Phi}_{k,ij} \\
D_{ijkl} &= \int_\Omega [(\frac{1}{r})_{,ij} + \frac{K'-K''}{K'+K''} M_{JJ}(\frac{1}{s})_{,ij}](x'_k - x^c_k)(x'_l - x^c_l)d\mathbf{x}' \\
&= \Phi_{kl,ij} + \frac{K'-K''}{K'+K''} M_{JJ} M_{KK} M_{LL} \overline{\Phi}_{kl,ij}
\end{aligned}
$$

$$
\Phi_{k,ij} = \begin{cases} \frac{4\pi a^5}{5r^4}(-\delta_{ij}n_k - \delta_{ik}n_j - \delta_{jk}n_i + 5n_i n_j n_k), & x \in (D-\Omega) \\ \frac{4\pi}{5}[-\delta_{ij}(x_k - x^C_k) - \delta_{ik}(x_j - x^C_j) - \delta_{jk}(x_i - x^C_i)], & x \in \Omega \end{cases}
$$

$$
\Phi_{kl,ij} = \begin{cases} \frac{16\pi a^7}{5r^5} n_i n_j n_k n_l - \frac{16\pi a^7}{35r^5}(\delta_{ij}n_k n_l + \delta_{kj}n_i n_l + \delta_{lj}n_i n_k) \\ -\frac{20\pi a^7}{35r^5}(\delta_{ki}n_j n_l + \delta_{li}n_j n_k) + \frac{4\pi a^7}{35r^5}(\delta_{lj}\delta_{ki} + \delta_{li}\delta_{kj}) \\ +\frac{\pi a^5}{105r^5}[(12a^2 - 28r^2)\delta_{ij}\delta_{kl} + (84r^2 - 60a^2)\delta_{kl}n_i n_j], \\ \text{for } x \in (D-\Omega) \\ -\frac{4\pi r^2}{7}(\delta_{ik}n_j n_l + \delta_{li}n_j n_k + \delta_{ij}n_l n_k + \delta_{kj}n_i n_l + \delta_{lj}n_l n_k) \\ +(\frac{2\pi a^2}{5} - \frac{2\pi r^2}{7})(\delta_{ik}\delta_{jl} + \delta_{il}\delta_{jk}) \\ +(\frac{4\pi r^2}{35} - \frac{4\pi a^2}{15})\delta_{ij}\delta_{kl} + \frac{8\pi r^2}{35} n_i n_j \delta_{kl}, \quad \text{for } x \in \Omega \end{cases}
\tag{7.36}
$$

All $\overline{\Phi}$ terms, as mentioned before, can be acquired by substituting r by s, n by $\overline{n}$ in Φ terms for the condition of $x \in (D-\Omega)$. Thus $\overline{\Phi}_{k,ij}$, $\overline{\Phi}_{kl,ij}$ can be derived from $\Phi_{k,ij}$, $\Phi_{kl,ij}$ here.

Extending the inclusion problem to the inhomogeneity one, we replace the inclusion Ω with an inhomogeneity of different thermal conductivity K^Ω. Meanwhile, we apply a uniform heat flux density q_3^0 in the far field. Due to material property mismatch, there will be a disturbance in the near field of inhomogeneity Ω. The concept of Eshelby's EIM [2,3] is applied to solve the problem. The inhomogeneity Ω is filled with the same material of K' and applied eigentemperature-gradient $T^*_{,i}(\mathbf{x})$, which will be determined later, to ensure equality of the heat flux inside Ω for two cases.

For one inhomogeneity in an infinite domain with uniform far-field heat flux density, the eigentemperature-gradient is uniform by analogy to the case in the elastic field [2,3]. Back to the case of uniform temperature distribution at the boundary of the semi-infinite domain, the ETG will not be uniform anymore. Instead, we can write it in terms of a polynomial form of $\mathbf{x}-\mathbf{x}^c$ with the origin at the center of the inhomogeneity as given in Eq. (7.26).

We may write the disturbed temperature gradient in Taylor's expansion with constant, linear, and quadratic terms as

$$\begin{aligned}-u'_{,i} &= \frac{1}{4\pi}(D_{ij}(\mathbf{x}^c)T_j^0 + D_{ijk}(\mathbf{x}^c)T_{jk}^1 + D_{ijkl}(\mathbf{x}^c)T_{jkl}^2) \\ &+ \frac{1}{4\pi}(D_{ij,r}(\mathbf{x}^c)T_j^0 + D_{ijk,r}(\mathbf{x}^c)T_{jk}^1 + D_{ijkl,r}(\mathbf{x}^c)T_{jkl}^2)(x_r - x_r^c) \\ &+ \frac{1}{8\pi}(D_{ij,pq}(\mathbf{x}^c)T_j^0 \\ &+ D_{ijk,pq}(\mathbf{x}^c)T_{jk}^1 + D_{ijkl,pq}(\mathbf{x}^c)T_{jkl}^2) * (x_p - x_p^c)(x_q - x_q^c)\end{aligned} \tag{7.37}$$

where $D_{ij,r}$, $D_{ijk,r}$, $D_{ijkl,r}$, $D_{ij,pq}$, $D_{ijk,pq}$, $D_{ijkl,pq}$ are also explicitly shown in the box below.

The remaining partial derivatives of D in relation with Φ for the analytical solutions are given below.

$$\begin{aligned} D_{ij,r} &= \Phi_{,ijr} + \frac{K'-K''}{K'+K''}M_{JJ}\overline{\Phi}_{,ijr} \\ D_{ijk,r} &= \Phi_{k,ijr} + \frac{K'-K''}{K'+K''}M_{JJ}M_{KK}\overline{\Phi}_{k,ijr} \\ D_{ijkl,r} &= \Phi_{kl,ijr} + \frac{K'-K''}{K'+K''}M_{JJ}M_{KK}M_{LL}\overline{\Phi}_{kl,ijr} \end{aligned} \tag{7.38}$$

$$\begin{aligned} D_{ij,pq} &= \Phi_{,ijpq} + \frac{K'-K''}{K'+K''}M_{JJ}\overline{\Phi}_{,ijpq} \\ D_{ijk,pq} &= \Phi_{k,ijpq} + \frac{K'-K''}{K'+K''}M_{JJ}M_{KK}\overline{\Phi}_{k,ijpq} \\ D_{ijkl,pq} &= \Phi_{kl,ijpq} + \frac{K'-K''}{K'+K''}M_{JJ}M_{KK}M_{LL}\overline{\Phi}_{kl,ijpq} \end{aligned} \tag{7.39}$$

When a uniform heat flux $q_i^0 = k'u_{,i}^0$ is applied on the boundary of the semi-infinite domain, the inhomogeneity is simulated with an inclusion. The heat flux on the inhomogeneity should be equivalent to that on the inclusion as

$$K^{\Omega}(u_{,i}^0 + u'_{,i}) = K'(u_{,i}^0 + u'_{,i} - T_i^*) \tag{7.40}$$

Substituting Eq. (7.37) into Eq. (7.40) and comparing the coefficients at each order of **x**, we can write

$$\begin{cases} (K^{\Omega} - K')\left(-4\pi u^0_{,i} + [D_{ij}(\mathbf{x}^c)T^0_j + D_{ijk}(\mathbf{x}^c)T^1_{jk} + D_{ijkl}(\mathbf{x}^c)T^2_{jkl}]\right) \\ \quad = 4\pi K' T^0_i \\ (K^{\Omega} - K')[D_{ij,r}(\mathbf{x}^c)T^0_j + D_{ijk,r}(\mathbf{x}^c)T^1_{jk} + D_{ijkl,r}(\mathbf{x}^c)T^2_{jkl}] = 4\pi K' T^1_{ir} \\ (K^{\Omega} - K')[D_{ij,pq}(\mathbf{x}^c)T^0_j + D_{ijk,pq}(\mathbf{x}^c)T^1_{jk} + D_{ijkl,pq}(\mathbf{x}^c)T^2_{jkl}] = 8\pi K' T^2_{ipq} \end{cases} \tag{7.41}$$

From the above, for the first set of equations, because the free index ranges $i = 1, 2, 3$, there are three equations. For the second set of equations, the free indices i and r form $3 \times 3 = 9$ equations. For the third set of equations with free indices i, p, and q, considering the symmetry of p and q, we have $3 \times 6 = 18$ equations. In total, we form 30 linear equations with 30 unknowns $(T^0_j, T^1_{jk}, T^2_{jkl})$. By solving the linear equations in Eq. (7.41) we can explicitly write the disturbed temperature gradient and thus the temperature gradient field.

Depending on the accuracy requirement, more or less order of the Taylor expansion of the ETG can be used. In this study case, most terms of $D(\mathbf{x}^c)$ in Φ part will be zero but nonzero in $\overline{\Phi}$ part. Although the present method does not provide an exact solution unless the variation of the temperature distribution on the particle can be described by a quadratic function, the previous work [10,36] shows that a linear function can provide high-fidelity results.

In addition, the elastic analysis for the arbitrary shape of inhomogeneity problems can be extended to the thermal analysis in a parallel way by using Eshelby's tensor and domain discretization.

Chapter 8

The iBEM for the Stokes flows

Incompressible vector potential

In this chapter, we introduce the iBEM application and implementation for the Stokes flows, which share a similar partial differential equation as the elastic problem but uses the velocity to replace displacement in the governing equation. A similar development process for elastic problems with displacement potentials is extended to the Stokes flows with incompressible vector potential of the velocity field. The shape effect on the particle motion is illustrated in a Stokes flow. The formulation is applicable to the quasistatic flow dynamics of the particle settlement and can be extended to multiphase flow with different liquid phases.

8.1 Equivalent inclusion method for the Stokes flows

Consider liquid subdomains Ω_i with viscosity μ_i $(i = 1, 2, \ldots, n)$ moving in a homogeneous fluid domain D with viscosity μ_0, as shown in Fig. 8.1. When we consider a highly viscous fluid flow with a small Reynolds number, the inertia effect of the fluid is negligible in the Navier–Stokes equation, and then it reduces to the Stokes equation. As discussed in Chapter 2, the mathematical partial differential equation of the Stokes flow exhibits the same form as the Navier equation for an elastic problem if we take Poisson's ratio 0.5 and replace the displacement field by the velocity field. When inhomogeneous drops or fluid particles move in a Stokes flow, they can induce disturbed velocity or stress field, and the EIM and iBEM can be extended to the multiphase Stokes flow.

When the radius of the fluid particles is small in our problem, the shape of the drops can be assumed to be ellipsoidal under the gravity, surface tension, and hydrodynamic forces. The body force induced by gravity is uniformly distributed on each particle. The body force in this system can be simply described as

$$f_i(x) = \begin{cases} f_i^0 & x \in \Omega_i \\ 0 & x \in D - \{\Omega_1, \Omega_2, \ldots, \Omega_i\} \end{cases} \tag{8.1}$$

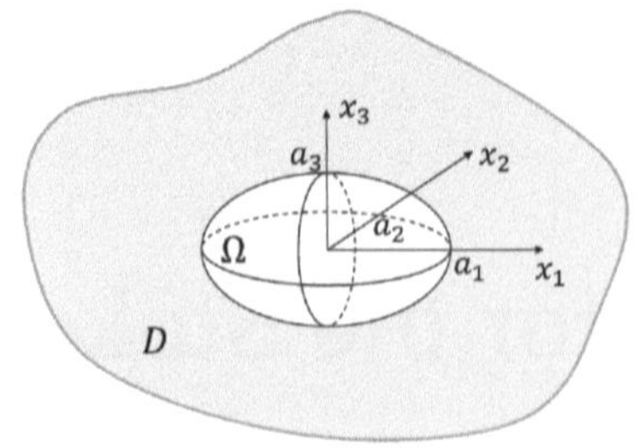

FIGURE 8.1 One ellipsoidal subdomain Ω embedded in an infinite domain D.

When the Reynolds number of flow behavior is not high, the equilibrium and motion of the fluid are described by the Stokes equation. Using the equivalent inclusion method, the material mismatch of different viscosities in fluid and particles can be represented by eigenstrain rate $e^*_{ij}(x)$. Then material properties of the whole domain D will keep as a homogeneous one with μ_0. The Stokes equation reads

$$2\mu^0 e_{ij,i} - p_{,j} = -f_i \quad v_{i,i} = 0, \quad x \in D - \Omega \tag{8.2}$$

$$2\mu^0 \left(e_{ij,i} - e^*_{ij,i}\right) - p_{,j} = -f_i \quad v_{i,i} = 0, \quad x \in \Omega \tag{8.3}$$

Compared with elastic analysis, the Stokes equation uses velocity as the basic unknown instead of displacement; the number of material constants reduces to one parameter of viscosity; because the liquid is incompressible, a pressure field is introduced into the equilibrium. Otherwise, the partial differential equation shares the same mathematical form. Based on the general EIM we have discussed before in the absence of far field strain rate [10], e^*_{ij} satisfies the stress equivalent conditions

$$\mu^0(e_{ij} - e^*_{ij}) = \mu^1 e_{ij}, \quad x \in \Omega \tag{8.4}$$

where the eigenstrain rate e^*_{ij} is a fictitious nonmechanical field, which makes the stress field in two problems equivalent, as shown in Fig. 8.2. Then the original problem with inhomogeneity can be treated as an inclusion problem with the eigenstrain rate satisfying Eq. (8.4). We assume that the eigenstrain rate follows the polynomial form up to the quadratic term:

$$e^*_{ij}(x) = E^0_{ij} + E^1_{ijk} x_k + E^2_{ijkl} x_k x_l \tag{8.5}$$

Because the eigenstrain rate is a continuous field over the particle, the above polynomial form function generally approaches the exact solution in an asymptotic manner. The more the terms are used, the higher the accuracy can be

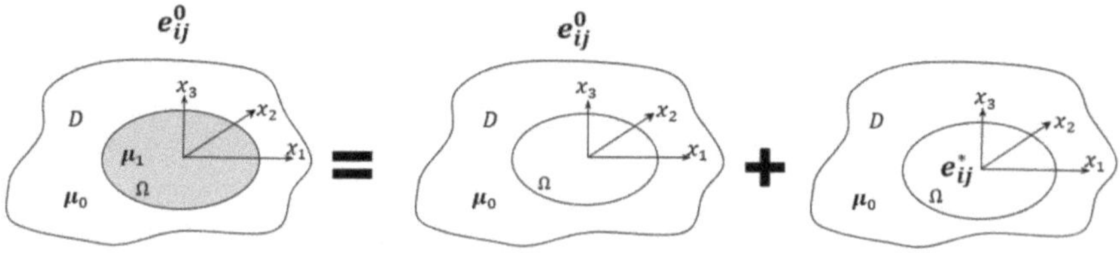

FIGURE 8.2 Equivalent condition for Stokes flow.

reached. However, for one ellipsoidal inclusion moving in a fluid driven by a constant body force, the strain rate over the inclusion is linearly distributed [95,105]. As shown in the next section, only one term $e_{ijk}^1 x_k$ may make the exact solution for one drop problem. For simplicity, we directly write

$$e_{ij}^*(x) = E_{ijk}^1 x_k, \quad x \in \Omega \tag{8.6}$$

where from the definition of e_{ij}^* we can write

$$E_{ijk}^1 = E_{jik}^1 \text{ and } E_{iik}^1 = 0 \tag{8.7}$$

Notice that for more than one drop, the eigenstrain rate should be more complex, and other terms in Eq. (8.5) will be nonzero. However, using constant and linear terms has produced very accurate results for the case of a pair of drops moving in a Stokes flow.

By the aid of the Green's function techniques the velocity field in the fluid can be expressed by the integral form of body force f and eigenstrain rate e^* in all particles as follows:

$$v_i = \int_\Omega (G_{ik} f_k - 2\mu^0 G_{im,n} e_{mn}^*) d\mathbf{x}' \tag{8.8}$$

where the Green's function is the following:

$$G_{ij} = \frac{1}{4\pi\mu^0}(\delta_{ij}\phi - \frac{1}{2}\frac{\partial^2\psi}{\partial x_i \partial x_j}), \quad \psi = |\mathbf{x} - \mathbf{x}'| \tag{8.9}$$

The function $G_{ij}(\mathbf{x}, \mathbf{x}')$ is the fundamental solution for the velocity response v_i at point $\mathbf{x}$ due to a unit point load δ_j acting at $\mathbf{x}'$ in the infinite domain in a Stokes flow. The integral in Eq. (8.8) can be analytically carried out in a spherical domain. Finally, the velocity can be obtained as follows (detailed expressions for Φ and Ψ can be found in Appendix 8.A):

$$v_i = \sum_{J=1}^{N} \left(L_{ik} f_k^J + M_{imn} E_{mn}^{J0} + N_{imnp} E_{mnp}^{J1} + Q_{imnpq} E_{mnpq}^{J2} \right) \tag{8.10}$$

where

$$L_{ik} = \frac{1}{4\pi\mu^0}\left[\delta_{ik}\Phi - \frac{1}{2}\Psi_{,ik}\right]$$
$$M_{imn} = -\frac{1}{4\pi}\left[\delta_{im}\Phi_{,n} + \delta_{in}\Phi_{,m} - \Psi_{,imn}\right]$$
$$N_{imnp} = -\frac{1}{4\pi}\left[\delta_{im}\Phi_{p,n} + \delta_{in}\Phi_{p,m} - \Psi_{p,imn}\right]$$
$$Q_{imnpq} = -\frac{1}{4\pi}\left[\delta_{im}\Phi_{pq,n} + \delta_{in}\Phi_{pq,m} - \Psi_{pq,imn}\right]$$

Based on the constitutive relation between strain rate and velocity field, the strain rate can also be straightforwardly derived in terms of the body force and eigenstrain rate:

$$\begin{aligned} e_{ij} &= \frac{1}{2}\left(v_{i,j} + v_{j,i}\right) \\ &= \frac{1}{8\pi\mu_0}\sum_{J=1}^{N}\left(F_{ijk}f_k^J + D^0_{ijmn}E^{J0}_{mn} + D^1_{ijmnp}E^{J1}_{mnp} + D^2_{ijmnpq}E^{J2}_{mnpq}\right) \end{aligned} \tag{8.11}$$

where

$$F_{ijk} = \delta_{ik}\Phi_{,j} + \delta_{jk}\Phi_{,i} - \Psi_{,ijk}$$

$$D^0_{ijmn} = -\mu_0\left[\delta_{im}\Phi_{,jn} + \delta_{in}\Phi_{,jm} + \delta_{jm}\Phi_{,in} + \delta_{jn}\Phi_{,im} - 2\Psi_{,ijmn}\right]$$

$$D^1_{ijmnp} = -\mu_0\left[\delta_{im}\Phi_{p,jn} + \delta_{in}\Phi_{p,jm} + \delta_{jm}\Phi_{p,in} + \delta_{jn}\Phi_{p,im} - 2\Psi_{p,ijmn}\right]$$

$$\begin{aligned} D^2_{ijmnpq} = &-\mu_0\left[\delta_{im}\Phi_{pq,jn} + \delta_{in}\Phi_{pq,jm} + \delta_{jm}\Phi_{pq,in} + \delta_{jn}\Phi_{pq,im}\right. \\ &\left. - 2\Psi_{pq,ijmn}\right] \end{aligned}$$

Substituting Eq. (8.11) into Eq. (8.4), we can get equivalent conditions for each particle:

$$\begin{aligned} E^{I0}_{ij} + E^{I1}_{ijr}x_r + E^{I2}_{ijrz}x_r x_z = \frac{\mu^0 - \mu^1}{\mu^0}\sum_{J=1}^{N}&(F_{ijk}f_k^J \\ + D^0_{ijmn}E^{J0}_{mn} + D^1_{ijmnp}E^{J1}_{mnp} + D^2_{ijmnpq}E^{J2}_{mnpq}) \quad &\text{for } I = 1, 2, \ldots, N \end{aligned} \tag{8.12}$$

Expanding the right side of Eq. (8.12) in a Taylor series of **x** at each center of the particle and comparing the coefficients of the terms **x** up to quadratic terms yields the following equations:

$$-\sum_{J=1}^{N} F_{ijk} f_k^J = -\frac{\mu^0}{\mu^0-\mu^1} E_{ij}^{I0} + \sum_{J=1}^{N} \left(D_{ijmn}^0 E_{mn}^{J0} + D_{ijmnp}^1 E_{mnp}^{J1} + D_{ijmnpq}^2 E_{mnpq}^{J2} \right) \quad \text{for } I = 1, 2, \ldots, N \tag{8.13}$$

$$-\sum_{J=1}^{N} F_{ijk,r} f_k^J = -\frac{\mu^0}{\mu^0-\mu^1} E_{ijr}^{I1} + \sum_{J=1}^{N} \left(D_{ijmn,r}^0 E_{mn}^{J0} + D_{ijmnp,r}^1 E_{mnp}^{J1} + D_{ijmnpq,r}^2 E_{mnpq}^{J2} \right) \quad \text{for } I = 1, 2, \ldots, N \tag{8.14}$$

$$-\sum_{J=1}^{N} F_{ijk,rz} f_k^J = -\frac{\mu^0}{\mu^0-\mu^1} E_{ijrz}^{I2} + \sum_{J=1}^{N} \left(D_{ijmn,rz}^0 E_{mn}^{J0} + D_{ijmnp,rz}^1 E_{mnp}^{J1} + D_{ijmnpq,rz}^2 E_{mnpq}^{J2} \right) \quad \text{for } I = 1, 2, \ldots, N \tag{8.15}$$

By solving the linear equation system (8.13)–(8.15) we can obtain all the unknown eigenstrain rate coefficients E^0, E^1, E^2 in all particles and evaluate the total velocity field and strain rate field by the body force and eigenstrain rate based on Eqs. (8.10) and (8.11), respectively.

8.2 Particle motion in a Stokes flow

In general, Eshelby's equivalent inclusion method (EIM) addresses the material inhomogeneity problem by replacing the inhomogeneity with the matrix but introducing a nonmechanical strain rate (eigenstrain rate) to make the constitutive law (8.6) equivalently satisfied. Therefore the boundary value problem is transferred to a homogeneous fluid with eigenstrain rate on the particle domains, which can be analytically solved with elliptic integrals for ellipsoidal particles. Ellipsoid is a versatile shape for engineering applications such that spherical voids, spheroidal drops, circular fibers, and penny shape disks can be particular cases of this general problem.

When a single drop moves in a fluid with $a_1 = a_2 = a_3 = a$, the ellipsoidal drop in Fig. 8.1 becomes a spherical drop. To simplify the calculation, only the term $E^1_{ijk} x_k$ is employed. Then, Eq. (8.11) can be simplified as

$$e_{ij} = \frac{1}{2}\left(v_{i,j} + v_{j,i}\right) = \frac{1}{8\pi\mu_0}\left[F_{ijk} f^0_k + D_{ijmnp} E^1_{mnp}\right] \tag{8.16}$$

When x is an interior point of Ω, F_{ijk} and D_{ijmnp} in Eq. (8.16) for the spherical drop can be explicitly derived as

$$F_{ijk} = -\frac{4\pi}{5}\delta_{ik}x_j - \frac{4\pi}{5}\delta_{jk}x_i + \frac{8\pi}{15}\delta_{ij}x_k \tag{8.17}$$

$$D_{ijmnp} = \mu_0 \begin{bmatrix} \left(\frac{16\pi}{35}\delta_{jp}\delta_{mn} - \frac{12\pi}{35}\delta_{jm}\delta_{np} - \frac{12\pi}{35}\delta_{jn}\delta_{mp}\right)x_i \\ +\left(\frac{16\pi}{35}\delta_{ip}\delta_{mn} - \frac{12\pi}{35}\delta_{im}\delta_{np} - \frac{12\pi}{35}\delta_{in}\delta_{mp}\right)x_j \\ +\left(\frac{16\pi}{35}\delta_{ij}\delta_{np} - \frac{12\pi}{35}\delta_{ip}\delta_{jn} - \frac{12\pi}{35}\delta_{in}\delta_{jp}\right)x_m \\ +\left(\frac{16\pi}{35}\delta_{ij}\delta_{mp} - \frac{12\pi}{35}\delta_{ip}\delta_{jm} - \frac{12\pi}{35}\delta_{im}\delta_{jp}\right)x_n \\ +\left(\frac{16\pi}{35}\delta_{ij}\delta_{mn} - \frac{8\pi}{7}\delta_{im}\delta_{jn} - \frac{8\pi}{7}\delta_{in}\delta_{jm}\right)x_p \end{bmatrix} \tag{8.18}$$

After e_{ij} is obtained, substituting Eq. (8.5) into Eq. (8.4) yields

$$e^*_{ij} = E^1_{ijr}x_r = \frac{\mu_0 - \mu_1}{\mu_0}e_{ij}, \quad x \in \Omega \tag{8.19}$$

Because Eq. (8.16) only includes the linear terms, the above equation could be rewritten as

$$E^1_{ijr} = \frac{\mu_0 - \mu_1}{\mu_0}e_{ij,r} \tag{8.20}$$

The substitution of Eq. (8.16) into Eq. (8.20) yields

$$E^1_{ijr} = \frac{\mu_0 - \mu_1}{8\pi\mu_0^2}\left(F_{ijk,r} f^0_k + D_{ijmnp,r} E^1_{mnp}\right) \tag{8.21}$$

The above relation between E^1_{ijr} and f^0_k can be rewritten as $E^1_{ijr} = S_{ijkr} f^0_k$. Then we can derive S_{ijkr} from Eqs. (8.19)–(8.21) and explicitly write as

$$S_{ijkr} = \gamma\delta_{ij}\delta_{kr} + \xi\left(\delta_{ik}\delta_{jr} + \delta_{jk}\delta_{ir}\right) \tag{8.22}$$

where

$$\gamma = \frac{\mu_0 - \mu_1}{3\mu_0 (2\mu_0 + 3\mu_1)}, \quad \xi = -\frac{\mu_0 - \mu_1}{2\mu_0 (2\mu_0 + 3\mu_1)} \tag{8.23}$$

Therefore the velocity field $v_i(\mathbf{x})$ from Eq. (8.10) can be rewritten as

$$v_i(\mathbf{x}) = \begin{cases} \frac{1}{3\mu_0(2\mu_0+3\mu_1)} \begin{bmatrix} \left(3a^2\mu_0 + 2a^2\mu_1 - 2x^2\mu_0\right) f_i^0 \\ +\mu_0 x_i x_k f_k^0 \end{bmatrix}, & \mathbf{x} \in \Omega \\ \frac{1}{6\mu_0(2\mu_0+3\mu_1)} \begin{bmatrix} a^2\rho\left(2\mu_0 + 3\mu_1 + \rho^2\mu_1\right) f_i^0 \\ +\rho^3\left(2\mu_0 + 3\mu_1 - 3\rho^2\mu_1\right) x_i x_k f_k^0 \end{bmatrix}, & \mathbf{x} \in D - \Omega \end{cases} \tag{8.24}$$

where

$$x^2 = x_i x_i; \ \rho = \frac{a}{x} \tag{8.25}$$

Similarly, the pressure field $p(\mathbf{x}) = -\frac{1}{4\pi}\int_\Omega (\Phi_{,k} f_k - 2\mu_0 \Phi_{,mn} e^*_{mn}) d\mathbf{x}'$ can be rewritten as

$$p(\mathbf{x}) = \begin{cases} \frac{6\mu_0 - \mu_1}{3(2\mu_0+3\mu_1)} x_k f_k^0, & \mathbf{x} \in \Omega \\ \frac{1}{3} a\rho^2 n_k f_k^0, & \mathbf{x} \in D - \Omega \end{cases} \tag{8.26}$$

Here the constant pressure term in the far field has not been included. Otherwise, a uniform pressure term should be superposed in the above equation.

With the known integrals

$$\int_\Omega a^2 dx = \frac{4\pi a^5}{3}, \quad \int_\Omega x^2 dx = \frac{4\pi a^5}{5}, \quad \int_\Omega x_i x_j dx = \frac{4\pi a^5}{15}\delta_{ij} \tag{8.27}$$

the average velocity field $\langle v_i \rangle$ over the drop can be determined as

$$\langle v_i \rangle = \frac{3}{4\pi a^3}\int_\Omega v_i(x) dx = \frac{2(\mu_0 + \mu_1) a^2}{3\mu_0 (2\mu_0 + 3\mu_1)} f_i^0 \tag{8.28}$$

For three different cases, such as spherical solid particle, drop, and air bubble, we obtain that

$$\begin{cases} \langle v_i \rangle_s = \frac{2a^2}{9\mu_0} f_i^0 & \text{for a solid sphere } \mu_1 = \infty \\ \langle v_i \rangle_a = \frac{a^2}{3\mu_0} f_i^0 & \text{for a spherical air void } \mu_1 = 0 \\ \langle v_i \rangle_d = \frac{2(\mu_0+\mu_1)a^2}{3\mu_0(2\mu_0+3\mu_1)} f_i^0 & \text{for a spherical drop with viscosity } \mu_1 \end{cases} \tag{8.29}$$

From $P_j = \int_{\partial\Omega} n_i \sigma_{ij} d\mathbf{x}$, we can determine the total drag force by substituting Eq. (8.29)

$$P_i = -\frac{4\pi a^3}{3} f_i^0$$
$$\Rightarrow \begin{cases} P_i^s = -6\pi a \mu_0 \langle v_i \rangle & \text{for a solid sphere } \mu_1 = \infty \\ P_i^a = -4\pi a \mu_0 \langle v_i \rangle & \text{for a spherical air void } \mu_1 = 0 \\ P_i^s = -2\pi a \mu_0 \langle v_i \rangle \frac{2\mu_0 + 3\mu_1}{\mu_0 + \mu_1} f_i^0 & \text{for a spherical drop with viscosity } \mu_1 \end{cases} \tag{8.30}$$

The above method shows exactly the same drag force as that derived by the Stokes law [111] for three different cases. Comparing the velocity and pressure fields in Eqs. (8.24) and (8.26) with the classic solutions [111], we found that they are also consistent as well. More detailed case studies for one single spheroidal drop and for multiple spherical particles moving in a fluid can be found in [10].

8.3 Boundary effect on the Stokes flow

To simulate particle sedimentation, consider that the plane S is a fixed boundary of a semi-infinite domain D filled with a fluid with viscosity μ_0 in Fig. 8.3. A cubic simulation box U is introduced within the domain D and a number N of particles are dispersed in U. Each particle domain is signified as Ω^I with diameters and densities written as d^I and ρ^I $(I = 1, 2, \ldots, N)$. The combination of all particle domains is written as Ω, so that the fluid domain is $D - \Omega$. The coordinate system is set up with the origin at the center of the bottom of U and the fluid in the domain $x_3 > 0$. Here U is only used to initialize the suspension at the beginning of the modeling; thus the particle can move out from the simulation box during the sedimentation. However, a particle becomes inactive once it makes contact with surface S.

The equilibrium equation is shown in Eqs. (8.2) and (8.3), and the continuity boundary conditions are as follows:

$$\begin{aligned} v_i^+ = v_i^-, \text{ and } n_j \sigma_{ji}^+ = n_j \sigma_{ji}^- \quad & \text{for } \mathbf{x} \in \partial\Omega \\ v_i = 0 \quad & \text{for } \mathbf{x} \in S \\ v_i = 0, \text{ and } p = 0, \quad & \text{for } \mathbf{x} \to \infty \end{aligned} \tag{8.31}$$

To solve the above boundary value problem for the particle velocity, we will use Eshelby's equivalent inclusion method (EIM). Following Mura's definition, we call a subdomain Ω in a matrix fluid an "inclusion," which exhibits the same viscosity as the matrix fluid, but on which an eigenstrain rate $e_{ij}^*(\mathbf{x})$ is applied as a fictitious nonmechanical strain rate. On the other hand, an "inhomogeneity" is a subdomain Ω in a homogeneous fluid with different material properties from

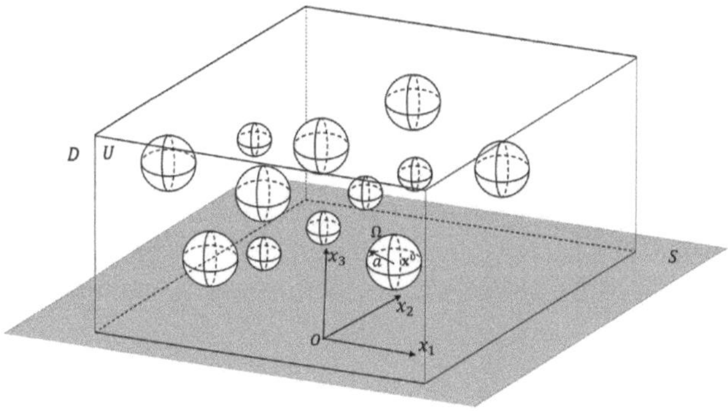

FIGURE 8.3 Many particles dispersed in a simulation box within a semi-infinite fluid with a fixed boundary.

the rest $D - \Omega$, but of which an existing eigenstrain rate is unnecessary in that subdomain. Therefore the particles are inhomogeneities instead of inclusions. However, when an inhomogeneity is subjected to a body force, the mismatch of the inhomogeneity from the matrix fluid can be simulated by an appropriately chosen eigenstrain rate on the drop, so that the velocity field can be obtained through solving the inclusion problem.

8.3.1 One particle moving in the semi-infinite fluid

To demonstrate the formulation, we start with the one particle case and then extend it to many particles. We first assume that other particles do not exist and that there is only a single particle in Fig. 8.3 with center x_0 and radius a. For the generality of formulation, we assume that the viscosity of the particle is written as μ_1, so that it can be generalized as a drop. In our case, $\mu_1 \to \infty$. Using Eshelby's equivalent inclusion method, we can assume that the particle Ω has the same viscosity as the fluid, but an eigenstrain rate $e_{ij}^*(\mathbf{x})$ is introduced in the particle Ω to represent the disturbance caused by the mismatch of the particle viscosity. Based on Eshelby's equivalent inclusion method, e_{ij}^* satisfies

$$\mu_0\left(e_{ij} - e_{ij}^*\right) = \mu_1 e_{ij}, \ \mathbf{x} \in \Omega \tag{8.32}$$

where $e_{ij}^*(\mathbf{x})$ also satisfies the mass conservation, i.e.,

$$e_{ii}^* = 0. \tag{8.33}$$

Therefore the governing equation (8.3) can be written as

$$2\mu_0 e_{ij,i} - p_{,j} = -f_i(\mathbf{x}) + 2\mu_0 e_{ij,i}^*(\mathbf{x}), \ v_{i,i} = 0, \ x \in \Omega \tag{8.34}$$

Comparing the above equation with Eq. (8.2), we can see that the two share the same form except that there are a body force and an eigenstrain rate on the particle. Here the eigenstrain rate is to be determined later by the stress equivalent condition. Therefore the overall velocity field would then be written in terms of the combination of the velocity field caused by the sources of both the body force and the eigenstrain rate on the particle.

The Green's function technique has been widely used to solve the above type of problem. For a concentrated force F_i in the interior point $\mathbf{x}'$ of a semi-infinite domain with fixed boundary, Rongved [112] has provided a fundamental solution, from which the velocity at any point $\mathbf{x}$ can be written in terms of the Green's function as

$$v_i(\mathbf{x}) = G_{ik}\left(\mathbf{x}, \mathbf{x}'\right) F_k\left(\mathbf{x}'\right) \tag{8.35}$$

where

$$\begin{aligned} G_{ij}\left(\mathbf{x}, \mathbf{x}'\right) = &\frac{1}{4\pi\mu^0}\delta_{ij}\left(\phi - \bar{\phi}\right) - \frac{1}{8\pi\mu^0}\left(\psi_{,ij} - Q_{im}Q_{jn}\bar{\psi}_{,mn}\right) \\ &+ \frac{x_3}{4\pi\mu^0}\left[Q_{jm}\bar{\psi}_{,im3} + 2\delta_{j3}\bar{\phi}_{,i} - Q_{jm}x_3\bar{\phi}_{,im}\right] \end{aligned} \tag{8.36}$$

in which $\psi = |\mathbf{x} - \mathbf{x}'|$, $\phi = \frac{1}{|\mathbf{x}-\mathbf{x}'|}$, $\bar{\psi} = |\mathbf{x} - \bar{\mathbf{x}}'|$, $\bar{\phi} = \frac{1}{|\mathbf{x}-\bar{\mathbf{x}}'|}$, $\bar{\mathbf{x}}' = \left(x_1', x_2', -x_3'\right)$ is the image point of $\mathbf{x}'$ through a mirror projection along plane S, and

$$Q = \begin{bmatrix} 1 & 0 & 0 \\ 0 & 1 & 0 \\ 0 & 0 & -1 \end{bmatrix}$$

is the mirror projection matrix, that is, $\bar{\mathbf{x}}' = Q.\mathbf{x}'$. Obviously, the Green's function satisfies the second and third boundary conditions in Eq. (8.31). Treating the left-hand side of Eq. (8.34) as a distributed body force in the inclusion, we can write

$$v_i(\mathbf{x}) = \int_D G_{ik}\left(\mathbf{x}, \mathbf{x}'\right)\left[f_i\left(\mathbf{x}'\right) - 2\mu_0 e^*_{ij,i}\left(\mathbf{x}'\right)\right] d\mathbf{x}' \tag{8.37}$$

Using the Gauss theorem and considering no eigenstrain rate on the boundary, we can rewrite the above equation as

$$v_i(\mathbf{x}) = \int_\Omega \left[G_{ik}\left(\mathbf{x}, \mathbf{x}'\right) f_k\left(\mathbf{x}'\right) + 2\mu_0 \Gamma^v_{imn}\left(\mathbf{x}, \mathbf{x}'\right) e^*_{mn}\left(\mathbf{x}'\right)\right] d\mathbf{x}' \tag{8.38}$$

where

$$\begin{aligned} \Gamma^v_{ijm} &= \frac{\partial G_{ij}}{\partial x'_m} \\ &= \frac{1}{4\pi\mu^0}\delta_{ij}\left(-\phi_{,m} + Q_{qm}\bar{\phi}_{,q}\right) + \frac{1}{8\pi\mu^0}\left(\psi_{,ijm} - Q_{qm}Q_{ik}Q_{jl}\bar{\psi}_{,klq}\right) \\ &\quad - \frac{x_3}{4\pi\mu^0}\left[Q_{qm}Q_{jk}\bar{\psi}_{,ik3q} + 2\delta_{j3}Q_{qm}\bar{\phi}_{,iq} - Q_{qm}Q_{jk}x_3\bar{\phi}_{,ikq}\right] \end{aligned}$$

Then the strain rate field can be written as

$$e_{ij} = \int_{\Omega} \left(\Gamma^{f}_{ijk} + 2\mu^{0} \Gamma^{e}_{ijmn} e^{*}_{mn} \right) dx' \tag{8.39}$$

where

$$\Gamma^{f}_{ijk} = \frac{1}{2} \left(G_{ik,j} + G_{jk,i} \right)$$

and

$$\Gamma^{e}_{ijmn} = \frac{1}{4} \left(\Gamma^{v}_{imn,j} + \Gamma^{v}_{jmn,i} + \Gamma^{v}_{inm,j} + \Gamma^{v}_{jnm,i} \right)$$

in which $e^{*}_{mn} = e^{*}_{nm}$ was used to make the tensor Γ^{e}_{ijmn} satisfy the major and minor symmetry of fourth-rank tensors.

Although the eigenstrain rate has not been obtained yet as a continuous tensor function, we can write it in a polynomial form of the coordinates as follows:

$$e^{*}_{ij}\left(\mathbf{x}'\right) = \begin{cases} e^{0}_{ij} + e^{1}_{ijk} x'_{k} + \cdots, & \mathbf{x}' \in \Omega \\ 0, & \mathbf{x}' \in D - \Omega \end{cases} \tag{8.40}$$

When the particle is embedded in an infinite fluid, the eigenstrain rate can be written in a single term of a linear function due to the body force [95]. Therefore, in this problem, the eigenstrain rate is also approximately linearly distributed when the particle is far from the boundary. However, when it is close to the fixed boundary, the eigenstrain rate distribution can be more complex, and it may include constant, linear, and even higher-order terms. We can asymptotically obtain more accurate results when higher-order terms of the eigenstrain rate are used. However, for simplicity, only constant and linear terms are used in the present section because it has shown very good agreement with the finite element results, which will be discussed in the next section. We can explicitly integrate Eq. (8.24) and write the velocity field as

$$e_{ij}(\mathbf{x}) = F_{ijk}\left(\mathbf{x}, \mathbf{x}^{0}\right) f_{k} + D_{ijmn}\left(\mathbf{x}, \mathbf{x}^{0}\right) e^{0}_{mn} + D_{ijmnp}\left(\mathbf{x}, \mathbf{x}^{0}\right) e^{1}_{mnp} + \cdots \tag{8.41}$$

where tensors F_{ijk}, D_{ijmn}, and D_{ijmnp} are explicitly provided in Appendix 8.A.

Expanding the strain field in the Taylor series referred to the particle center $\mathbf{x}^{0}$, the strain rate on the inclusion can be written as

$$\begin{aligned} e_{ij}(\mathbf{x}) = {} & F_{ijk}\left(\mathbf{x}^{0}, \mathbf{x}^{0}\right) f_{k} + D_{ijmn}\left(\mathbf{x}^{0}, \mathbf{x}^{0}\right) e^{0}_{mn} + D_{ijmnp}\left(\mathbf{x}^{0}, \mathbf{x}^{0}\right) e^{1}_{mnp} \\ & + \left[\begin{array}{c} F_{ijk,r}\left(\mathbf{x}^{0}, \mathbf{x}^{0}\right) f_{k} + D_{ijmn,r}\left(\mathbf{x}^{0}, \mathbf{x}^{0}\right) e^{0}_{mn} \\ + D_{ijmnp,r}\left(\mathbf{x}^{0}, \mathbf{x}^{0}\right) e^{1}_{mnp} \end{array} \right] \left(x_{r} - x^{0}_{r} \right) + \cdots \end{aligned} \tag{8.42}$$

Inside Ω, the equivalent inclusion condition of Eq. (8.32) must be satisfied, i.e.,

$$e_{ij}^{*} = \frac{\mu^0 - \mu^1}{\mu^0} e_{ij} \tag{8.43}$$

Then we obtain

$$\begin{aligned} & e_{ij}^0 + e_{ijr}^1 \left(x_r - x_r^0\right) + \cdots \\ & = \frac{\mu^0 - \mu^1}{\mu^0} \left[\begin{array}{c} F_{ijk}\left(\mathbf{x}^0, \mathbf{x}^0\right) f_k + D_{ijmn}\left(\mathbf{x}^0, \mathbf{x}^0\right) e_{mn}^0 + D_{ijmnp}\left(\mathbf{x}^0, \mathbf{x}^0\right) e_{mnp}^1 \\ + \left[\begin{array}{c} F_{ijk,r}\left(\mathbf{x}^0, \mathbf{x}^0\right) f_k \\ + D_{ijmn,r}\left(\mathbf{x}^0, \mathbf{x}^0\right) e_{mn}^0 \\ + D_{ijmnp,r}\left(\mathbf{x}^0, \mathbf{x}^0\right) e_{mnp}^1 \end{array} \right] \left(x_r - x_r^0\right) + \cdots \end{array} \right] \end{aligned} \tag{8.44}$$

Comparing the coefficients for different orders of $\left(x_r - x_r^0\right)$ terms, we obtain

$$\left(D_{ijkl} - \frac{\mu^0}{\mu^0 - \mu^1} I_{ijkl}\right) e_{kl}^0 + D_{ijklp} e_{klp}^1 = -F_{ijk} f_k \tag{8.45}$$

$$D_{ijkl,r} e_{kl}^0 + \left(D_{ijklp,r} - \frac{\mu^0}{\mu^0 - \mu^1} I_{ijklpr}\right) e_{klp}^1 = -F_{ijk,r} f_k \tag{8.46}$$

where I_{ijkl} and I_{ijklpr} are the unit tensors, which map the corresponding tensors to themselves through the multiplications. Notice that if a higher order of polynomial form of eigenstrain rate is used, then this method can be correspondingly used with more equations. Solving this linear equation system, we can find the final eigenstrain rate terms. Substituting them into Eq. (8.24) yields the velocity

$$\begin{aligned} v_i(\mathbf{x}) &= \int_\Omega \left(G_{ik} f_k + 2\mu_0 \Gamma_{imn}^v e_{mn}^*\right) d\mathbf{x}' \\ &= \int_\Omega \left(G_{ik} f_k + \mu_0 \Gamma_{imn} \left(e_{mn}^0 + e_{mnp}^1 x_p'\right)\right) d\mathbf{x}' \\ &= L_{ik} f_k + M_{imn} e_{mn}^0 + N_{imnp} e_{mnp}^1 \end{aligned} \tag{8.47}$$

where

$$\begin{aligned} L_{ik} = & \frac{1}{4\pi\mu^0} \delta_{ik} \left(\Phi - \bar{\Phi}\right) - \frac{1}{8\pi\mu^0} \left(\Psi_{,ik} - Q_{im} Q_{kn} \bar{\Psi}_{,mn}\right) \\ & + \frac{x_3}{4\pi\mu^0} \left[Q_{km} \bar{\Psi}_{,im3} + 2\delta_{k3} \bar{\Phi}_{,i} - Q_{km} x_3 \bar{\Phi}_{,im}\right] \\ M_{imn} = & \frac{1}{2\pi} \delta_{ij} \left(-\Phi_{,m} + Q_{qm} \bar{\Phi}_{,q}\right) + \frac{1}{4\pi} \left(\Psi_{,ijm} - Q_{qm} Q_{ik} Q_{jl} \bar{\Psi}_{,klq}\right) \\ & - \frac{x_3}{2\pi} \left[Q_{qm} Q_{jk} \bar{\Psi}_{,ik3q} + 2\delta_{j3} Q_{qm} \bar{\Phi}_{,iq} - Q_{qm} Q_{jk} x_3 \bar{\Phi}_{,ikq}\right] \end{aligned}$$

$$
\begin{aligned}
N_{ijmp} = & \frac{1}{2\pi}\delta_{ij}\left(-\Phi_{p,m} + Q_{pt}Q_{qm}\bar{\Phi}_{t,q}\right) \\
& + \frac{1}{4\pi}\left(\Psi_{,ijm} - Q_{pt}Q_{qm}Q_{ik}Q_{jl}\bar{\Psi}_{t,klq}\right) \\
& - \frac{x_3}{2\pi}\left[Q_{pt}Q_{qm}Q_{jk}\bar{\Psi}_{t,ik3q} + 2\delta_{j3}Q_{pt}Q_{qm}\bar{\Phi}_{t,iq}\right. \\
& \left. - Q_{pt}Q_{qm}Q_{jk}x_3\bar{\Phi}_{t,ikq}\right]
\end{aligned}
$$

8.3.2 Multiple particles moving in a semi-infinite fluid

The above formulation can be extended from a single particle to multiple particles in a very straightforward way. When many particles are moving in the viscous fluid, their interactions significantly affect the velocity field of the whole simulation domain. Fig. 8.3 illustrates the whole system with many particles, say Al and HDPE particles. In the Stokes flow, all particles are assumed to shortly reach their equilibrium velocities. To formulate this problem, a global coordinate system has been set up in Fig. 8.3, and each particle, say Ω^I, is centered at $\mathbf{x}^I = \left(x_1^I, x_2^I, x_3^I\right)$. All particles being treated as the same viscosity as the fluid, the eigenstrain rate can be introduced with a polynomial form again as

$$
e_{ij}^* = e_{ij}^{0,I} + e_{ijk}^{1,I}\left(x_k' - x_k^I\right) + \cdots \quad \text{for } \mathbf{x}' \in \Omega^I \tag{8.48}
$$

where no summation is applied to the superscript, whereas repeated subscripts still mean summation. As discussed before, we compute up to linear terms. Then the equivalent condition for one of the particles, say Ω^J, reads

$$
\mu^0\left(e_{ij} - e_{ij}^*\right) = \mu^J e_{ij} \quad \text{for } \mathbf{x}' \in \Omega^J \tag{8.49}
$$

The velocity field of any point in the fluid domain is caused by the eigenstrain rate and body force of all particles, which can be obtained by the superposition of the field induced by these two sources as follows:

$$
v_i(\mathbf{x}) = \frac{1}{4\pi\mu_0}\int_{\sum_{I=1}^N \Omega^I}\left[G_{ik}\left(\mathbf{x},\mathbf{x}'\right)f_k\left(\mathbf{x}'\right) + \mu_0\Gamma_{imn}^v\left(\mathbf{x},\mathbf{x}'\right)e_{mn}^*\left(\mathbf{x}'\right)\right]\mathrm{d}\mathbf{x}' \tag{8.50}
$$

Using the same procedure as in the case of one particle, we can write the strain rate on domain Ω^J in terms of the Taylor expansion as

$$
\begin{aligned}
e_{ij}(\mathbf{x}) = & \sum_{I=1}^{N} F_{ijk}^I\left(\mathbf{x}^J,\mathbf{x}^I\right)f_k^I + D_{ijmn}^I\left(\mathbf{x}^J,\mathbf{x}^I\right)e_{mn}^{0,I} + D_{ijmnp}^I\left(\mathbf{x}^J,\mathbf{x}^I\right)e_{mnp}^{1,I} \\
& + \left[\begin{array}{c} F_{ijk,r}^I\left(\mathbf{x}^J,\mathbf{x}^I\right)f_k^I + D_{ijmn,r}^I\left(\mathbf{x}^J,\mathbf{x}^I\right)e_{mn}^{0,I} \\ +D_{ijmnp,r}^I\left(\mathbf{x}^J,\mathbf{x}^I\right)e_{mnp}^{1,I}\end{array}\right]\left(x_r - x_r^J\right) + \cdots
\end{aligned} \tag{8.51}
$$

where F^I_{ijk}, D^I_{ijmn}, ..., and $D^I_{ijmnp,r}$ are the corresponding integrals on the domain Ω^I. Comparing the coefficients at different orders of the terms $\left(x_r - x^J_r\right)$ of the equivalent inclusion condition (8.49), we can also obtain

$$\begin{aligned}
&\sum_{I=1}^{N} D^I_{ijkl} e^{0,I}_{kl} - \frac{\mu^0}{\mu^0 - \mu^J} e^{0,J}_{ij} + \sum_{I=1}^{N} D^I_{ijklp} e^{1,I}_{klp} = -\sum_{I=1}^{N} F^I_{ijk} f^I_k \\
&\sum_{I=1}^{N} D^I_{ijkl,r} e^{0,I}_{kl} + \sum_{I=1}^{N} D^I_{ijklp,r} e^{1,I}_{klp} - \frac{\mu^0}{\mu^0 - \mu^J} e^{1,J}_{ijr} = -\sum_{I=1}^{N} F^I_{ijk,r} f^I_k
\end{aligned} \tag{8.52}$$

Note that there are 36 components of the eigenstrain rate tensorial function in each particle, Eq. (8.52) can be rewritten in a matrix notation as

$$[A]^J_{36\times 36\mathrm{n}}[E]_{36\mathrm{n}} = [F]^J_{36} \tag{8.53}$$

where the eigenstrain rate terms of all the particles are represented by a list as $[E]$ whose length is 36 n, $[A]^J$ is the local coefficient matrix of dimensions $36 \times 36\mathrm{n}$ whose elements are the coefficients in Eq. (8.52), and $[F]^J$ is the local force list of length 36, whose elements are the sums of the body force multiplied by the corresponding coefficients.

For each particle, we can write one set of equations like above. When a higher rank of the polynomial form of eigenstrain rate is used, more equations can be obtained as above. Rewriting all the equations in the matrix notation, assembling all the $[A]^J$ into the global coefficient matrix and all the $[F]^J$ into global force vector, we obtain

$$[A]_{36\mathrm{n}\times 36\mathrm{n}}[E]_{36n} = [F]_{36\mathrm{n}} \tag{8.54}$$

Solving this global linear system of equations, we can obtain the eigenstrain rate terms of all particles. Substituting them into Eq. (8.50), we can also obtain the velocity field.

Because we consider the boundary value problem through a continuum mechanics-based approach, the particle interactions have been implicitly included in the analysis. For example, the velocity field on particle Ω^J is affected by the body force and eigenstrain rate acting on particle Ω^I. Thus the effect of particle Ω^J has been transferred to particle Ω^I by the fluid. It is similar to the pairwise interaction of inhomogeneities through the matrix [113,114]. This kind of pairwise interaction exists between any pair of particles, and each particle is subject to the effects of all other particles, which rapidly decay with the particle–particle distance.

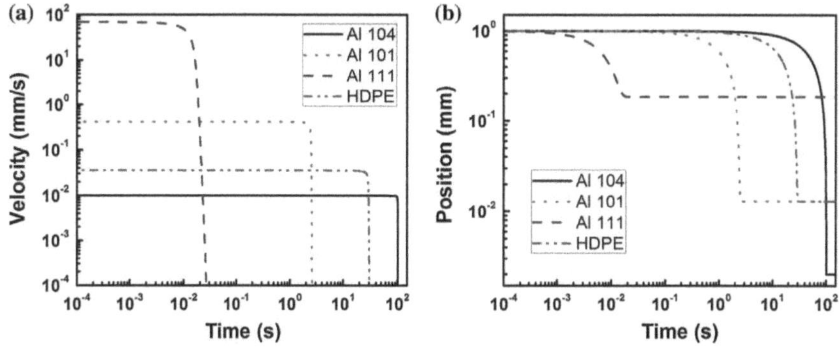

FIGURE 8.4 A single Al or HDPE particle moving in a semi-infinite fluid of ethanol. (a) Velocity versus time and (b) position versus time.

*8.4 Virtual experiments of particle settlement in a viscous fluid

In this section, we use the EIM to simulate the sedimentation of particles and demonstrate the formation of a graded mix. First, we simulate the motion of a single particle. The boundary effect has been illustrated for HDPE particles and the three types of aluminum particles, namely Al104, Al101, and Al111. Then a many-particle system moving in a viscous fluid is simulated. A random microstructure has been generated with the quartet structure generation set (QSGS) method for the simulation of sedimentation. The motion of particles has been illustrated and a graded mix is obtained.

*8.4.1 Sedimentation of one particle

The sedimentation of three types of aluminum particles and HDPE particle is first simulated for the case of a single particle moving in the ethanol fluid. The particle's center is located at $(0, 0, 1\,\text{mm})$ and the time step is 5×10^{-4} s. Fig. 8.4a illustrates the velocities changing with time. It is observed that the velocity of Al111 is the highest because of its large size and high density. The size of Al101 and HDPE is similar, but the velocity of Al101 is higher than that of HDPE because the density of aluminum is much higher than that of HDPE. Though having a higher density, the velocity of Al104 is lower than that of HDPE due to it having the smallest size. Because of the difference in their sedimentation velocity, the gradient distribution can be observed in the experiment when they are put into the same container. Additionally, we can find that the velocity decreases slowly at the beginning but drops rapidly when it is close to the boundary. This indicates that the effect of the boundary is significant near the boundary but decreases fast with the distance from the boundary.

Fig. 8.4b illustrates the particle position changing with time. Notice that the final position is not zero because the position of the particle center is $(0, 0, r)$

when it reaches the boundary. Here r indicates the radius of the corresponding particle. We can observe that it takes only about 0.011 s for Al111 particles to reach the boundary due to their high velocity; Al 101 spends more than 1 s to reach the boundary; HDPE spends more than 10 s, whereas Al104 spends more than 100 s to finish the sedimentation process. This means that the main time cost is caused by the sedimentation of HDPE and small aluminum particles. This observation can explain the phenomenon that the velocity of slower particles needs more time to decrease to zero. Actually, the smaller the velocity is, the more the time it will take to reach the boundary. Then the boundary effect increases more slowly, and the velocity takes longer to decrease to zero.

*8.4.2 Sedimentation of multiple particles toward FGM manufacturing

During the process of FGM fabrication, the aluminum particles and HDPE particles are mixed together and then placed in a container. Their interactions have a significant effect on their velocities. They interact with each other, which leads to a higher velocity. However, when a particle is close to the boundary, other particles and the fixed boundary will affect its velocity. To investigate this complex and interesting phenomenon, the sedimentation process of many Al 101 and HDPE particles of similar size is considered. A simulation box of 220 μm × 220 μm × 660 μm is placed at a fixed boundary. The length of each cell is 44 μm, and there are in total 375 cells in the simulation box. Using the QSGS microstructure generator, 75 Al 101 particles and 75 HDPE particles are randomly placed in this box. Fig. 8.5a illustrates the initial positions of all particles. The red particles are Al 101, and the green ones are HDPE. Then all particles begin to fall slowly forming a Stokes flow. When one particle reaches the bottom, it is inactivated and does not contribute to the sedimentation of the rest. When the bottom layer of cells is filled with particles, the bottom layer moves up 22 μm. A particle becomes inactive when it contacts any inactive particles. It takes 5.84 s for all the particles to reach the fixed boundary. Fig. 8.5 illustrates the positions of particles at different times. The sedimentation of Al 101 particles is faster than that of HDPE ones because of their higher densities. Fig. 8.5c illustrates that almost all Al 101 particles and some HDPE ones have settled in 0.5 s. However, the remaining particles move very slowly because the effect of interaction of particles significantly decreases. Fig. 8.5f illustrates the final configuration of the particle system, in which the volume fraction of Al 101 is the highest at the bottom and continuously decreases from the bottom to the top, which exhibits a graded distribution.

As the velocities of particles keep changing during the sedimentation process, the average velocity of all particles changing with time is illustrated in Fig. 8.6. The average velocity decreases with time and is close to zero at the

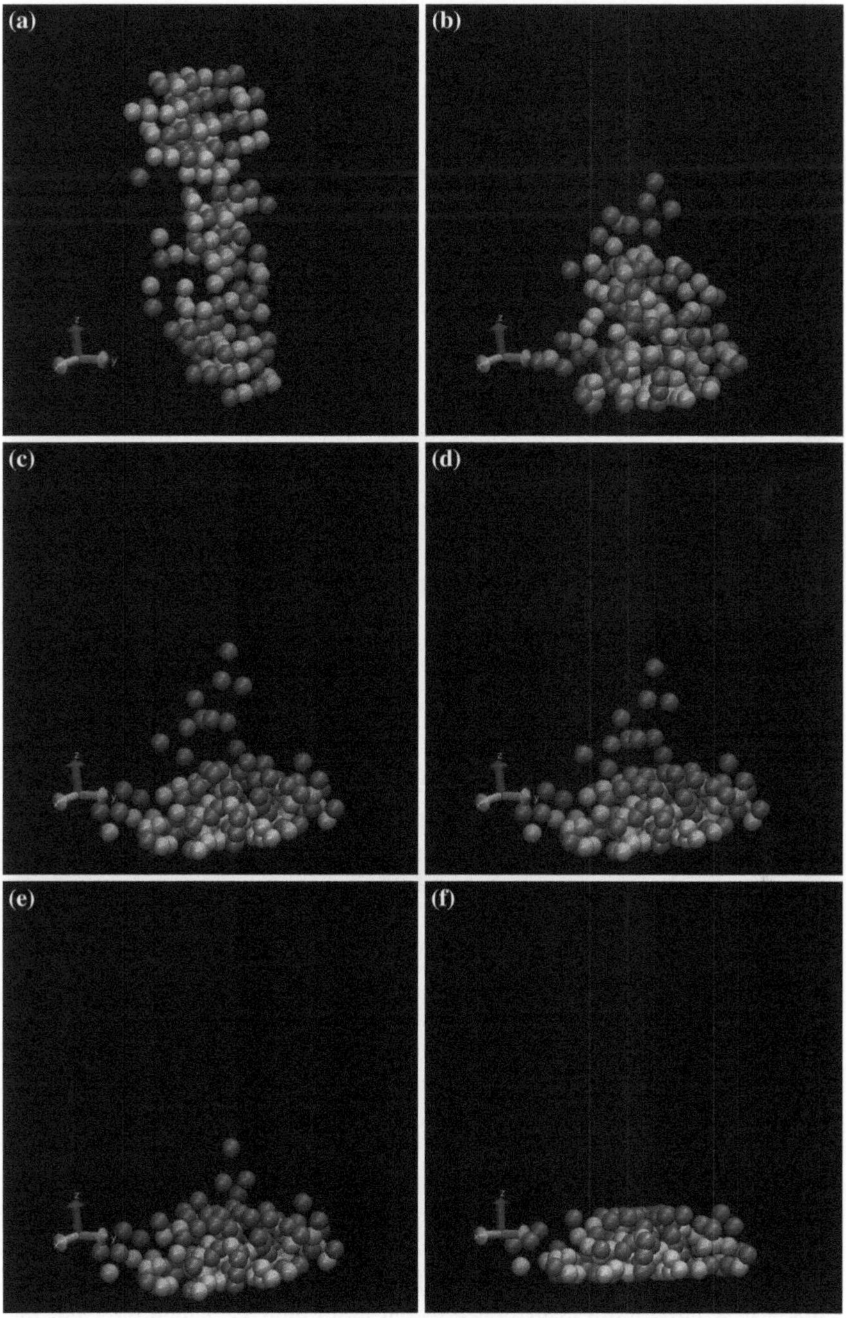

FIGURE 8.5 EIM-based simulation of the sedimentation process of Al 101-HDPE in ethanol: (a) t = 0; (b) t = 0.25 s; (c) t = 0.50 s; (d) t = 1 s; (e) t = 2.50 s; (f) t = 5.84 s.

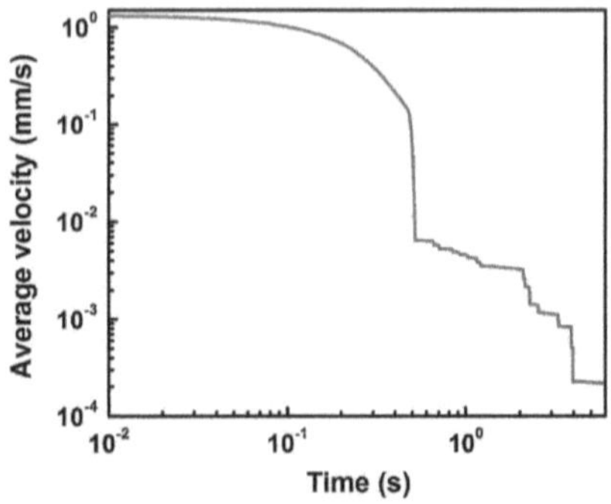

FIGURE 8.6 Average velocity of Al 101-HDPE changing with time.

end of the simulation, because the distance between the particles and the fixed boundary is low, and therefore the boundary effect becomes more significant. In addition, the velocities of HDPE particles are always small due to their smaller densities, although their size is similar to that of Al 101. The average velocity changes fast at first, but the changing rate reduces, which is much different from the sedimentation of one particle. When the Al 101 particles are close to the boundary, the change of average velocity is mainly contributed by the change of HDPE velocities. Due to their small velocities, the distance between HDPE particles and the boundary changes slowly, which lowers the changing rate of their velocities as well. Notice that log-coordinates are used and the changing rate may not be straightforwardly visible in the figure. We may observe that there are some jumps in the curve. Each jump means that some particles reached the fixed boundary, which will make a contribution to the sudden change of the average velocity. A large jump is observed at about $t = 0.5$ s. It is caused by many Al 101 particles reaching the boundary.

When Al 104 particles are mixed with HDPE for the sedimentation test, a different phenomenon is reported due to its smallest size. A simulation box of 177 μm × 177 μm × 531 μm is divided into 192 cells whose length is still 44 μm. Fifty Al 104 particles and 50 HDPE ones initially are randomly distributed in the box. The time step is 0.01 s, and in total 37.9 s are needed to finish the whole sedimentation process. Fig. 8.7 illustrates the average velocity changing with time. As expected, the average velocity is much smaller compared with the case using Al 101 particles. It is observed that the velocity decreases rapidly at the beginning, and then the changing rate lowers down. Compared with Fig. 8.6, we can find that it takes much more time for the average velocity to decrease to zero, although it is smaller. The reason is that it takes a longer time to reach the boundary. A large jump is also observed at more than 1 s, which indicates that many HDPE particles become inactive. Then some small jumps exist, and each one indicates that some particles reach the boundary.

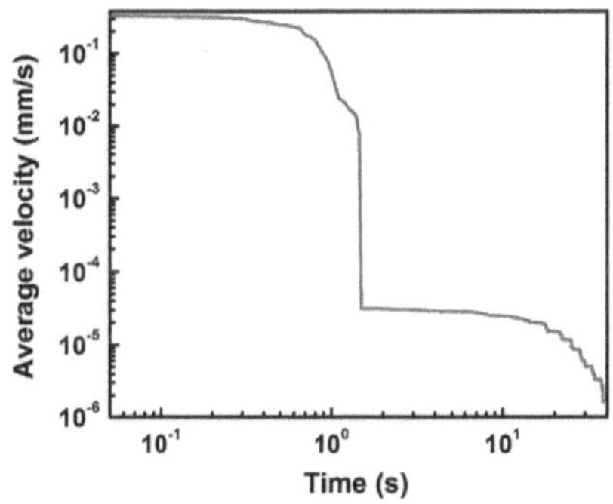

FIGURE 8.7 Average velocity of Al 104-HDPE changing with time.

8.5 Formulation of iBEM for the Stokes flow containing elliptical particles

Consider a flow domain D with viscosity μ_0 containing multiple elliptical particles with different viscosity μ_1 in subdomains Ω_I. The behavior of the flow is governed by the Stokes equation. In the matrix domain $D - \sum_{I=1}^{n} \Omega_I$, in the absence of body forces, the governing equations and associated boundary conditions are as follows:

$$\mu_0 u_{i,jj} - p_{,i} = 0 \qquad u_{i,i} = 0 \quad \text{in } D - \sum_{I=1}^{n} \Omega_I \tag{8.55}$$

$$\sigma_{ij}(x) n_j = t_i'(x) \quad \text{on } \partial D^t \tag{8.56}$$

$$u_i(x) = u_i'(x) \quad \text{on } \partial D^u \tag{8.57}$$

In each particle domain Ω_I, due to the material mismatch, separated Stokes equations should be established, and the continuity on the interfaces should also be satisfied as well:

$$\mu_1 u_{i,jj}(\mathbf{x}) - p_{,i}(\mathbf{x}) = 0 \quad \text{in } \Omega_I \tag{8.58}$$

$$\sigma_{ij}^{in}(\mathbf{x}) n_j^{in} = \sigma_{ij}^{out}(\mathbf{x}) n_j^{out} \quad \text{on } \partial \Omega_I \tag{8.59}$$

$$u_i^{in}(\mathbf{x}) = u_i^{out}(\mathbf{x}) \quad \text{on } \partial \Omega_I \tag{8.60}$$

The above equations form a complete and well-posed boundary value problem. Finding the analytical solutions for such a complicated problem is very challenging. As shown in the following sections, iBEM can be deployed to solve this system numerically and efficiently.

Basically, the inner velocity field in this colloidal system is expressed by iBEM as follows:

$$u_i\left(\mathbf{x}'\right)=\int_{\partial D} G_{ji}\left(\mathbf{x}',\mathbf{x}\right)t_j\left(\mathbf{x}\right)\mathrm{d}\mathbf{x}-\int_{\partial D} T_{ji}\left(\mathbf{x}',\mathbf{x}\right)u_j\left(\mathbf{x}\right)\mathrm{d}\mathbf{x}+\sum_{I=1}^{n} D^I\left(\mathbf{x}'\right)\varepsilon_{mn}^{I*} \tag{8.61}$$

where the Green's functions G and T are written as

$$G_{ij}\left(\mathbf{x}',\mathbf{x}\right)=\frac{1}{4\pi\mu}\left(\frac{\delta_{ij}}{|\mathbf{x}-\mathbf{x}'|}-\frac{1}{2}\frac{\partial\,|\mathbf{x}-\mathbf{x}'|}{\partial x_i\partial x_j}\right) \tag{8.62}$$

$$P_m\left(\mathbf{x}',\mathbf{x}\right)=-\frac{1}{4\pi}\frac{\partial}{\partial x_m}\left(\frac{1}{|\mathbf{x}-\mathbf{x}'|}\right) \tag{8.63}$$

$$T_{im}\left(\mathbf{x}',\mathbf{x}\right)=\left(\mu\left(G_{im,j}\left(\mathbf{x}',\mathbf{x}\right)+G_{jm,i}\left(\mathbf{x}',\mathbf{x}\right)\right)-P_m\left(\mathbf{x}',\mathbf{x}\right)\delta_{ij}\right)n_j \tag{8.64}$$

The first two terms in Eq. (8.61) describe the effect from boundary conditions to the inner velocity field. The last term takes the material mismatch into account at each particle by introducing eigenstrain rate ε^*. The eigenstrain rate is typically a nonmechanical strain rate. The effect of eigenstrain rate is described by Green's function technique as follows. In individual particle domain the viscosity is forced to be the same as the matrix but assigned with eigenstrain rate ε^*. The Stokes equation becomes

$$2\mu_0\left(\frac{1}{2}u_{i,j}+\frac{1}{2}u_{j,i}-\varepsilon_{ij}^*\right)_{,j}-p_{,i}=0 \tag{8.65}$$

After rearranging the above equation and applying $u_{i,i}=0$ for the impressible flow, we arrive at

$$\mu_0 u_{i,jj}-p_{,i}-2\mu_0\varepsilon_{ij,j}^*=0 \tag{8.66}$$

If we consider a new force vector $f_i=-2\mu_0\varepsilon_{i,jj}^*$ and apply the Green's function technique, the disturbed field due to the existence of eigenstrain rates is

$$u_m=-2\mu_0\int_{\Omega} G_{mi}\varepsilon_{ij,j'}^*d\mathbf{x}' \tag{8.67}$$

Usually, the following polynomial distribution of eigenstrain rates in each particle is assumed:

$$\varepsilon_{ij}^{I*}=\varepsilon_{ij}^{I0}+\varepsilon_{ijp}^{I1}x_p+\cdots,\qquad \mathbf{x}\in\Omega_I \tag{8.68}$$

Substituting Eqs. (8.62) and (8.68) into Eq. (8.67) and applying the divergence theorem, we can write the last term in Eq. (8.61) as follows:

$$D^I \varepsilon_{mn}^{I*} = D_{imn}^0 \varepsilon_{mn}^{I0} + D_{imnp}^1 \varepsilon_{ijp}^{I1} \tag{8.69}$$

where

$$D_{imn}^0 = -2\mu_0 \int_\Omega G_{im,n}\left(\mathbf{x}', \mathbf{x}\right) \mathrm{d}\mathbf{x}' = -\frac{1}{2\pi}\left(\delta_{im}\Phi_{,n} - \frac{1}{2}\Psi_{,imn}\right) \tag{8.70}$$

$$D_{imnp}^1 = -2\mu_0 \int_\Omega G_{i,mn}\left(\mathbf{x}', \mathbf{x}\right) x_p' \mathrm{d}\mathbf{x}' = -\frac{1}{2\pi}\left(\delta_{im}\Phi_{p,n} - \frac{1}{2}\Psi_{p,imn}\right) \tag{8.71}$$

The explicit definition and expression for Ψ and Φ can be found in Appendix 8.A.

From the relation between the strain rates and velocity $\varepsilon_{ij} = \frac{1}{2}\left(u_{i,j} + u_{j,i}\right)$ the strain rates anywhere inside the domain D are expressed as

$$\begin{aligned}
\varepsilon_{ij}\left(\mathbf{x}'\right) = &-\int_{\partial D} \frac{1}{2}\left(G_{mi,j}\left(\mathbf{x}', \mathbf{x}\right) + G_{mj,i}\left(\mathbf{x}', \mathbf{x}\right)\right) t_m(\mathbf{x})\, \mathrm{d}\mathbf{x} \\
&+ \int_{\partial D} \frac{1}{2}\left(T_{mi,j}\left(\mathbf{x}', \mathbf{x}\right) + T_{mj,i}\left(\mathbf{x}', \mathbf{x}\right)\right) u_m(\mathbf{x})\, \mathrm{d}\mathbf{x} \\
&+ \sum_{I=1}^{n} \frac{1}{2}\left(D_{ijmn}^{I0}\left(\mathbf{x}'\right) + D_{jimn}^{I0}\left(\mathbf{x}'\right)\right) \varepsilon_{mn}^{I0} \\
&+ \sum_{I=1}^{n} \frac{1}{2}\left(D_{ijmnp}^{I1}\left(\mathbf{x}'\right) + D_{jimnp}^{I1}\left(\mathbf{x}'\right)\right) \varepsilon_{mnp}^{I1}
\end{aligned} \tag{8.72}$$

where the properties $G_{,i} = -G_{,i'}$ and $T_{,i} = -T_{,i'}$ are used. The detailed expressions of tensor D can be found in Appendix 8.A.

To resolve the additional eigenstrain rate on each particle, stress equivalent conditions are developed inside each particle,

$$\mu_0\left(\varepsilon_{ij} - \varepsilon_{ij}^{I*}\right) = \mu_1 \varepsilon_{ij} \quad \text{for } I = 1, \ldots, n \tag{8.73}$$

since the local coordinates of elliptical particles differ from each other. Let $\mathbf{x^I}$ be the local coordinate of particle I consistent with its own axes, and let $\mathbf{x^0}$ be the global coordinate associated with the whole composite as shown in Fig. 8.8. Based on this configuration, the coordinate transformation tensor is defined as $a_{ij}^{IJ} = x_i \cdot x_j$. By expanding the ε_{ij} in Eq. (8.73) as the Taylor series at the center of the local coordinate of particle J for the Jth equation and matching the coefficients of $\mathbf{x^J}$, we can obtain the following equivalent conditions after

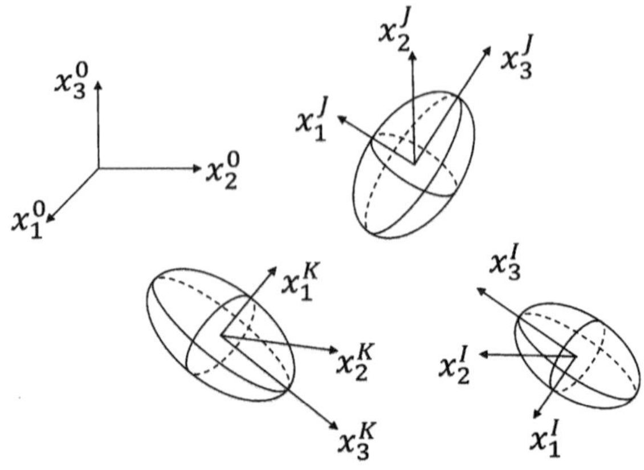

FIGURE 8.8 Global coordinate and local coordinates for elliptical particles I, J, K.

considering the coordinate transformations:

$$(\mu_0-\mu_1)\left[\begin{array}{c} a_{si}^{J0}a_{tj}^{J0}\int_{\partial D}-\frac{1}{2}\left(G_{mi,j}(\mathbf{0})+G_{mj,i}(\mathbf{0})\right)t_m \\ +a_{si}^{J0}a_{tj}^{J0}\int_{\partial D}\frac{1}{2}\left(T_{mi,j}(\mathbf{0})+T_{mj,i}(\mathbf{0})\right)u_m \\ +\sum_{I=1}^{n}\frac{1}{2}a_{si}^{JI}a_{tj}^{JI}\left(D_{ijmn}^{I0}(\mathbf{0})+D_{jimn}^{I0}(\mathbf{0})\right)\varepsilon_{mn}^{I0} \\ +\sum_{I=1}^{n}\frac{1}{2}a_{si}^{JI}a_{tj}^{JI}\left(D_{ijmnp}^{I1}(\mathbf{0})+D_{jimnp}^{I1}(\mathbf{0})\right)\varepsilon_{mnp}^{I1} \end{array}\right]=\mu_0\varepsilon_{st}^{J0}$$

$$\text{for } J=1,\ldots,n \tag{8.74}$$

$$(\mu_0-\mu_1)\left[\begin{array}{c} a_{si}^{J0}a_{tj}^{J0}a_{rq}^{J0}\int_{\partial D}\frac{1}{2}\left(G_{mi,jq}(\mathbf{0})+G_{mj,iq}(\mathbf{0})\right)t_m \\ +a_{si}^{J0}a_{tj}^{J0}a_{rq}^{J0}\int_{\partial D}-\frac{1}{2}\left(T_{mi,jq}(\mathbf{0})+T_{mj,iq}(\mathbf{0})\right)u_m \\ +\sum_{I=1}^{n}\frac{1}{2}a_{si}^{JI}a_{tj}^{JI}a_{rq}^{JI}\left(D_{ijmn,q}^{I0}(\mathbf{0})+D_{jimn,q}^{I0}(\mathbf{0})\right)\varepsilon_{mn}^{I0} \\ +\sum_{I=1}^{n}\frac{1}{2}a_{si}^{JI}a_{tj}^{JI}a_{rq}^{JI}\left(D_{ijmnp,q}^{I1}(\mathbf{0})+D_{jimnp,q}^{I1}(\mathbf{0})\right)\varepsilon_{mnp}^{I1} \end{array}\right]$$

$$=\mu_0\varepsilon_{str}^{J0}\quad\text{for } J=1,\ldots,n \tag{8.75}$$

where $\mathbf{0}$ denotes that this formula should be evaluated at the center of the local coordinate of particle J for the Jth equation.

In practical implementation, Eqs. (8.61), (8.74), and (8.75) are constructed and assembled into a linear equation system. The unknown quantities are unknown boundary responses ($3N$) and eigenstrain rates ($24n$), where N and n are the numbers of boundary nodes and elliptical particles. After obtaining the boundary values and eigenstrain rates, the total velocity field and strain rates can be derived from its equivalent inclusion problem with appropriate eigenstrain rates as Eqs. (8.61) and (8.68).

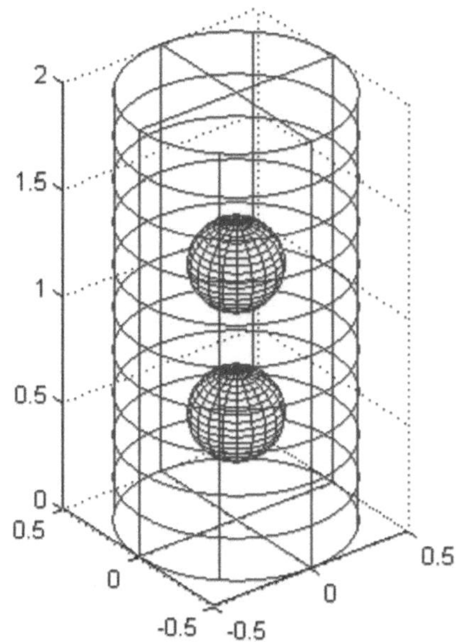

FIGURE 8.9 The problem configuration for numerical verification.

8.5.1 Verification of iBEM

In this section, we perform numerical verification of iBEM for three cases, two prolates, two oblates, and one prolate and one oblate. The same configurations of Stokes flow containing hard elliptical particles are implemented in ABAQUS as reference FEM results. All the quantities follow the SI units. The configuration is illustrated as in Fig. 8.9. One cylinder is built containing two elliptical particles at the center axis. The bottom of the cylinder is fixed in all directions, and a uniform tensile stress $t_z = 1$ is applied to the top surface. To simulate hard particles, the viscosity of particles is much larger than the viscosity of the fluid matrix. The velocity profiles u_z along the center axis are recorded. The comparisons between iBEM and FEM are provided in Figs. 8.10, 8.11, and 8.12 for three different cases.

From Figs. 8.10, 8.11, and 8.12 we observe a good agreement between iBEM and FEM results. The velocities in the particle domain are almost constant, which implies that the particles are subjected to rigid-body motions. It is demonstrated that the iBEM indeed can capture the underlying physics in this multiphase flow with particle interactions and boundary effects taken into account. It is worth noticing that the FEM simulation requires about one hundred thousand elements to obtain results with acceptable accuracy for this 3-D problem, whereas only one thousand elements and 48 more degrees of freedom are desired in the implementation of iBEM.

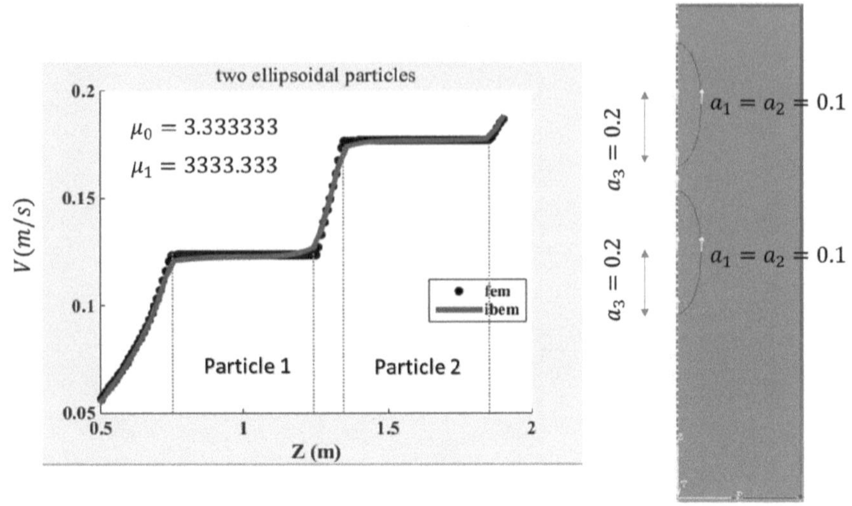

FIGURE 8.10 Comparison of iBEM and FEM for two-prolate case.

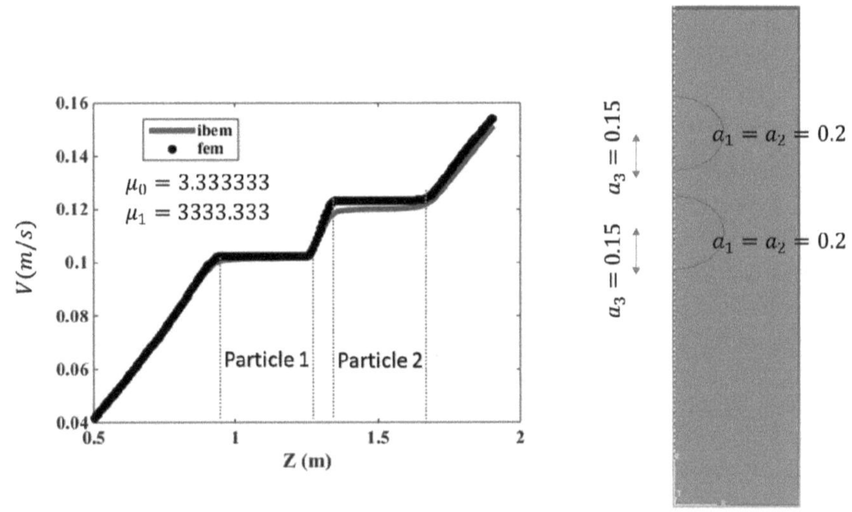

FIGURE 8.11 Comparison of iBEM and FEM for two-oblate case.

8.5.2 Simulation of the colloidal system with rods and disks like particles under shear

The advantage of iBEM is that it is convenient to simulate the multiphase flow system. In a Stokes fluid, one typical and challenging problem is to study the colloidal system where a number of particles are dispersed in a Stokes fluid. Applications of the colloidal system are widespread problems of sedimenta-

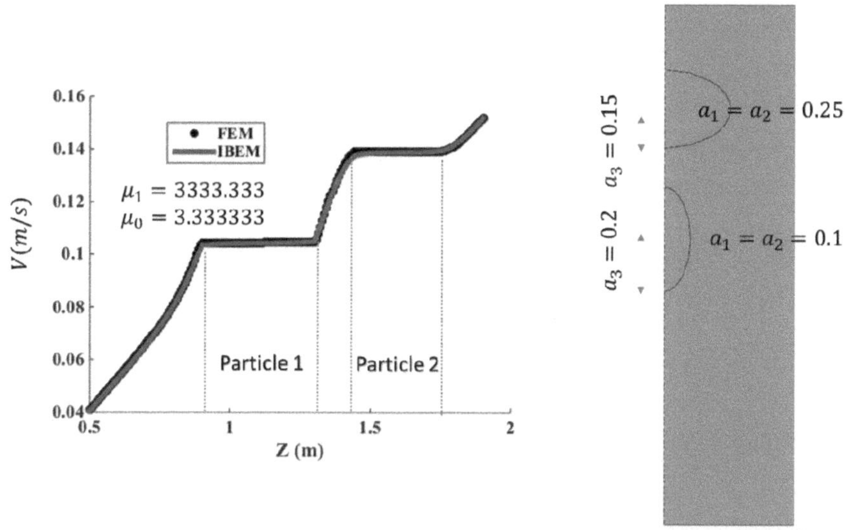

FIGURE 8.12 Comparison of iBEM and FEM for one-prolate–one-oblate case.

tion, flocculation, diffusion, polymer rheology, and transport in porous media, all falling within its domain. The difficulty is to characterize the material discontinuity in the system. One way to approach this problem is using a molecular-dynamics-like approach. However, it is hard to describe the interaction force between liquid and particles and the boundary effect to the particles. Little progress has been made beyond the two-particle limit model. In our method, we can easily study the colloidal system with a high solid load and do not need to make any assumptions and simplifications for modeling the interaction forces between particles and liquids. All the motions of particles are directly governed by Stokes equations. Mathematically speaking, solid particles are materials with infinite large viscosity, i.e., $\mu = \infty$. In our practical simulation the ratio between the viscosity of particles and fluid is set to be $\mu_1/\mu_0 = 10^5$. To investigate the evolution of microstructure of asphalt binders under the rheometer, the colloidal system with rod- and disk-shape asphaltene under sinusoidal shear is simulated by iBEM. The viscosities of fluid matrix and particles are 1000 and 100000000, respectively. The three axes for rod-like asphaltene are $a_1 = a_2 = 0.03$ and $a_3 = 0.12$, and the three axes for disk-like asphaltene are $a_1 = a_2 = 0.12$ and $a_3 = 0.03$. Displacement control is applied as a virtual DSR experiment for measuring viscosity. The magnitude of the resulting sinusoidal velocity is applied as a function $u = 0.15r \sin\left(\frac{\pi}{3}t\right)$ at the top, where r is the distance from the applied point to the center of the top circular surface. The direction of the shear is tangential to the r direction. The bottom surface is fixed in all directions. The configuration is shown as in Fig. 8.13. 200 particles are generated randomly to simulate this 3-D problem. The simulation results are shown

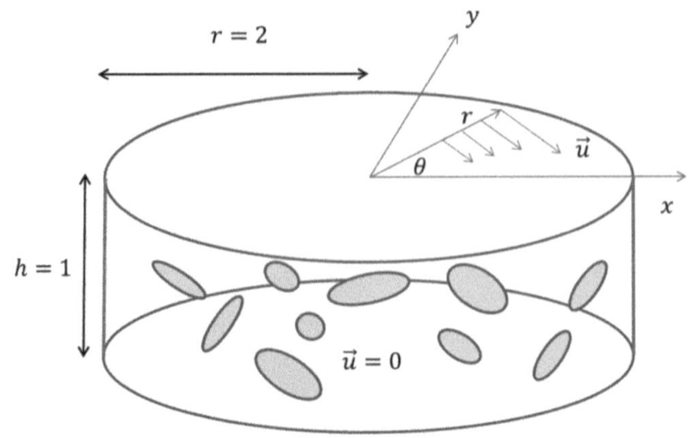

FIGURE 8.13 Schematic plot of configuration of the colloidal system under sinusoidal shear.

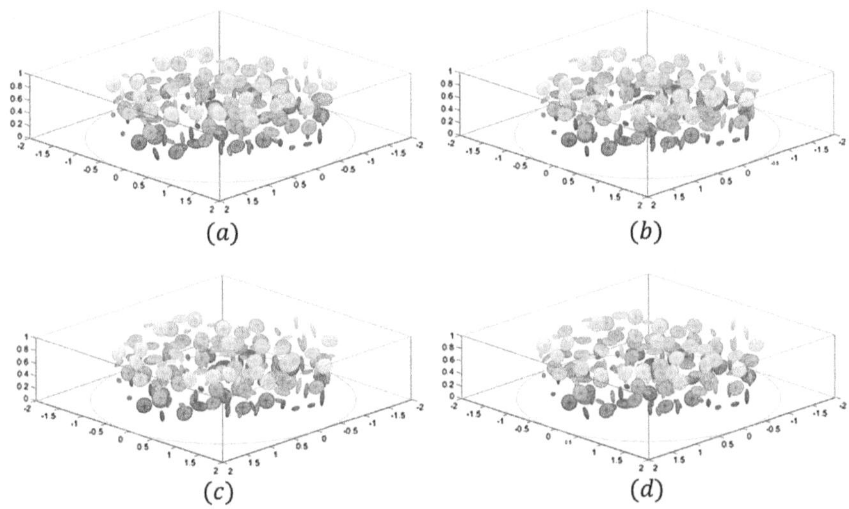

FIGURE 8.14 Snapshots of microstructure of the colloidal system during one period: (a) $t = 0$, (b) $t = T/3$, (c) $t = 2T/3$, (d) $t = T$.

in Fig. 8.14. Since the microstructure of asphalt binder with asphaltene evolves in time, the fluctuation of measured viscosity is observed as shown in Fig. 8.15. The trend of the curve is decreasing gradually, which also matches the common observation results from real experiments.

In this chapter, we present an application of iBEM for modeling and simulation of the colloidal system with rod and disk particles in asphalt binders. We derive an explicit formulation of iBEM considering particle–particle and particle–boundary interaction for elliptical particles with arbitrary orientations. The proposed formulas are verified with finite element results for two prolate,

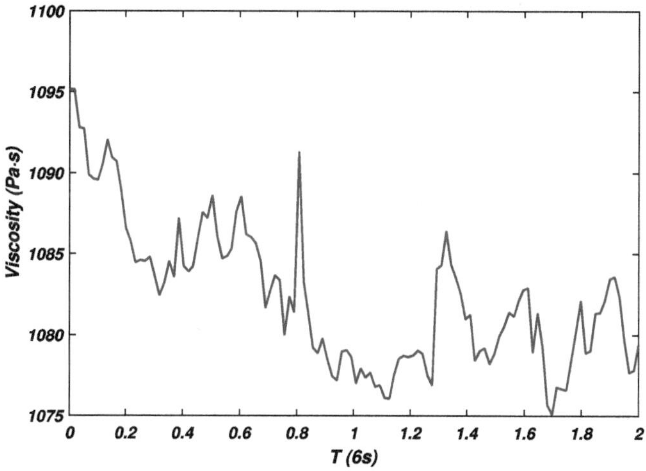

FIGURE 8.15 Simulated viscosity changing with time.

two oblate and one prolate, and one oblate cases. To investigate the evolution of microstructure of asphalt binders under the rheometer, the colloidal system with rod- and disk-shape asphaltene under sinusoidal shear is successfully efficiently simulated by iBEM. In the future, more experiments could be done to measure the effective viscosities of asphalt binders with respect to the different content of asphaltene. The microstructure evolution and homogenized viscosity predicted by iBEM could be validated with those experimental results.

Appendix 8.A Derivation and explicit expression of the tensors

In Section 8.3.1, the tensors F_{ijk}, D_{ijmn}, and D_{ijmnp} can be derived from the integrals of the following Green's function and its derivative:

$$\begin{aligned} G_{ij}\left(\mathbf{x},\mathbf{x}'\right) &= \frac{1}{4\pi\mu^0}\delta_{ij}\left(\phi-\bar{\phi}\right)-\frac{1}{8\pi\mu^0}\left(\psi_{,ij}-Q_{im}Q_{jn}\bar{\psi}_{,mn}\right) \\ &+\frac{x_3}{4\pi\mu^0}\left[Q_{jm}\bar{\psi}_{,im3}+2\delta_{j3}\bar{\phi}_{,i}-Q_{jm}x_3\bar{\phi}_{,im}\right] \end{aligned} \tag{8.76}$$

and

$$\begin{aligned} \Gamma^{v}_{ijm} &= \frac{1}{4\pi\mu^0}\delta_{ij}\left(-\phi_{,m}+Q_{qm}\bar{\phi}_{,q}\right)+\frac{1}{8\pi\mu^0}\left(\psi_{,ijm}-Q_{qm}Q_{ik}Q_{jl}\bar{\psi}_{,klq}\right) \\ &-\frac{x_3}{4\pi\mu^0}\left[Q_{qm}Q_{jk}\bar{\psi}_{,ik3q}+2\delta_{j3}Q_{qm}\bar{\phi}_{,iq}-Q_{qm}Q_{jk}x_3\bar{\phi}_{,ikq}\right] \end{aligned} \tag{8.77}$$

To explicitly derive the integrals, we used Φ and Ψ to denote the integrals of the functions ϕ and ψ over an ellipsoidal inclusion Ω:

$$\Phi = \int_\Omega \phi d\mathbf{x}', \ \Psi = \int_\Omega \psi d\mathbf{x}' \tag{8.78}$$

where ϕ and ψ are defined in Section 8.3.1. In the following, Φ_p and Ψ_p denote the integrals of $\phi x'_p$ and $\psi x'_p$. For a spherical domain with radius a, we can explicitly obtain the integral terms of Φ, Ψ, Φ_p, and Ψ_p and their derivative terms.

The tensors F_{ijk}, D_{ijmn}, and D_{ijmnp} are given as

$$F_{ijk} = \int_\Omega \Gamma^f_{ijk} d\mathbf{x}' = \frac{1}{2}\int_\Omega \left(G_{ik,j} + G_{jk,i}\right) d\mathbf{x}' = \frac{1}{2}\left(H_{ikj} + H_{jki}\right) \tag{8.79}$$

$$\begin{aligned} D_{ijmn} &= \int_\Omega \left(2\mu^0 \Gamma^e_{ijmn}\right) d\mathbf{x}' \\ &= \frac{\mu^0}{2}\int_\Omega \left(\Gamma^v_{imn,j} + \Gamma^v_{jmn,i} + \Gamma^v_{inm,j} + \Gamma^v_{jnm,i}\right) d\mathbf{x}' \\ &= \frac{1}{4}\left(P_{ijmn} + P_{ijnm} + P_{jimn} + P_{ijnm}\right) \end{aligned} \tag{8.80}$$

and

$$\begin{aligned} D_{ijmnp} &= \int_\Omega \left(2\mu^0 \Gamma^e_{ijmn} x'_p\right) d\mathbf{x}' \\ &= \frac{\mu^0}{2}\int_\Omega \left(\Gamma^v_{imn,j} + \Gamma^v_{jmn,i} + \Gamma^v_{inm,j} + \Gamma^v_{jnm,i}\right) x'_p d\mathbf{x}' \\ &= \frac{1}{4}\left(K_{pijmn} + K_{pijnm} + K_{pjimn} + K_{pijnm}\right) \end{aligned} \tag{8.81}$$

where the tensors H_{ikj}, P_{ijmn}, and K_{pijmn} are derived as follows. Considering that

$$\begin{aligned} G_{ik,j} &= \frac{1}{4\pi\mu^0}\delta_{ik}\left(\phi_{,j} - \bar{\phi}_{,j}\right) - \frac{1}{8\pi\mu^0}\left(\psi_{,ikj} - Q_{im}Q_{kn}\bar{\psi}_{,mnj}\right) \\ &+ \frac{x_3}{4\pi\mu^0}\left[Q_{jm}\bar{\psi}_{,im3j} + 2\delta_{k3}\bar{\phi}_{,ij} - Q_{km}x_3\bar{\phi}_{,imj} - \delta_{3j}Q_{km}\bar{\phi}_{,im}\right] \\ &+ \frac{\delta_{3j}}{4\pi\mu^0}\left[Q_{km}\bar{\psi}_{,im3} + 2\delta_{k3}\bar{\phi}_{,i} - Q_{km}x_3\bar{\phi}_{,im}\right] \end{aligned} \tag{8.82}$$

its integral is calculated as follows:

$$
\begin{aligned}
H_{ikj} = {} & \frac{1}{4\pi\mu^0}\delta_{ik}\left(\Phi_{,j} - \bar{\Phi}_{,j}\right) - \frac{1}{8\pi\mu^0}\left(\Psi_{,ikj} - Q_{im}Q_{kn}\bar{\Psi}_{,mnj}\right) \\
& + \frac{x_3}{4\pi\mu^0}\left[Q_{jm}\bar{\Psi}_{,im3j} + 2\delta_{k3}\bar{\Phi}_{,ij} - Q_{km}x_3\bar{\Phi}_{,imj} - \delta_{3j}Q_{km}\bar{\Phi}_{,im}\right] \\
& + \frac{\delta_{3j}}{4\pi\mu^0}\left[Q_{km}\bar{\Psi}_{,im3} + 2\delta_{k3}\bar{\Phi}_{,i} - Q_{km}x_3\bar{\Phi}_{,im}\right]
\end{aligned} \tag{8.83}
$$

Considering Eq. (8.77), we get

$$
\begin{aligned}
\Gamma^{v}_{imn,j} = {} & \frac{1}{4\pi\mu^0}\delta_{im}\left(-\phi_{,nj} + Q_{qm}\bar{\phi}_{,qj}\right) + \frac{1}{8\pi\mu^0}\left(\psi_{,imnj} - Q_{qn}Q_{ik}Q_{ml}\bar{\psi}_{,klqj}\right) \\
& - \frac{x_3}{4\pi\mu^0}\left[Q_{qn}Q_{mk}\bar{\psi}_{,ik3qj} + 2\delta_{m3}Q_{qn}\bar{\phi}_{,iqj} - Q_{qn}Q_{mk}x_3\bar{\phi}_{,ikqj}\right. \\
& \left. - \delta_{3j}Q_{qn}Q_{mk}\bar{\phi}_{,ikq}\right] \\
& - \frac{\delta_{3j}}{4\pi\mu^0}\left[Q_{qn}Q_{mk}\bar{\psi}_{,ik3q} + 2\delta_{m3}Q_{qn}\bar{\phi}_{,iq} - Q_{qn}Q_{mk}x_3\bar{\phi}_{,ikq}\right]
\end{aligned} \tag{8.84}
$$

We obtain the integrals as

$$
\begin{aligned}
\Gamma^{v}_{ijm} = {} & \frac{1}{4\pi\mu^0}\delta_{ij}\left(-\Phi_{,m} + Q_{qm}\bar{\Phi}_{,q}\right) + \frac{1}{8\pi\mu^0}\left(\Psi_{,ijm} - Q_{qm}Q_{ik}Q_{jl}\bar{\Psi}_{,klq}\right) \\
& - \frac{x_3}{4\pi\mu^0}\left[Q_{qm}Q_{jk}\bar{\Psi}_{,ik3q} + 2\delta_{j3}Q_{qm}\bar{\Phi}_{,iq} - Q_{qm}Q_{jk}x_3\bar{\Phi}_{,ikq}\right]
\end{aligned} \tag{8.85}
$$

and

$$
\begin{aligned}
\Gamma^{v}_{imn,j} = {} & \frac{1}{4\pi\mu^0}\delta_{im}\left(-\Phi_{,nj} + Q_{qm}\bar{\Phi}_{,qj}\right) \\
& + \frac{1}{8\pi\mu^0}\left(\Psi_{,imnj} - Q_{qn}Q_{ik}Q_{ml}\bar{\Psi}_{,klqj}\right) \\
& - \frac{x_3}{4\pi\mu^0}\left[Q_{qn}Q_{mk}\bar{\Psi}_{,ik3qj} + 2\delta_{m3}Q_{qn}\bar{\Phi}_{,iqj} - Q_{qn}Q_{mk}x_3\bar{\Phi}_{,ikqj}\right. \\
& \left. - \delta_{3j}Q_{qn}Q_{mk}\bar{\Phi}_{,ikq}\right] \\
& - \frac{\delta_{3j}}{4\pi\mu^0}\left[Q_{qn}Q_{mk}\bar{\Psi}_{,ik3q} + 2\delta_{m3}Q_{qn}\bar{\Phi}_{,iq} - Q_{qn}Q_{mk}x_3\bar{\Phi}_{,ikq}\right]
\end{aligned} \tag{8.86}
$$

Chapter 9

The iBEM for time-dependent loads and material behavior

Dynamics, transient heat conduction, and viscoelasticity

In this chapter, we extend the iBEM applications to time-dependent loads and material properties, including transient or dynamic problems and viscoelastic material behavior, in which the scalar or vector potential changes with temporal and spatial variables. For time-dependent loads, the time-dependent Green's functions are introduced in the iBEM to consider their variations with time. For harmonic variation with time, the method of separation of variables can be used, and a periodic solution is obtained. An elastic problem has been used to demonstrate the process, which can be applied to periodic heat conduction problems as well. For general transient or dynamic problems, both time and spatial integrals are needed. Numerical implementation has been demonstrated for phase change material (PCM) based building envelopes. For viscoelastic material behavior, the Fourier or Laplace transforms can be used to transform the time-dependent governing equations to the same form of elastic equations through the elastic–viscoelastic correspondence principle. Therefore the iBEM can be extended to time-dependent loads and materials.

9.1 Harmonic vibration with time

When composite materials are in working condition under a harmonic load, such as vibration in mechanical systems or daily temperature change in thermal systems, the local fields exhibit periodic behavior. Due to the existence of the inhomogeneities, the local field distribution is disturbed by the material mismatch. However, it is still periodic and can be represented by the method of separation of variables. This section focuses on the elastic analysis, which will be extended to thermal analysis subsequently.

9.1.1 Fundamental solution for harmonic excitation

Consider a domain D subjected to harmonic surface tractions $\bar{\mathbf{t}}$ and prescribed displacements $\bar{\mathbf{u}}$ as in Fig. 9.1. There are multiple subdomains Ω^I with different material properties λ^I, μ^I, and ρ^I, in which λ^I and μ^I are the Lame constants,

The Inclusion-Based Boundary Element Method (iBEM). https://doi.org/10.1016/B978-0-12-819384-6.00017-6

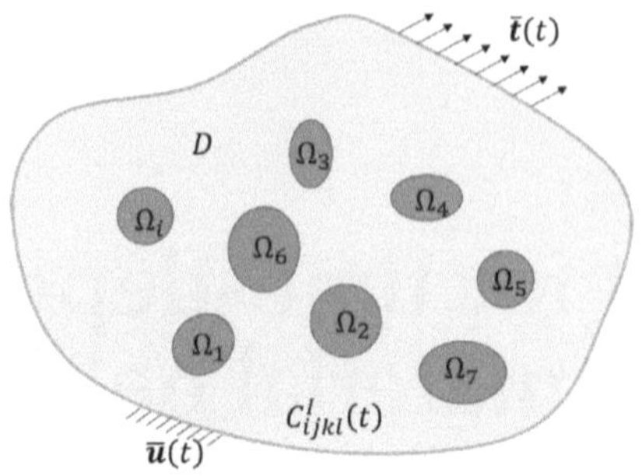

FIGURE 9.1 Schematic illustration of multiple inhomogeneities embedded in a domain D under harmonic surface traction and displacements.

ρ^I is the density, and $I = 0, \ldots, n$ denote the material phases. The material is assumed to be isotropic. The stiffness tensor C^I_{ijkl} can be expressed by the Lame constants, i.e., $C^I_{ijkl} = \lambda^I \delta_{ij}\delta_{kl} + \mu^I(\delta_{ik}\delta_{jl} + \delta_{il}\delta_{jk})$. For an elastic material system, the harmonic surface traction $\bar{\mathbf{t}}$ and prescribed displacements $\bar{\mathbf{u}}$ can be written in the forms

$$\bar{t}_i(\mathbf{x}, t) = t_i(\mathbf{x}) \exp(-i\omega t), \quad \bar{u}_i(\mathbf{x}, t) = u_i(\mathbf{x}) \exp(-i\omega t) \tag{9.1}$$

where the dynamic response can be decoupled by the spatial dependent term and the harmonic term with frequency ω of the harmonic load. The latter part can be solved by the governing equation in the absence of body force as follows:

$$C^I_{ijkl} u_{k,li}(\mathbf{x}) = \rho \ddot{u}_j \tag{9.2}$$

where $\ddot{u}$ represents the second time derivative of u or acceleration. The periodic solution of $\mathbf{u}$ can be written as $u_i = u_i(\mathbf{x})\exp(-i\omega t)$, so that the above governing equation is written as

$$(\lambda^I + \mu^I)u_{k,kj}(\mathbf{x}) + \mu^I u_{j,kk}(\mathbf{x}) = -\rho^I \omega^2 u_j \tag{9.3}$$

By using Eshelby's EIM the material inhomogeneities can be simulated by harmonic eigenstrain of $\varepsilon^*_{ij}(\mathbf{x})\exp(-i\omega t)$ as well. The inertia force caused by the density difference can be simulated by a fictitious force $f^*_i = (\rho^0 - \rho^I)\ddot{u}_i(\mathbf{0})$, where the acceleration at the particle center is used to represent the overall particle acceleration. Therefore the above governing equation can be written with homogeneous stiffness of C^0_{ijkl} and density ρ^0 as follows:

$$(\lambda^0 + \mu^0)u_{k,kj}(\mathbf{x}) + \mu^0 u_{j,kk}(\mathbf{x}) + \rho^0\omega^2 u_j(\mathbf{x}) + f'_j(\mathbf{x}) = 0 \tag{9.4}$$

where the source field $f'_j(\mathbf{x})$ includes the eigenstrain and the fictitious force on the particle domain, which can be determined by stress equivalent conditions for each point later as

$$f_j'(\mathbf{x}) = -\left[\lambda^0 \varepsilon^*_{kk,j}(\mathbf{x}) + 2\mu^0 \varepsilon^*_{kj,k}(\mathbf{x})\right] + f_j^*(\mathbf{x}) \tag{9.5}$$

Eq. (9.4) can be rewritten as

$$(\lambda^0 + 2\mu^0)\nabla(\nabla \cdot \mathbf{u}) - \mu^0 \nabla \times \nabla \times \mathbf{u} + \rho^0 \omega^2 \mathbf{u} + \mathbf{f}' = 0 \tag{9.6}$$

The Green's function for the new governing equation with unit harmonic point force $\mathbf{f}' = \mathbf{f}^1 \delta(\mathbf{x} - \mathbf{x}')$ with $|\mathbf{f}^1| = 1$ can be written as

$$G_{ij} = \frac{1}{4\pi\rho\omega^2}\left[\beta^2 \delta_{ij} \frac{e^{i\beta r}}{r} - \frac{\partial^2}{\partial x_i \partial x_j}\left(\frac{e^{i\alpha r}}{r} - \frac{e^{i\beta r}}{r}\right)\right] \tag{9.7}$$

where $r = |\mathbf{x} - \mathbf{x}'|$, the distance between the field point $\mathbf{x}$ and source point $\mathbf{x}'$, $\alpha = \sqrt{\frac{\rho^0}{\lambda^0 + 2\mu^0}}\omega$, and $\beta = \sqrt{\frac{\rho^0}{\mu^0}}\omega$. The derivation of the elastodynamic fundamental solution is elaborated in Appendix 9.A. Therefore the displacement caused by the load on the boundary and source on the particles can be written as

$$\begin{aligned} u_i\left(\mathbf{x}'\right) &= \sum_{I=1}^{N} \int_{\Omega^I} G_{ji}\left(\mathbf{x}, \mathbf{x}'\right) f_j'(\mathbf{x}) d\mathbf{x} \\ &\quad + \int_{\partial D} G_{ji}\left(\mathbf{x}, \mathbf{x}'\right) \bar{t}_j(\mathbf{x}) - T_{ji}\left(\mathbf{x}, \mathbf{x}'\right) \bar{u}_j(\mathbf{x})\, \mathrm{d}\mathbf{x} \end{aligned} \tag{9.8}$$

where $T_{ji}\left(\mathbf{x}, \mathbf{x}'\right)$ is the traction tensor, which can be directly derived from Green's function with the aid of constitutive law as follows:

$$T_{ij} = \frac{1}{2} n_m C_{imkl}\left(G_{kj,l} + G_{lj,k}\right) \tag{9.9}$$

which is also the fundamental solution of traction in the ith direction at the boundary point x for time-harmonic elastodynamic equation with a point force $\delta\left(x'\right)$ acting at point x' in the jth direction.

9.1.2 Inclusion problem for the infinite domain

Eq. (9.8) includes the integral from both the particles and the boundary. Consider an infinite domain D containing a nonmechanical strain, or eigenstrain ϵ^*_{ij}, and a fictitious force f_i^* in subdomain Ω, which is treated as a new source of a body force f_i' distributed in Ω, in Eq. (9.5).

By the Green's function techniques the displacement vector at any spatial point $\mathbf{x}$ in the infinite domain D is

$$
\begin{aligned}
u_m(\mathbf{x}) &= \int_{\Omega} \left(-C_{ijkl}\epsilon^*_{kl,i'} G_{mj}(\mathbf{x},\mathbf{x}') + f^*_j G_{mj}(\mathbf{x},\mathbf{x}')\right) \mathrm{d}\mathbf{x}' \\
&= \int_{\partial\Omega} -C_{ijkl}\epsilon^*_{kl} G_{mj}(\mathbf{x},\mathbf{x}') n'_i \mathrm{d}\mathbf{x}' + \int_{\Omega} C_{ijkl}\epsilon^*_{kl} G_{mj,i'}(\mathbf{x},\mathbf{x}') \mathrm{d}\mathbf{x}' \\
&+ \int_{\Omega} f^*_j G_{mj}(\mathbf{x},\mathbf{x}') \mathrm{d}\mathbf{x}' \\
&= \int C_{ijkl}\epsilon^*_{kl} G_{mj,i'}(\mathbf{x},\mathbf{x}') \mathrm{d}\mathbf{x}' + \int_{\Omega} f^*_j G_{mj}(\mathbf{x},\mathbf{x}') \mathrm{d}\mathbf{x}' \\
&= \frac{\partial}{\partial x_i} \int -C_{ijkl}\epsilon^*_{kl} G_{mj}(\mathbf{x},\mathbf{x}') \mathrm{d}\mathbf{x}' + \int_{\Omega} f^*_j G_{mj}(\mathbf{x},\mathbf{x}') \mathrm{d}\mathbf{x}' \\
&= J_{mkl}\epsilon^*_{kl} + L_{ij} f^*_j
\end{aligned}
\tag{9.10}
$$

Here we applied Gauss's theorem and also two facts, $\epsilon^*_{kl} = 0$ on ∂D and $\frac{\partial G_{ij}(x-x')}{\partial x'} = -\frac{\partial G_{ij}(x-x')}{\partial x}$. Define $\Psi(\alpha, x) = \int_{\Omega} \frac{e^{i\alpha r}}{r} dx'$, which is the integral of Helmholtz potential [115]. We can obtain the explicit forms of tensors J_{mkl} and L_{ij}:

$$
J_{mkl} = -C_{ijkl} \frac{1}{4\pi\rho\omega^2} \left(\beta^2 \delta_{mj} \Psi(\beta, x)_{,i} - \Psi(\alpha, x)_{,mji} + \Psi(\beta, x)_{,mji}\right)
\tag{9.11}
$$

$$
L_{mj} = \frac{1}{4\pi\rho\omega^2} \left(\beta^2 \delta_{mj} \Psi(\beta, x) - \Psi(\alpha, x)_{,mj} + \Psi(\beta, x)_{,mj}\right)
\tag{9.12}
$$

The integral of the Helmholtz potential is closely related to the dynamic Eshelby tensor [115]. Following Mikata and Nemat-Nasser [116], the explicit form of $\Psi(\alpha, x)$ over a spherical domain Ω with radius a is

$$
\Psi(\alpha, x) = \begin{cases} \frac{4\pi}{\alpha^2}\left[e^{i\alpha r}(1 - i\alpha a)\frac{\sin(\alpha r)}{\alpha r} - 1\right], & x \in \Omega \\ \frac{4\pi e^{i\alpha r}}{\alpha^3 r}\left[\sin(\alpha a) - \alpha a \cos(\alpha a)\right], & x \notin \Omega \end{cases}
\tag{9.13}
$$

By combining Eqs. (9.10)–(9.13) can obtain the total disturbed displacement field by the eigenstrain and fictitious force for the inclusion problem.

9.1.3 Inclusion problem for a bounded domain (time harmonic)

For a bounded domain, the boundary effect is calculated by the boundary integral. To numerically solve this boundary value problem by BEM, we introduce the shape function N_{ij} to discretize the surface traction and displacements by $t_j = N_{jm} t^d_m$, $u_j = N_{jm} u^d_m$. The superscript d denotes the value of tractions and displacements on nodes in that element e. So after discretization, we can write the boundary integral in Eq. (9.8) in the discretized form

$$u_i^{BEM} = \sum_{e=1}^{Ne} U_{ij} t_j^d - \sum_{e=1}^{Ne} H_{ij} u_j^d \tag{9.14}$$

where

$$U_{ij} = \int_{\partial D^e} G_{mi}\left(\mathbf{x}, \mathbf{x}'\right) N_{mj}\left(\mathbf{x}\right) \mathrm{d}\mathbf{x} \tag{9.15}$$

and

$$H_{ij} = \int_{\partial D^e} T_{mi}\left(\mathbf{x}, \mathbf{x}'\right) N_{mj}\left(\mathbf{x}\right) \mathrm{d}\mathbf{x} \tag{9.16}$$

Basically, the matrices U and H in each element are first formed and then assembled into the global matrix. They are calculated as the integrals of the kernel shape function product [21] in each element. Ne is the total number of elements. Notice that the Green's function in Eq. (9.7) is only applicable to a homogeneous material. Following the conventional BEM, the microvoid surface should be considered as a part of the boundary, which may require a fairly large number of elements to describe the geometry of the void surfaces. However, I-BEM will treat the microvoid as the same material as the matrix but introduce eigenstrain and fictitious force to represent the material mismatch. Consider a bounded domain D containing n multiple inclusions Ω_i with continuously distributed eigenstrain ε_{ij}^I and fictitious force f^I, also subjected to prescribed boundary conditions u and t. By superposition the total displacement at any point inside D can be presented as

$$u_i = \sum_{e=1}^{Ne} U_{ij} t_j^d - \sum_{e=1}^{Ne} H_{ij} u_j^d + \sum_{I=1}^{n} J_{mkl} \epsilon_{kl}^{*I} + \sum_{I=1}^{n} L_{mj} f_j^{*I} \tag{9.17}$$

By the relation between displacement and strain for the small deformation the strain can be obtained from the displacement field as

$$\epsilon_{im} = \sum_{e=1}^{Ne} U'_{imj} t_j^d - \sum_{e=1}^{Ne} H'_{imj} u_j^d + \sum_{I=0}^{n} M_{imkl} \epsilon_{kl}^{*I} + \sum_{I=0}^{n} N_{imj} f_j^{*I} \tag{9.18}$$

where

$$U'_{imj} = \int_{\partial D^e} -\frac{1}{2}\left(G_{ni,m} + G_{nm,i}\right) N_{nj}\left(\mathbf{x}\right) \mathrm{d}\mathbf{x} \tag{9.19}$$

$$H'_{imj} = \int_{\partial D^e} -\frac{1}{2}\left(T_{ni,m} + T_{nm,i}\right) N_{nj}\left(\mathbf{x}\right) \mathrm{d}\mathbf{x} \tag{9.20}$$

$$M_{mnkl} = \frac{-C_{ijkl}}{8\pi\rho\omega^2}\Big[\beta^2\delta_{mj}\Psi(\beta,x)_{,in} + \beta^2\delta_{nj}\Psi(\beta,x)_{,im} - 2\Psi(\alpha,x)_{,mjin} + 2\Psi(\beta,x)_{,mjin}\Big] \tag{9.21}$$

and

$$N_{ijm} = \frac{1}{8\pi\rho\omega^2}\Big[\beta^2\delta_{im}\Psi(\beta,x)_{,j} + \beta^2\delta_{jm}\Psi(\beta,x)_{,i} - 2\Psi(\alpha,x)_{,ijm} + 2\Psi(\beta,x)_{,ijm}\Big] \tag{9.22}$$

9.1.4 Equivalent inclusion method for the bounded domain (time harmonic)

The equivalent inclusion method [2,3] can convert the original inhomogeneity problem to inclusion problem by introducing an eigenstrain and fictitious force, which can be solved by the equivalent inclusion condition [1]. On each inhomogeneity, the stress equivalence and force equivalence are provided as follows:

$$C^I_{ijkl}\epsilon_{kl}(\mathbf{0}) = C^0_{ijkl}\left(\epsilon_{kl}(\mathbf{0}) - \epsilon^{*I}_{kl}\right) \quad \text{for } I = 1:n \tag{9.23}$$

$$\rho^I \ddot{u}_i(\mathbf{0}) = \rho^0 \ddot{u}_i(\mathbf{0}) - f^{*I}_i \quad \text{for } I = 1:n \tag{9.24}$$

where C^I_{ijkl} and C^0_{ijkl} are the elastic tensors for the Ith inhomogeneity and matrix, respectively, and ρ^I and ρ^0 are the densities for the Ith inhomogeneity and matrix, respectively. The equivalent inclusion condition for the static problem can be obtained directly by making the density ρ equal to zero. The global equations for each collocation node and each inclusion can be formed based on EIM and BEM. The linear system can be solved in one shot. After obtaining the eigenstrain, fictitious forces, and unknown boundary conditions, the displacement and strain field can be calculated by Eqs. (9.17) and (9.18). Substituting Eqs. (9.17) and (9.18) into Eqs. (9.23) and (9.24), we obtain

$$\Delta C^M_{ijkl}\left(\sum_{e=1}^{Ne} U'_{klm}t^d_m - \sum_{e=1}^{Ne} H'_{kln}u^d_m + \sum_{I=1}^{n} M^I_{klpq}\epsilon^{*I}_{pq} + \sum_{I=1}^{n} N^I_{kls}f^{*I}_s\right) + C^0_{ijrz}\epsilon^{*M}_{rz} = 0 \tag{9.25}$$

$$-\omega^2\Delta\rho^M\left(\sum_{e=1}^{Ne} U_{im}t^d_m - \sum_{e=1}^{Ne} H_{im}u^d_m + \sum_{I=1}^{n} J^I_{ipq}\epsilon^{*I}_{pq} + \sum_{I=1}^{n} L^I_{is}f^{*I}_s\right) + f^{*M}_i = 0 \tag{9.26}$$

where $\Delta C^M_{ijkl} = C^M_{ijkl} - C^0_{ijkl}$ and $\Delta\rho^M = \rho^M - \rho^0$ with $M = 1, 2, \ldots, N$. In addition, the displacement on each node [21] can be written as

$$cu_i^d - \left(\sum_{e=1}^{Ne} U_{im} t_m^d - \sum_{e=1}^{Ne} H_{im} u_m^d + \sum_{I=1}^{n} J_{ipq}^I \epsilon_{pq}^{*I} + \sum_{I=1}^{n} L_{is}^I f_s^{*I}\right) = 0 \quad (9.27)$$

where $d = 1, 2, \ldots, N$.

We can rewrite the above three equations in the matrix form by listing the tensors into vectors as follows:

$$\begin{bmatrix} c+H & -J & -L \\ -\Delta CH' & \Delta CM & \Delta CN \\ \omega^2\Delta\rho H & -\omega^2\Delta\rho J & -\omega^2\Delta\rho L \end{bmatrix} \begin{bmatrix} u \\ \epsilon^* \\ f^* \end{bmatrix} - \begin{bmatrix} U \\ -\Delta CU' \\ \omega^2\Delta\rho U \end{bmatrix} [t] = 0 \quad (9.28)$$

From the matrix form it is clear that besides the traditional BEM and EIM terms on the diagonal blocks, the new coupling terms $-J$ and $-L$ describe the effect of the inclusions on the boundary terms, and the terms $-\Delta CH'$, $-\Delta CU'$, $\omega^2\Delta\rho H$, and $\omega^2\Delta\rho U$ are critical for incorporating the boundary effect into the EIM formula. After applying the boundary conditions, a linear equation system can be solved. Then using Eqs. (9.17) and (9.18), the displacement and strain field can be derived. By the constitutive relation the stress field can be obtained as well from the constitutive law. Some numerical implementation and results can be found in Song and Yin [39].

Note that for arbitrarily shaped particles, a similar process of the static elastic analysis can be applicable, and domain discretization can provide tailorable accuracy to the solution under harmonic excitation.

9.1.5 Extension to heterogeneous materials under harmonic temperature variation

The temperature change with time in nature often exhibits a repetitive pattern with a period in day or year, which can be approximated by a harmonic temperature variation with the average temperature superposing a sinusoidal function. When the material system is heterogeneous, the iBEM can be a powerful tool to solve the temperature distribution. For example, the ground temperature profile changing with depth and seasons is a good application.

A similar procedure of the above elastic analysis under harmonic excitation can be extended to a heterogeneous material system under harmonic temperature variation of the boundary points. We provide a brief description as follows.

The governing equation for the transient heat conduction problem can be written as

$$\frac{\partial u}{\partial t} = \alpha^I u_{,ii} + \frac{q}{c_p} \quad (9.29)$$

where α^I is the heat diffusivity of the material phase I, which depends on the thermal conductivity k^I and the specific heat c_p^I based on the volume as $\alpha = \frac{k}{c_p}$.

Consider a homogeneous material with thermal diffusivity α under a harmonic thermal loading $q(\mathbf{x})\exp(-i\omega t)$. The temperature field u can be written in terms of the spatial and time parts by the method of separation of variables as

$$u(\mathbf{x}, t) = u^x(\mathbf{x})\exp(-i\omega t) \tag{9.30}$$

and we can rewrite the governing equation as

$$\alpha^I u^x_{,ii}(\mathbf{x}) + i\omega u^x(\mathbf{x}) + \frac{q(\mathbf{x})}{c_p} = 0 \tag{9.31}$$

Given a harmonic point heat source at **x**', say $\delta(\mathbf{x} - \mathbf{x}')\exp(-i\omega t)$, the Green's function can be written as follows (Appendix 9.B):

$$G(\mathbf{x},\mathbf{x}') = \frac{\exp\left[-(1-i)\sqrt{\frac{\omega}{2\alpha}}r\right]}{4\pi c_p r} \tag{9.32}$$

where $r = |\mathbf{x} - \mathbf{x}'|$. The temperature field compared with a reference temperature can be indicated by the Green's function as

$$\begin{aligned} u(\mathbf{x}, t) &= Re\left(\int_{-\infty}^{\infty} G(\mathbf{x},\mathbf{x}')q(\mathbf{x}')d\mathbf{x}' \exp(-i\omega t)\right) \\ &= Re\left(\int_{-\infty}^{\infty} \frac{\exp\left[-i\omega t - (1-i)\sqrt{\frac{\omega}{2\alpha}}r\right]}{4\pi c_p r} q(\mathbf{x}')d\mathbf{x}'\right) \end{aligned} \tag{9.33}$$

where $Re(\cdot)$ means the real part of a function. Therefore, for a composite material system under a harmonic thermal loading condition, the problem can be solved in the same fashion as the elastic problem with the EIM and iBEM.

In addition, for 2D or 1D problems, the Green's functions have been provided in Appendix 9.B. The same procedure can be used for iBEM implementation.

9.2 Transient heat conduction problems

For harmonic loading, the solution is harmonic and can be expressed in terms of the two parts with explicitly given time-dependent behavior. Therefore the governing equation can be reduced to solving a spatial variable. When the load is not periodic in time, the spatial and temporal variables will be coupled. The problem becomes a general dynamic elastic problem or a transient heat conduction problem, which can still be solved by the Green's function technique. The Green's function for an impulse point source can be obtained by the Fourier integral of the Green's function for the harmonic loading over ω. In this section, we use the transient heat conduction problem to demonstrate the application of iBEM to general time-dependent loading conditions. The dynamic problem can be solved by iBEM in a similar fashion with the Green's function provided in Appendix 9.A.

9.2.1 Boundary element method for transient heat conduction

Consider a spatial domain D subjected to boundary conditions such as a heat flux of q_{bc} and a prescribed temperature of u_{bc} during time domain Ω with given initial condition (Fig. 9.2).

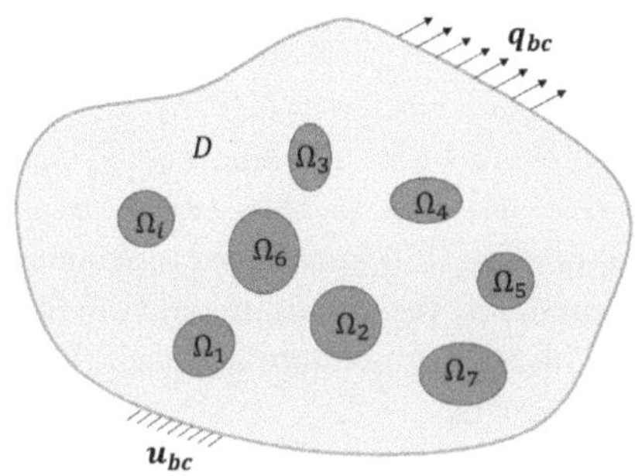

FIGURE 9.2 Schematic illustration of multiple inhomogeneities embedded in a domain $\mathcal{D}$ under heat flux and temperature.

Using the Green's function technique, the temperature field at $\mathbf{x}$ and t can be expressed by the boundary integral over the time domain as [21,35,117]

$$\begin{aligned} u(\mathbf{x}, t) = & \int_D G(\mathbf{r}, t) u^0 \, d\mathbf{x}' + \frac{1}{C_p} \int_\Lambda \int_\Omega G(\mathbf{r}, \tau) Q(\mathbf{x}', t') \, dt' \, d\mathbf{x}' \\ & + \alpha \int_{\partial D} \int_\Omega G(\mathbf{r}, \tau) q_{bc} dt' \, d\mathbf{x}' - \alpha \int_{\partial D} \int_\Omega T(\mathbf{r}, \tau) u_{bc} \, dt' \, d\mathbf{x}' \end{aligned} \tag{9.34}$$

where the transient Green's function in the infinite domain and its derivative are written as

$$\begin{aligned} G(\mathbf{r}, \tau) &= \frac{1}{(4\pi\alpha\tau)^{\frac{3}{2}}} e^{\frac{-r^2}{4\alpha\tau}} H(\tau) \\ T(\mathbf{r}, \tau) &= -\frac{2\pi r_i n_i}{(4\pi\alpha\tau)^{\frac{5}{2}}} e^{\frac{-r^2}{4\alpha\tau}} H(\tau) \end{aligned} \tag{9.35}$$

where $\tau = t - t'$, $\alpha = \frac{k}{C_p}$, and C_p and k denote the specific heat on unit volume and thermal conductivity, respectively. The superscript 0 represents the initial condition, Q and Λ are the heat source and its domain, $H(\tau)$ is the Heaviside function, the Green's function $G(\mathbf{r}, \tau)$ is also called the fundamental solution due to an impulse heat source at $\mathbf{x}'$ and time t', $T(\mathbf{r}, \tau)$ is the derivative of the

Green's function with respect to the surface normal **n** [117], and the first term in Eq. (9.34) involves the initial temperature field.

To numerically evaluate the boundary integral in Eq. (9.34), Gupta [117] interchanged the integral sequence and derived the analytical form with respect to time Ω from t_0 to t_F, where t_0 and t_F are the start time and ending time (or the observing time). Discretize Ω by N_F time steps with time coordinates t_i, $i = 0, 1, 2, \ldots, N_F$, and the time intervals $t_i - t_{i-1}$ $(i = 1, 2, \ldots, N_F)$ are not necessarily the same, although in this section, we use equal time intervals for convenience. Assume that the thermal responses remain the same over one time step, and thus the influences from thermal source during the time interval from t_{f-1} to t_f and location $\mathbf{x}'$ are handled by the time integral of the Green's function over the time interval of $t_f - t_{f-1}$. The time integrals of G and T defined in Eq. (9.35) for each time step are written as $\overline{G}(\mathbf{r}, t_\Omega, \Delta t)$ and $\overline{T}(\mathbf{r}, t_\Omega, \Delta t)$, where t_Ω is the duration of the overall time domain, i.e., $t_\Omega = t_F - t_f$, and the time interval is defined as $\Delta t = t_f - t_{f-1}$. They can be explicitly expressed with the incomplete gamma function (IGF) presented in Appendix 9.C.1. Eq. (9.117) can be numerically integrated by the approximation of temperature and heat flux fields of each element with the temperature and heat flux field provided by the corresponding shape function as

$$
\begin{aligned}
u(\mathbf{x}, t) = & \int_D G(\mathbf{r}, t) u^0 \, d\mathbf{x}' + \frac{1}{C_p} \int_\Lambda \int_\Omega G(\mathbf{r}, \tau) Q(\mathbf{x}', t') \, dt \, d\mathbf{x}' \\
& + \sum_{f=1}^{N_F} \sum_{l=1}^{N_e} \sum_{i=1}^{N_n} (U_i q_i^{bc})_f - \sum_{f=1}^{N_F} \sum_{l=1}^{N_e} \sum_{i=1}^{N_n} (H_i u_i^{bc})_f
\end{aligned} \tag{9.36}
$$

where N_e and N_n are the numbers of elements and nodes, respectively, in the corresponding element, U_i and H_i are the boundary integrals with the specific shape functions, i.e., $(U_i)_f = \alpha \int_{\partial D} \overline{G}(\mathbf{r}, t_\Omega, \Delta t) N_i \, ds$ and $(H_i)_f = \alpha \int_{\partial D} \overline{T}(\mathbf{r}, t_\Omega, \Delta t) N_i \, ds$, where N_i is the shape function, and the subscript f denotes the thermal source active time interval during time t_{f-1} and t_f. Based on the concept of Green's function, the heat fields at any future time are dependent on the past heat quantities. It is necessary to store the previous matrices and boundary solutions.

9.2.2 Inclusion problem for infinite and finite domain

Without the loss of generality, consider an infinite homogeneous domain, D_{inf}, containing an eigentemperature gradient ϵ_i^* and heat source Q^* at the subdomain Λ_I. Here the eigentemperature gradient denotes a prescribed temperature gradient that does not contribute to any heat flux, i.e., $q_i = -k(u_{,i} - \epsilon_i^*)$ [60,118]. Now introduce a fictitious quantity $Q' = Q^* - k\epsilon_{i,i}^*$, which is an equivalent heat source caused by ϵ_i^* and Q^*. The temperature field caused by the source Q' in Eq. (9.34) can be rewritten as [35,39]

$$u(\mathbf{x},t) = \frac{\alpha}{k}\sum_{f=1}^{N_F}\Big[\Big[\int_{\Lambda_I}\overline{G}(\mathbf{r},t_\Omega,\Delta t)_f\,d\mathbf{x}'(Q^*)_f - \int_{\partial\Lambda_I}\overline{G}(\mathbf{r},t_\Omega,\Delta t)_f k n_i'\,dS\,(\epsilon_i^*)_f - \frac{\partial}{\partial x_i}\int_{\Lambda_I}\overline{G}(\mathbf{r},t_\Omega,\Delta t)_f\,k\,d\mathbf{x}'\,(\epsilon_i^*)_f\Big] \quad (9.37)$$

where the subscript f represents the time step of the thermal quantities existing during time from t_{f-1} and t_f. Here only the second term in Eq. (9.34) exists, and the Gauss theorem is used to transfer the domain integral to the boundary of the subdomain Λ_I. In Eq. (9.37) the fictitious quantity only exists in the subdomain Λ_I, and the contour integral at the outer surface yields zero. Therefore only the domain integral terms of the first and third terms in Eq. (9.37) exist. Introduce the domain integral terms as

$$\begin{aligned} M_f &= \frac{\alpha}{k}\int_{\Lambda_I}\overline{G}(\mathbf{r},t_\Omega,\Delta t)_f\,d\mathbf{x}' \\ J_{f,i} &= -\frac{\partial}{\partial x_i}\alpha\int_{\Lambda_I}\overline{G}(\mathbf{r},t_\Omega,\Delta t)_f\,d\mathbf{x}' \end{aligned} \quad (9.38)$$

where M_f and $J_{f,i}$ denote the influence of the eigentemperature gradient and heat source of the inclusion (Λ_I) existing during time t_{f-1} and t_f in an infinite domain. Combining the ordinary BEM (9.36) and the disturbed field by fictitious quantity (9.37), the temperature field at $\mathbf{x}$ ($x \in D$) can be presented as

$$u(\mathbf{x},t) = \sum_{f=1}^{N_F}\Big(\sum_{l=1}^{N_e}\sum_{i=1}^{N_n}[(U_i\,q_i^{bc}) - (H_i\,u_i^{bc})]_f + \sum_{I=1}^{N_S}(M_f\,Q^{I*} + J_{f,i}\,\epsilon_i^{I*})\Big) + \int_D G(\mathbf{r},t)u^0\,d\mathbf{x}' \quad (9.39)$$

where N_S denotes the number of subdomains. The above equation is based on the homogeneous material with boundary effects and heat sources on a number of subdomains or inclusions. However, actual composites contain phase change material (PCM) particles that are different from the matrix. The same Green's function cannot be used for both PCM particles and the matrix due to the different governing equations. Traditional BEM uses boundary integrals over each material domain with boundary continuities, which requires numerical boundary integrals on the subdomains with high computational costs. In the following, iBEM is introduced to use Eq. (9.39) to avoid the numerical boundary integral of the subdomains.

9.2.3 Phase change process

Following Eshelby [60], PCM capsules can be treated as inclusions with eigentemperature gradient and heat source, which can be determined by the equivalence of heat flux to simulate the material mismatch. One challenge involved in the modeling is that PCM specific heat increases during phase change. Hsiao [119] proposed a piecewise specific heat function in terms of temperature for numerical simulation of the phase change process. For a pure material, it has a sharp freezing/melting temperature, and thus a small temperature interval is desirable and vice versa. In general, the PCM capsules used in the building envelopes feature a comparatively larger temperature interval, measured by differential scanning calorimetry (DSC) [120]:

$$C_p(u) = \begin{cases} C_{p,s} & \text{if } u < u_s \\ \frac{\Delta H_m}{u_h - u_l} + \frac{C_{p,s}+C_{p,l}}{2} & \text{if } u_s \le u \le u_e \\ C_{p,l} & \text{if } u_e < u \end{cases} \tag{9.40}$$

where $C_{p,s}$ and $C_{p,l}$ are specific heats for the solid and liquid phases of PCM, u_s and u_e are the temperatures at which the phase changing starts and ends, respectively, and ΔH_m is the latent heat fusion per unit mass of PCM. Since the volume of each PCM capsule is small, one assumption was made that the entire status of the particle can be described by the spatial-averaged temperature of sampling points. Therefore the heat capacity of the entire particle can be considered uniform.

9.2.4 Inclusion-based boundary element method for transient heat conduction with phase change process

In Section 9.2.2, we built an equation for the bounded inclusion problem with eigentemperature gradient and fictitious heat source. EIM treats each particle with different thermal conductivity the same material as the matrix, but an eigentemperature gradient and a fictitious heat source are introduced to simulate the effect of material difference on the heat transfer. Therefore the boundary value problem for a finite domain with many inhomogeneities can be solved with a uniform material with the corresponding inclusions of source field of eigentemperature gradient and heat source. The iBEM formula combines BEM and EIM using the equivalent flux condition within the inhomogeneity domain Λ_I. For each inhomogeneity, two additional thermal quantities, eigentemperature gradient and fictitious heat source, are introduced to describe the material mismatch and phase change process.

9.2.5 Implementation of EIM

Consider a bounded domain D with N_S embedded PCM inhomogeneities Λ_I, $I = 1, 2, 3, \ldots, N_S$. The material properties of matrix and inhomogeneities are written as k_0, C_p^0 and k_I, C_p^I, respectively, where k and C_p are the thermal conductivity and specific heat, respectively. Notice that here the specific heat is based on a unit volume and can be transferred from mass-based specific heat by multiplying density. The equivalent conditions are built upon the heat flux field within the inhomogeneity to match the thermal conductivity difference and product of density and specific heat at each time step as

$$\begin{aligned} -k_0(u_{,i} + \epsilon_i' - \epsilon_i^*) &= -k_I(u_{,i} + \epsilon_i') \quad \text{in } \Lambda_I \\ C_p^0 \frac{\partial u}{\partial t} &= C_p^I \frac{\partial u}{\partial t} + Q_i^* \quad \text{in } \Lambda_I \end{aligned} \tag{9.41}$$

where the temperature gradient $u_{,i}$ can be obtained by taking the derivative with respect to $x_{,i}$ in Eq. (9.39). Since the boundary response remains the same at each time step, the derivatives are only taken with the coefficients of U, H, M, J, and G. As for the temperature changing rate $\frac{\partial u}{\partial t}$, it can be derived similarly as the temperature gradient. Eq. (9.41) can be simplified as

$$\begin{aligned} \Delta k u_{,i} &= k_0 \epsilon_i^* \\ \Delta C_p^I \frac{\partial u}{\partial t} &= Q_i^* \end{aligned} \tag{9.42}$$

Written the temperature and its changing rate in terms of the boundary integral and volume integral over particles with ϵ_i^* and Q_i^* on each particle, we can solve for ϵ_i^* and Q_i^*.

9.2.6 Equivalent heat flux condition with the heat capacity method

The physics of PCM is that when the temperature is in the range of phase change temperature window, the heat capacity C_p, i.e., the product of density and specific heat, is larger than that in other temperature ranges. By modifying the heat equation the left-hand side heat capacity can be expressed as $C_p(u)$, which is a function of temperature. As assumed in Section 9.2.3, a spatial-averaged specific heat is considered for the entire PCM particle, which is a piecewise function of temperature. Therefore a heat source term Q_{rc}^I is used to match the heat capacity difference, which is similar to the inclusion problem with fictitious heat source. Although Q_{rc}^I and Q_I^* are introduced for PCM particle in the mushy state with solid/liquid coexisting, respectively, they both serve to match the heat capacity difference. Therefore it is natural to consider a heat source $Q_I^*(u)$ as a general form for heat capacity mismatch in Eqs. (9.40) and (9.42). In Eq. (9.39) the temperature at any point and time can be expressed. The heat source term Q_I^* depends on the temperature changing rate and the heat capacity difference ΔC_p^I. Employing a backward finite difference scheme [121], we write the temperature changing rate at time step n as

$$\frac{\partial u}{\partial t} = \frac{u^n - u^{n-1}}{t^n - t^{n-1}} \tag{9.43}$$

Since the time interval and temperature at the previous time step are known, the key of the scheme is to express the temperature at the current time step by Eq. (9.39). According to Hsiao [119], the heat capacity of the PCM particle is computed at the previous time step. The explicit scheme will be presented in Eq. (9.47) in the next subsection.

9.2.7 Assembly of the global matrix of iBEM

To solve the problem, the global matrix of iBEM can be assembled as follows. Firstly, assemble the boundary element equation related to boundary node temperature and surface element fluxes. Secondly, assemble EIM equations and phase change process constraint equations (9.42). In Sections 9.2.5 and 9.2.6 the equations will be solved for the material properties mismatch and phase change process simultaneously with modified coefficients of the variables ϵ^* and Q^*. The initial conditions are implemented in Eq. (9.39), where they are multiplied by Green's function and then integrated in the spatial domain. The implementation of boundary conditions is the same as BEM, which involves the computed coefficient matrix multiplying the boundary variable vectors. The superscript ($'$) denotes the spatial derivative, and the spatial integral of Green's function is written as G_{int}. Following the pioneering work on BEM [117], below we show a direct iBEM for the phase change problem. Based on Eq. (9.41), we write the modified EIM equation as

$$
\begin{aligned}
(\epsilon_i^{I*})_f &= \frac{\Delta k}{k_0}\Bigg\{\sum_{c=1}^{f}\sum_{l=1}^{N_e}\sum_{i=1}^{N_n}[(U_i' q_i^{bc}) - (H_i' u_i^{bc})]_c \\
&\quad + \sum_{I=1}^{N_S}(M' Q^{I*} + J_i' \epsilon_i^I)_c + G_{int}'(\mathbf{r},t)u^0\Bigg\} \\
&= \frac{\Delta k}{k_0}\Bigg\{\sum_{l=1}^{N_e}\sum_{i=1}^{N_n}[(U_i' q_i^{bc})_f - (H_i' u_i^{bc})_f] \\
&\quad + \sum_{I=1}^{N_S}[(J_i'\epsilon_i^{I*})_f + (M' Q^{I*})_f]\Bigg\} \\
&\quad + \frac{\Delta k}{k_0}\Bigg\{\sum_{c=1}^{f-1}\sum_{l=1}^{N_e}\sum_{i=1}^{N_n}[(U_i' q_i^{bc}) - (H_i' u_i^{bc})]_c \\
&\quad + \sum_{I=1}^{N_S}(M' Q^{I*} + J_i' \epsilon_i^I)_c + G_{int}'(\mathbf{r},t)u^0\Bigg\}
\end{aligned} \tag{9.44}
$$

where the subscripts f and c represent the thermal quantities existing during times t_{f-1}, t_f and t_{c-1}, t_c, respectively. Combining the coefficients of ϵ_i^{I*} yields

$$\sum_{l=1}^{N_e}\sum_{i=1}^{N_n}[(U_i' q_i^{bc})_f - (H_i' u_i^{bc})_f] + \sum_{I=1}^{N_S}[((J_i' - \frac{k_0}{\Delta k})\epsilon_i^{I*})_f + (M' Q^{I*})_f]$$
$$+\sum_{c=1}^{f-1}\sum_{l=1}^{N_e}\sum_{i=1}^{N_n}[(U_i' q_i^{bc}) - (H_i' u_i^{bc})]_c + \sum_{I=1}^{N_S}(M' Q^{I*} + J_i' \epsilon_i^I)_c$$
$$+ G'_{int}(\mathbf{r}, t) u^0 = 0 \tag{9.45}$$

where the modified coefficient of $(\epsilon_i^{I*})_f$ is $\frac{\Delta k^I}{k_0} J'_{f,i} - 1$, and $\Delta k^I = k_0 - k_I$. Note that the modified coefficient does not exist in the domain of dependence except the current time step matrix. Similarly, the modified coefficients for the heat source can be obtained as $M_f - \frac{t_f - t_{f-1}}{\Delta C_p{}^I}$ and $\Delta C_p^I = C_p^0 - C_p^I$. Finally, the modified constraint equation of PCM particle temperature can be expressed. Note that the heat capacity C_p^I is a function of u_I^n, which is the spatial-average temperature of the Ith subdomain at the previous time step. The constraint equation at the Ith subdomain is expressed by the equation

$$\sum_{l=1}^{N_e}\sum_{i=1}^{N_n}[(U_i q_i^{bc})_f - (H_i u_i^{bc})_f] + \sum_{I=1}^{N_S}\left\{(J_i \epsilon_i^{I*})_f + [(M - \frac{t_f - t_{f-1}}{\Delta (C_p)^I}) Q^{I*}]_f\right\}$$
$$+\sum_{c=1}^{f-1}\sum_{l=1}^{N_e}\sum_{i=1}^{N_n}[(U_i q_i^{bc}) - (H_i u_i^{bc})]_c + \sum_{I=1}^{N_S}(M Q^{I*} + J_i \epsilon_i^{I*})_c$$
$$+ G_{int}(\mathbf{r}, t) u^0 - u_{f-1}^I = 0 \tag{9.46}$$

Combining the above equivalent equations with BEM equations, we can form the main matrix and variable vectors for iBEM. To simplify the notation of the main matrix, denote by $\overline{U}, \overline{H}, \overline{J}, \overline{M}$ the coefficient matrices for the variable vectors $\overline{q_{bc}}, \overline{u_{bc}}, \overline{\epsilon^*}, \overline{Q^*}$, and denote the coefficients of initial conditions as the vector $\overline{G_u^0}$. Thus we obtain the direct form of the iBEM as

$$\begin{bmatrix} C + \overline{H_f} & -\overline{J_f} & -\overline{M_f} \\ -\overline{H_f'} & \overline{J_f'} - \frac{k_0}{\Delta k} & \overline{M_f'} \\ -\overline{H_f} & \overline{J_f} & \overline{M_f} - \frac{t_f - t_{f-1}}{\Delta C_p{}^I} \end{bmatrix} \begin{bmatrix} \overline{u_{bc}^f} \\ \overline{\epsilon_*^f} \\ \overline{Q_*^f} \end{bmatrix} - \begin{bmatrix} \overline{U_f} \\ -\overline{U_f}' \\ -\overline{U_f} \end{bmatrix} \begin{bmatrix} \overline{q_{bc}^f} \end{bmatrix}$$
$$= -\sum_{c=1}^{f-1} \begin{bmatrix} \overline{H_c} & -\overline{J_c} & -\overline{M_c} \\ -\overline{H_c'} & \overline{J_c'} & \overline{M_c'} \\ -\overline{H_c} & \overline{J_c} & \overline{M_c} \end{bmatrix} \begin{bmatrix} \overline{u_{bc}^c} \\ \overline{\epsilon_*^c} \\ \overline{Q_*^c} \end{bmatrix} + \begin{bmatrix} \overline{U_c} \\ -\overline{U_c} \\ -\overline{U_c} \end{bmatrix} \begin{bmatrix} \overline{q_{bc}^c} \end{bmatrix}$$
$$+ \begin{bmatrix} \overline{G_u^0} \\ -\overline{G_u^0}' \\ -\overline{G_u^0} \end{bmatrix} \begin{bmatrix} \overline{u_0} \end{bmatrix} + \begin{bmatrix} 0 \\ 0 \\ u_f^{I-1} \end{bmatrix} \tag{9.47}$$

To reduce the computational cost, keeping the same time intervals, there is only one coefficient matrix to be solved and stored at each time step, which brings the influence of the initial thermal field.

9.3 Time-dependent material behavior of composites

For elastic materials under a static load, the mechanical response is unique and does not change with time. When a dynamic load is applied, the Green's function or fundamental solution is still uniquely determined by the elastic constants and the relative time and space with reference to the source. Viscoelastic materials exhibit time-dependent material behavior. Therefore, even when static loading is applied, the mechanical response still changes with time. This section introduces the iBEM applications to viscoelastic materials.

When the mechanical constants of a viscoelastic material are given, a boundary value problem can be solved by the elastic–viscoelastic correspondence principle [122,123] through either the Laplace transform (LT) or Fourier transform (FT). Therefore Eshelby's [2] theory can still be applicable for both inclusion problems and the EIM. As a consequence, the iBEM can still be extended to viscoelastic material behavior of composites. Here we provide the main difference in comparison with the elastic material behavior in Chapters 3 to 6.

For a composite containing isotropic viscoelastic particles and matrix under a certain load on the boundary, the constitutive law will be different from Hooke's law of the elastic materials. It can be given in terms of the master curves of either creep compliance or relaxation modulus in the time domain of $0 \sim \infty$, or complex modulus in the frequency domain of $0 \sim \infty$. The stress–strain relationship can be written either in the integral form with time or in differential form with stress or strain rates. LT and FT have been often used to transfer the viscoelastic problem to a similar elastic problem in the Laplace or Fourier domain and then to inversely transform the solution to the time domain. In addition, because many material characterizations or applications have been conducted under harmonic loading, the results in the frequency domain or Fourier domain carry the physical meaning as well. In the following, we show the cases with LT and FT, respectively.

9.3.1 Laplace transform-based correspondence to elastic problems

Consider a material system in Fig. 1.1. The elastic BVP problem is formulated in Eqs. (1.1)–(1.5). For a general viscoelastic material system, we can also use the superscript $I = 0, 1$ to show the materials of inhomogeneity and matrix, respectively. In fact, all those equations in elastic BVP apply to the viscoelastic material system except Eq. (1.2), which can be written in either the integral or differential form as follows.

The integral form of constitutive relation is written

$$\sigma_{ij}(\mathbf{x},t)=\int_{-\infty}^{t}C_{ijkl}^{I}(t-\tau)\frac{d\varepsilon_{kl}(\mathbf{x},\tau)}{d\tau}d\tau \tag{9.48}$$

where $C_{ijkl}(t)^I$ is the relaxation modulus changing with time, which can be derived from the master curve of creep compliance as well. For isotropic materials, we can also use two Lame constants to represent $C_{ijkl}(t)=\lambda(t)\delta_{ij}\delta_{kl}+\mu(t)\left(\delta_{ik}\delta_{jl}+\delta_{il}\delta_{jk}\right)$. Notice that in actual calculation, bulk and shear moduli are often used instead of the two Lame constants to show the relationship of the volumetric and deviatoric components of the stress and strain. Here to illustrate the correspondence between the elastic and viscoelastic formulations, we first keep using the two Lame constants and later show the case of using bulk modulus in the differential form.

Applying the Laplace transform (LT) to the convolution (9.48) yields an algebraic relation in the Laplace domain as

$$\hat{\sigma}_{ij}(\mathbf{x},s)=s\hat{C}_{ijkl}^{I}\hat{\varepsilon}_{kl}(\mathbf{x},s) \tag{9.49}$$

where the LT of a function $f(t)$ is shown with $\hat{f}(s)=\int_0^\infty f(t)\exp(-st)dt$, $s>0$, as well.

Similarly, other equations can be transferred to the Laplace domain as follows:

$$\hat{\sigma}_{ij,i}+\hat{b}_j=0 \tag{9.50}$$

and

$$\hat{\varepsilon}_{ij}=\frac{1}{2}\left(\hat{u}_{i,j}+\hat{u}_{j,i}\right) \tag{9.51}$$

The boundary condition and interfacial continuity can be transformed in the same fashion

$$\begin{aligned}\hat{t}_j^0&=\hat{\sigma}_{ij}n_i \quad \text{for } x\in\partial\mathcal{D}^t\\ \hat{u}_i^0&=\hat{u}_i \quad \text{for } x\in\partial\mathcal{D}^u\end{aligned} \tag{9.52}$$

and

$$\hat{\sigma}_{ij}^{+}(\mathbf{x})n_j^{+}=\hat{\sigma}_{ij}^{-}(\mathbf{x})n_j^{-},\quad \hat{u}_i^{+}(\mathbf{x})=\hat{u}_i^{-}(\mathbf{x}) \quad \text{on } \partial\Omega_I \tag{9.53}$$

In the Laplace domain, all elastic fields depend on s instead of t. From the boundary conditions and material properties we can obtain $\hat{b}_j$, $\hat{C}_{ijkl}$, $\hat{t}_j^0$, and $\hat{u}_i^0$. Comparing Eqs. (9.49)–(9.53) with Eqs. (1.1)–(1.5), we can see that they share the same mathematical form except that the stiffness tensor is replaced by $s\hat{C}_{ijkl}$. Therefore, given a value of s, the above equations can be exactly solved by the iBEM with the equivalent inclusion method. Taking the inverse Laplace transform of stress or displacement field [124,125], we can obtain the mechanical response in the time domain. Notice that the inertia force caused by acceleration is not considered in the above process as the effect is typically small under a static load. Otherwise, if the load is time-dependent or viscous effect is very large, then the elastodynamic equilibrium equation can be used, so that the elastic–viscoelastic correspondence principle needs to be updated in the same fashion.

When the constitutive relationship is given in the differential form, the time derivatives of the stress and strain appear in the constitutive relation. For isotropic material behavior, the stress and strain relationship can be written in volumetric and deviatoric cases, and each case exhibits a set of material constants as follows:

$$P_d s_{ij} = Q_d e_{ij}; \quad P_v \sigma^v = Q_v \varepsilon^v \tag{9.54}$$

where the volumetric stress $\sigma^v = \frac{1}{3}\sigma_{ii}$ and strain $\varepsilon^v = \varepsilon_{ii}$, the deviatoric stress $s_{ij} = \sigma_{ij} - \sigma^v \delta_{ij}$ and strain $e_{ij} = \varepsilon_{ij} - \frac{1}{3}\varepsilon^v$, and the differential operators with respect to time are written as

$$P_x = \Sigma_{r=0}^{N} p_r^x \frac{\partial^r}{\partial t^r}; \quad Q_x = \Sigma_{r=0}^{N} q_r^x \frac{\partial^r}{\partial t^r} \tag{9.55}$$

where the subscript x can be either d or v to show the deviatoric or volumetric cases, $p_0^x = 1$ to show the stress, and the summation upper limit N can be different for each case of P, Q, d or v, depending on viscoelastic material models. For simplicity, we can take the maximum of them and make the parameters at zero for the unnecessary terms.

Taking the Laplace transform of Eq. (9.54), we can write

$$\hat{s}_{ij}(\mathbf{x}, s) = \frac{\hat{Q}_d(s)}{\hat{P}_d(s)} \hat{e}_{ij}(\mathbf{x}, s); \quad \hat{\sigma}^v(\mathbf{x}, s) = \frac{\hat{Q}_v(s)}{\hat{P}_v(s)} \hat{\varepsilon}^v(\mathbf{x}, s) \tag{9.56}$$

where

$$\hat{P}_x(s) = \Sigma_{r=0}^{N} p_r^x s^r; \quad \hat{Q}_x(s) = \Sigma_{r=0}^{N} q_r^x s^r \tag{9.57}$$

Eq. (9.56) is similar to Eq. (9.49) in the Laplace domain. Therefore by combining with the other equations the problem can be solved by iBEM in the same fashion in the Laplace domain and inversely transformed to the time domain.

9.3.2 Fourier transform-based correspondence to elastic problems

Consider a viscoelastic material is under a harmonic mechanical load, say $F = \overline{F}\cos(\omega t)$ or $\overline{F}\exp(i\omega t)$. The displacement, stress, and strain will also exhibit a harmonic variation with time as

$$\begin{aligned} \sigma_{ij}(\mathbf{x}, t) &= \overline{\sigma}_{ij}(\mathbf{x}) \exp(i\omega t) \\ \varepsilon_{ij}(\mathbf{x}, t) &= \overline{\varepsilon}_{ij}(\mathbf{x}) \exp(i\omega t) \\ u_i(\mathbf{x}, t) &= \overline{u}_i(\mathbf{x}) \exp(i\omega t) \end{aligned} \tag{9.58}$$

where the actual numbers are taken as the real parts of complex quantities, but the operations are conducted with complex variables. Substituting the above relation into Eq. (9.48) yields

$$\overline{\sigma}_{ij}(\mathbf{x})\exp(i\omega t) = i\omega \int_{-\infty}^{t} C_{ijkl}(t-\tau)\overline{\varepsilon}_{kl}(\mathbf{x})\exp(i\omega\tau)d\tau \tag{9.59}$$

By replacing the variable $t' = t - \tau$ we rewrite thsi equation as

$$\overline{\sigma}_{ij}(\mathbf{x}) = C^{*}_{ijkl}\overline{\varepsilon}_{kl}(\mathbf{x}) \tag{9.60}$$

where $C^{*}_{ijkl} = i\omega\int_0^\infty C_{ijkl}(t')\exp(-i\omega t')dt'$ is also called the complex modulus at ω, which includes both real and imaginary parts, namely storage and loss moduli, and can be further illustrated by its absolute value, say dynamic modulus and a phase angle. Therefore we obtain a linear algebraic stress–strain relationship at domain ω.

Similarly, for the differential form of constitutive relation, the volumetric and deviatoric parts of the stress and strain can also be written in the form of Eq. (9.58). Using that in Eq. (9.54) yields

$$\overline{s}_{ij}(\mathbf{x}) = \frac{\overline{Q}_d}{\overline{P}_d}\overline{e}_{ij}(\mathbf{x}); \quad \overline{\sigma}^{v}(\mathbf{x}) = \frac{\overline{Q}_v}{\hat{P}_v}\overline{\varepsilon}^{v}(\mathbf{x}) \tag{9.61}$$

where

$$\overline{P}_x = \Sigma_{r=0}^{N} p_r^x (i\omega)^r; \quad \overline{Q}_x = \Sigma_{r=0}^{N} q_r^x (i\omega)^r \tag{9.62}$$

Similarly, at the frequency domain ω, the other equations can also be obtained in the same fashion as Eqs. (9.50)–(9.53). Therefore the equations can be solved with iBEM in the frequency domain by iBEM and then inversely transformed back to the time domain.

Appendix 9.A Elastodynamic Green's functions for an isotropic infinite domain

As shown in Eq. (9.7), the fundamental solution to the elastodynamic equation with a time-harmonic load is present. As a well established issue, the derivation of it can be simplified as the extension to the celebrated Helmholtz equation [126–132], which relates the study of acoustics as well.

Without loss of generality, consider an infinite homogeneous elastic material with the Lame constants λ and μ and density ρ. The elastodynamic equation (9.2) with distributed body force $f_i(\mathbf{x}, t)$, with i representing the direction and $\mathbf{x}$ the location of the load, can be rewritten as follows:

$$\ddot{u}_i = \frac{(\lambda + \mu)}{\rho} u_{j,ji} + \frac{\mu}{\rho} u_{i,jj} + \frac{f_i}{\rho} \tag{9.63}$$

Note that this equation for three-dimensional (3D) problem can be reduced to a 2D plane strain problem in the same form with the Einstein notation, which can be further reduced to a 1D problem. Here we focus on the 3D problem and the derivation for 2D and 1D can be similarly conducted.

9.A.1 Derivation of the elastodynamic Green's function for a harmonic point force

When the body force is given in the form of a harmonic single wave as

$$f_i(\mathbf{x}, t) = \overline{f}_i \exp(i\boldsymbol{\xi} \cdot \mathbf{x}) \exp(-i\omega t) = \overline{f}_i \exp(i(\boldsymbol{\xi} \cdot \mathbf{x} - \omega t)) \tag{9.64}$$

the displacement field should also be a harmonic periodic function

$$u_i(\mathbf{x}, t) = \overline{u}_i \exp(i(\boldsymbol{\xi} \cdot \mathbf{x} - \omega t)) \tag{9.65}$$

Therefore Eq. (9.63) can be written as

$$\frac{(\lambda + \mu)}{\rho} \xi_i \xi_j \overline{u}_j + \frac{\mu}{\rho} \xi^2 \overline{u}_i - \omega^2 \overline{u}_i = \frac{\overline{f}_i}{\rho} \tag{9.66}$$

where $\xi = \sqrt{\xi_i \xi_i}$. Define the matrix $\mathbf{K}$ as

$$K_{ij} = \frac{\lambda + \mu}{\rho} \xi_i \xi_j + \left(\frac{\mu}{\rho} - \frac{\omega^2}{\xi^2} \right) \xi^2 \delta_{ij} \tag{9.67}$$

Therefore we obtain

$$\overline{u}_i = K_{ij}^{-1} \frac{\overline{f}_j}{\rho} \tag{9.68}$$

where K_{ij}^{-1} is the inverse of $\mathbf{K}$:

$$K_{ij}^{-1} = \frac{\left(v_p^2 - \frac{\omega^2}{\xi^2} \right) \delta_{ij} - (v_p^2 - v_s^2) s_i s_j}{\xi^2 \left(v_s^2 - \frac{\omega^2}{\xi^2} \right) \left(v_p^2 - \frac{\omega^2}{\xi^2} \right)} \tag{9.69}$$

in which $v_p = \sqrt{\frac{\lambda+2\mu}{\rho}}$ and $v_s = \sqrt{\frac{\mu}{\rho}}$ are the speeds of the primary and secondary waves in the elastic medium, respectively, and $\mathbf{s} = \boldsymbol{\xi}/\xi$ is the directional vector of $\boldsymbol{\xi}$.

The derivation of K_{ij}^{-1} is as follows. Due to the closeness of the multiplication operation of $\xi_i\xi_j$ and δ_{ij}, we can assume that K_{ij}^{-1} also is of the form

$$K_{ij}^{-1} = K_1\xi_i\xi_j + K_2\xi^2\delta_{ij} \tag{9.70}$$

We can write

$$K_{ij}^{-1}K_{jk} = \left[K_1\left(\frac{\lambda+2\mu}{\rho} - \frac{\omega^2}{\xi^2}\right) + K_2\frac{\lambda+\mu}{\rho}\right]\xi^2\xi_i\xi_k + K_2\xi^4\left(\frac{\mu}{\rho} - \frac{\omega^2}{\xi^2}\right)\delta_{ik} \tag{9.71}$$

Using $K_{ij}^{-1}K_{jk} = \delta_{ik}$, we can determine

$$K_1 = -\frac{\frac{\lambda+\mu}{\rho}}{\xi^4\left(\frac{\mu}{\rho} - \frac{\omega^2}{\xi^2}\right)\left(\frac{\lambda+2\mu}{\rho} - \frac{\omega^2}{\xi^2}\right)}; \quad K_2 = \frac{1}{\xi^4\left(\frac{\mu}{\rho} - \frac{\omega^2}{\xi^2}\right)} \tag{9.72}$$

Therefore we can rewrite K_{ij}^{-1} as

$$K_{ij}^{-1} = -\frac{\frac{\lambda+\mu}{\rho}\xi_i\xi_j}{\xi^4\left(\frac{\mu}{\rho} - \frac{\omega^2}{\xi^2}\right)\left(\frac{\lambda+2\mu}{\rho} - \frac{\omega^2}{\xi^2}\right)} + \frac{\delta_{ij}}{\xi^2\left(\frac{\mu}{\rho} - \frac{\omega^2}{\xi^2}\right)} \tag{9.73}$$

$$= \frac{\left(v_p^2 - \frac{\omega^2}{\xi^2}\right)\delta_{ij} - (v_p^2 - v_s^2)s_i s_j}{\xi^2\left(v_s^2 - \frac{\omega^2}{\xi^2}\right)\left(v_p^2 - \frac{\omega^2}{\xi^2}\right)} \tag{9.74}$$

as is given in Eq. (9.69).

Therefore by Eq. (9.65) the displacement can be written as

$$u_i(\mathbf{x}, t) = \frac{K_{ij}^{-1}\overline{f}_j}{\rho}\exp\left[i(\boldsymbol{\xi}\cdot\mathbf{x} - \omega t)\right] \tag{9.75}$$

For a harmonic body force with arbitrary spatial distribution, it can be separated by the spatial and time components and then written in the wave form by the Fourier integral as

$$f_i(\mathbf{x}, t) = f_i^x(\mathbf{x})\exp(-i\omega t) = \int_{-\infty}^{\infty}\overline{f}_i(\boldsymbol{\xi})\exp(i\boldsymbol{\xi}\cdot\mathbf{x})d\boldsymbol{\xi}\exp(-i\omega t) \tag{9.76}$$

where $\overline{f}_i(\boldsymbol{\xi})$ is the Fourier transform of the spatial component $f_i^x(\mathbf{x})$,

$$\overline{f}_i(\boldsymbol{\xi}) = \frac{1}{(2\pi)^3}\int_{-\infty}^{\infty} f_i^x(\mathbf{x}')\exp(-i\boldsymbol{\xi}\cdot\mathbf{x}')d\mathbf{x}' \tag{9.77}$$

in which

$$\int_{-\infty}^{\infty} d\boldsymbol{\xi} = \int_{-\infty}^{\infty} d\xi_1 \int_{-\infty}^{\infty} d\xi_2 \int_{-\infty}^{\infty} d\xi_3$$

and

$$\int_{-\infty}^{\infty} d\mathbf{x}' = \int_{-\infty}^{\infty} dx_1' \int_{-\infty}^{\infty} dx_2' \int_{-\infty}^{\infty} dx_3'$$

for 3D integrals. Substituting the source of Eq. (9.76) into Eq. (9.75), we obtain

$$u_i(\mathbf{x}, t) = \int_{-\infty}^{\infty} \frac{K_{ij}^{-1} \overline{f}_j}{\rho} \exp i(\boldsymbol{\xi} \cdot \mathbf{x} - \omega t) d\boldsymbol{\xi} \tag{9.78}$$

Using Eq. (9.77) in Eq. (9.78) yields

$$u_i(\mathbf{x}, t) = \frac{1}{(2\pi)^3} \int_{-\infty}^{\infty} \int_{-\infty}^{\infty} \frac{K_{ij}^{-1} f_j^x(\mathbf{x}')}{\rho} \exp i(\boldsymbol{\xi} \cdot (\mathbf{x} - \mathbf{x}') - \omega t) d\mathbf{x}' d\boldsymbol{\xi} \tag{9.79}$$

The Green's function caused by a harmonic point source at $\mathbf{x}'$ can be written as

$$G_{ij}(\mathbf{x}, \mathbf{x}') = \frac{1}{(2\pi)^3} \int_{-\infty}^{\infty} \frac{K_{ij}^{-1}}{\rho} \exp[i\boldsymbol{\xi} \cdot (\mathbf{x} - \mathbf{x}')] d\boldsymbol{\xi} \tag{9.80}$$

Using Eq. (9.74) in the above equation yields

$$G_{ij}(\mathbf{x}, \mathbf{x}') = \frac{1}{(2\pi)^3} \int_{-\infty}^{\infty} \frac{\left(v_p^2 - \frac{\omega^2}{\xi^2}\right) \delta_{ij} - (v_p^2 - v_s^2) s_i s_j}{\rho \xi^2 \left(v_s^2 - \frac{\omega^2}{\xi^2}\right)\left(v_p^2 - \frac{\omega^2}{\xi^2}\right)} \exp[i\boldsymbol{\xi} \cdot (\mathbf{x} - \mathbf{x}')] d\boldsymbol{\xi} \tag{9.81}$$

After the triple integration of $d\boldsymbol{\xi} = d\xi_1 d\xi_2 d\xi_3$, we obtain the explicit Green's function

$$G_{ij}(\mathbf{x}, \mathbf{x}') = \frac{1}{4\pi\rho\omega^2} \left[\beta^2 \delta_{ij} \frac{e^{i\beta r}}{r} - \frac{\partial^2}{\partial x_i \partial x_j} \left(\frac{e^{i\alpha r}}{r} - \frac{e^{i\beta r}}{r} \right) \right] \tag{9.82}$$

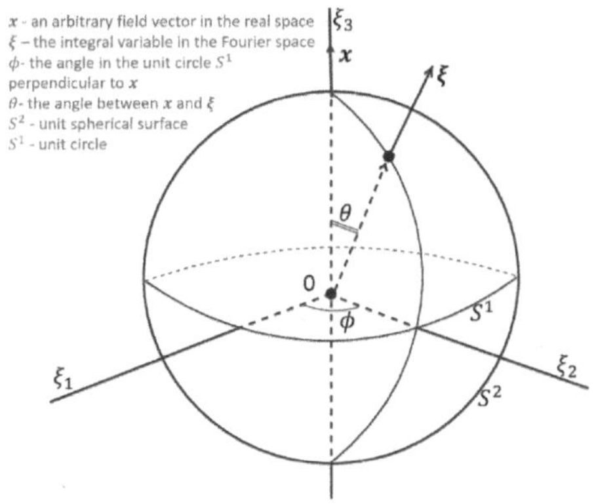

FIGURE 9.3 The integral for Green's function at point **x** over $\boldsymbol{\xi}$ through a spherical coordinate.

where $r = |\mathbf{x} - \mathbf{x}'|$ is the distance between the field point **x** and source point $\mathbf{x}'$, $\alpha = \omega/v_p$, and $\beta = \omega/v_s$.

The derivation of Eq. (9.82) is as follows.
Similarly to the derivation of the static elastic Green's function in pages 23–25 of Mura [1], $d\boldsymbol{\xi} = \xi^2 d\xi dS$, where dS is the surface element on the unit sphere S^2 in the $\boldsymbol{\xi}$ space, and the unit vector **s** measures the surface given the orientation with the starting point at the origin of the $\boldsymbol{\xi}$ space (see Fig. 9.3). Let the source be at the origin in the **x** space, i.e., $\mathbf{x}' = \mathbf{0}$, which is the same as defining $\mathbf{t} = \mathbf{x} - \mathbf{x}'$ and using **t** for the following derivation in the place of $\dot{\mathbf{x}}$. Eq. (9.81) is rewritten as

$$G_{ij}(\mathbf{x}) = \frac{1}{(2\pi)^3 \rho} \int_0^{\infty} d\xi \int_{S^2} \frac{\left(v_p^2 - \frac{\omega^2}{\xi^2}\right)\delta_{ij} - (v_p^2 - v_s^2) s_i s_j}{\left(v_s^2 - \frac{\omega^2}{\xi^2}\right)\left(v_p^2 - \frac{\omega^2}{\xi^2}\right)} \exp(i\boldsymbol{\xi} \cdot \mathbf{x}) dS \tag{9.83}$$

Considering the symmetry of $\boldsymbol{\xi}$ and $-\boldsymbol{\xi}$ [1], we can write

$$G_{ij}(\mathbf{x}) = \frac{1}{2(2\pi)^3 \rho} \int_{-\infty}^{\infty} \int_{S^2} \frac{\left(v_p^2 - \frac{\omega^2}{\xi^2}\right)\delta_{ij} - (v_p^2 - v_s^2) s_i s_j}{\left(v_s^2 - \frac{\omega^2}{\xi^2}\right)\left(v_p^2 - \frac{\omega^2}{\xi^2}\right)} \exp(i\boldsymbol{\xi} \cdot \mathbf{x}) d\xi dS \tag{9.84}$$

where $i\boldsymbol{\xi} \cdot \mathbf{x} = i\xi x s_i n_i$ with $x = |\mathbf{x}|$ and $n_i = x_i/x$. Notice that the integral function includes three parts as follows:

$$\frac{\left(v_p^2 - \frac{\omega^2}{\xi^2}\right)\delta_{ij} - (v_p^2 - v_s^2) s_i s_j}{\left(v_s^2 - \frac{\omega^2}{\xi^2}\right)\left(v_p^2 - \frac{\omega^2}{\xi^2}\right)} = \frac{\delta_{ij}}{v_s^2 - \frac{\omega^2}{\xi^2}} - \frac{s_i s_j}{v_s^2 - \frac{\omega^2}{\xi^2}} + \frac{s_i s_j}{v_p^2 - \frac{\omega^2}{\xi^2}} \tag{9.85}$$

The integral of the first part with respect to ξ can be written as

$$\int_{-\infty}^{\infty} \frac{\delta_{ij}}{v_s^2 - \frac{\omega^2}{\xi^2}} \exp(i\boldsymbol{\xi} \cdot \mathbf{x}) d\xi = \frac{\delta_{ij}}{v_s^2} \int_{-\infty}^{\infty} \left(1 + \frac{\beta^2}{\xi^2 - \beta^2}\right) \exp(i\xi x s_k n_k d\xi$$
$$= 2\pi \frac{\delta_{ij}}{v_s^2} \left[\delta(x s_k n_k) - \frac{\beta}{2} i \exp(-i\beta |x s_k n_k|)\right] \tag{9.86}$$

The integral of the second part with respect to ξ can be written as

$$\int_{-\infty}^{\infty} \frac{s_i s_j}{v_s^2 - \frac{\omega^2}{\xi^2}} \exp(i\boldsymbol{\xi} \cdot \mathbf{x}) d\xi = -\frac{1}{v_s^2} \frac{\partial^2}{\partial x_i \partial x_j} \int_{-\infty}^{\infty} \frac{1}{\xi^2 - \beta^2} \exp(i\boldsymbol{\xi} \cdot \mathbf{x}) d\xi$$
$$= \frac{2\pi}{v_s^2} \frac{\partial^2}{\partial x_i \partial x_j} \left[\frac{i}{2\beta} \exp(-i\beta |x s_k n_k|)\right] \tag{9.87}$$

The third part can also be integrated in the same fashion, so that Eq. (9.84) can be rewritten as

$$G_{ij}(\mathbf{x}) = \frac{1}{8\pi^2 \rho} \int_{S^2} \frac{\delta_{ij} \left[\delta(x s_k n_k) - \frac{\beta}{2} i \exp(-i\beta |x s_k n_k|)\right]}{v_s^2}$$
$$- \frac{1}{v_s^2} \frac{\partial^2}{\partial x_i \partial x_j} \left[\frac{i}{2\beta} \exp(-i\beta |x s_k n_k|)\right] + \frac{1}{v_p^2} \frac{\partial^2}{\partial x_i \partial x_j} \left[\frac{i}{2\alpha} \exp(-i\alpha |x s_k n_k|)\right] dS \tag{9.88}$$

Define $s_i n_i = \cos\theta$ with θ being the angle between $\boldsymbol{\xi}$ and $\mathbf{x}$. Use ϕ to indicate the angle of a unit circle on the plane perpendicular to $\mathbf{x}$. Then $dS = \sin\theta d\theta d\phi = -d(s_i n_i) d\phi$ with θ from $-\pi/2$ to $\pi/2$ or $s_i n_i$ from -1 to 1 and ϕ from 0 to 2π. Using the identities

$$\int_{-1}^{1} \delta(x s_i n_i) d(s_i n_i) = \frac{1}{x}$$
$$\frac{\beta}{2} i \int_{-1}^{1} \exp(-i\beta |x s_k n_k|) d(\mathbf{s} \cdot \mathbf{n}) = \frac{1 - \exp(-i\beta x)}{x} \tag{9.89}$$

we can rewrite Eq. (9.88) as

$$G_{ij}(\mathbf{x}, \mathbf{x}') = \frac{1}{4\pi\rho\omega^2} \left[\beta^2 \delta_{ij} \frac{e^{-i\beta r}}{r} - \frac{\partial^2}{\partial x_i \partial x_j} \left(\frac{e^{-i\alpha r}}{r} - \frac{e^{-i\beta r}}{r}\right)\right] \tag{9.90}$$

Notice that the Green's function takes the real part of the complex variable, $\exp(-i\beta x)$ can be replaced by $\exp(i\beta x)$ and the same for $\exp(-i\alpha x)$, so that the above equation is the same as Eq. (9.82).

For the 2D or 1D problems under a harmonic load, the above procedure can be similarly conducted and is provided in Section 9.A.3, which should be obtained by the integral of a line source under the plane strain assumption in the same fashion as the static Green's function.

9.A.2 Derivation of the elastic Green's function for an impulse point force

The above derivation in Appendix 9.A can be extended to general dynamic problems. The 3D dynamic elastic Green's function can be considered as the field caused by an impulse point force. To derive it, replace Eq. (9.76) by

$$f_i(\mathbf{x},t)=\int_{-\infty}^{\infty}d\omega\int_{-\infty}^{\infty}\overline{f}_i(\boldsymbol{\xi},\omega)\exp(i\boldsymbol{\xi}\cdot\mathbf{x}-i\omega t)d\boldsymbol{\xi} \tag{9.91}$$

where ω is extended to $(-\infty\sim\infty)$ by an even function, and

$$\overline{f}_i(\boldsymbol{\xi},\omega)=\frac{1}{(2\pi)^4}\int_{-\infty}^{\infty}dt'\int_{-\infty}^{\infty}f_i(\mathbf{x}',t')\exp(-i\boldsymbol{\xi}\cdot\mathbf{x}')d\mathbf{x}' \tag{9.92}$$

Therefore we obtain the displacement as

$$u_i(\mathbf{x},t)=\int_{-\infty}^{\infty}d\omega\int_{-\infty}^{\infty}\frac{K_{ij}^{-1}\overline{f}_j(\boldsymbol{\xi},\omega)}{\rho}\exp(i\boldsymbol{\xi}\cdot\mathbf{x}-i\omega t)d\boldsymbol{\xi} \tag{9.93}$$

Substituting Eq. (9.92) into the above equation yields

$$u_i(\mathbf{x},t)=\int_{-\infty}^{\infty}dt'\int_{-\infty}^{\infty}G_{ij}(\mathbf{x}-\mathbf{x}',t-t')f_i(\mathbf{x}',t')d\mathbf{x}' \tag{9.94}$$

where

$$G_{ij}(\mathbf{x},t)=\frac{1}{(2\pi)^4}\int_{-\infty}^{\infty}d\omega\int_{-\infty}^{\infty}\frac{K_{ij}^{-1}}{\rho}\exp(i\boldsymbol{\xi}\cdot(\mathbf{x}-\mathbf{x}')-i\omega(t-t'))d\boldsymbol{\xi} \tag{9.95}$$

Using Eqs. (9.80) and (9.82) in the above equation yields

$$\begin{aligned}
&G_{ij}(\mathbf{x},t)\\
&=\frac{1}{8\pi^2\rho}\int_{-\infty}^{\infty}\left[\frac{\delta_{ij}}{v_s^2 r}e^{i\omega r/v_s}-\frac{\partial^2}{\partial x_i\partial x_j}\left(\frac{e^{i\omega r/v_p}}{\omega^2 r}-\frac{e^{i\omega r/v_s}}{\omega^2 r}\right)\right]e^{-i\omega(t-t')}d\omega\\
&=\frac{1}{8\pi^2\rho}\int_{-\infty}^{\infty}\frac{\delta_{ij}}{v_s^2 r}e^{i\omega\left(\frac{r}{v_s}-(t-t')\right)}\\
&\quad-\frac{\partial^2}{\partial x_i\partial x_j}\left(\frac{e^{i\omega\left(\frac{r}{v_p}-(t-t')\right)}}{\omega^2 r}-\frac{e^{i\omega\left(\frac{r}{v_s}-(t-t')\right)}}{\omega^2 r}\right)d\omega
\end{aligned} \tag{9.96}$$

We can explicitly write the above inverse Fourier transform as

$$G_{ij}(\mathbf{x},t)=\frac{1}{4\pi\rho}\left[\frac{\delta_{ij}}{v_s^2 r}\delta\left(t-t'-\frac{r}{v_s}\right)\right.\\ \left.+\frac{\partial^2}{\partial x_i\partial x_j}\frac{R\left(t-t'-\frac{r}{v_p}\right)-R\left(t-t'-\frac{r}{v_s}\right)}{r}\right] \tag{9.97}$$

where $R(x)=xH(x)$ is the ramp function defined in Appendix 1.B.

9.A.3 Summary of elastodynamic Green's functions

The above Green's functions can be extended to lower dimensions, say 2D or 1D cases [132], systemically with the identities of Fourier transform of a scalar potential in the Helmoholtz equation. Using the Fourier transform technique, Eq. (9.100) becomes

$$\mathbf{u}(\mathbf{x},t)=\int_{-\infty}^{\infty}\frac{d\omega}{2\pi}\int_{-\infty}^{\infty}\frac{1}{(2\pi)^n}e^{-i(\boldsymbol{\xi}\cdot\mathbf{x}-\omega t)}\bar{\mathbf{u}}(\boldsymbol{\xi},\omega)d\boldsymbol{\xi} \tag{9.98}$$

where ω is the frequency in the Fourier domain, $\boldsymbol{\xi}$ is the wave vector transforming from the Cartesian coordinate to the Fourier space, and its dimension equals n ($n=1,2,3$), and $^-$ represents the quantities after the Fourier transform. In the Fourier domain, the partial differentiation process is transformed as the multiplication of vector $\boldsymbol{\xi}$, and time differentiation changes as $i\omega$:

$$\dot{\mathbf{u}}\leftrightarrow -i\omega\bar{\mathbf{u}} \tag{9.99a}$$

$$\ddot{\mathbf{u}}\leftrightarrow -\omega^2\bar{\mathbf{u}} \tag{9.99b}$$

$$\nabla\mathbf{u}(\mathbf{x},t)\leftrightarrow i\boldsymbol{\xi}\bar{\mathbf{u}} \tag{9.99c}$$

$$\Delta\mathbf{u}(\mathbf{x},t)\leftrightarrow -(\boldsymbol{\xi}\cdot\boldsymbol{\xi})\bar{\mathbf{u}}=-\xi^2\bar{\mathbf{u}} \tag{9.99d}$$

where $\Delta=\nabla\cdot\nabla$ is the Laplace operator, and $\leftrightarrow$ represents the Fourier transform and its inverse transform.

Using the 3D case as an example, the governing equation (9.2) can be written in the vector form as

$$\ddot{\mathbf{u}}=v_p^2\nabla(\nabla\cdot\mathbf{u})-v_s^2\nabla\times(\nabla\times\mathbf{u})+\frac{f_i}{\rho} \tag{9.100}$$

With the aid of Eq. (9.99), Eq. (9.100) can be transformed into the Fourier domain of the simplified form

$$-\omega^2\bar{\mathbf{u}}(\boldsymbol{\xi},\omega)=\frac{\bar{\mathbf{f}}}{\rho}-v_p^2\boldsymbol{\xi}(\boldsymbol{\xi}\cdot\bar{\mathbf{u}})+v_s^2\boldsymbol{\xi}\times\boldsymbol{\xi}\times\bar{\mathbf{u}} \tag{9.101}$$

Multiplying both sides of Eq. (9.101) by $\boldsymbol{\xi}\cdot$ and applying the identity of $\boldsymbol{\xi}\cdot\boldsymbol{\xi}\times(\boldsymbol{\xi}\times\bar{\mathbf{u}})=\mathbf{0}$, we have

$$-\omega^2\boldsymbol{\xi}\cdot\bar{\mathbf{u}}=\frac{\boldsymbol{\xi}\cdot\bar{\mathbf{f}}}{\rho}-v_p^2\xi^2(\boldsymbol{\xi}\cdot\bar{\mathbf{u}}) \tag{9.102}$$

where $\boldsymbol{\xi}\cdot\bar{\mathbf{u}}=\frac{\boldsymbol{\xi}\cdot\bar{\mathbf{f}}}{\rho(v_p^2\xi^2-\omega^2)}$. For the first term on the right-hand side of Eq. (9.102), applying the identity $\boldsymbol{\xi}\times\boldsymbol{\xi}\times\bar{\bar{\mathbf{f}}}=\boldsymbol{\xi}(\boldsymbol{\xi}\cdot\bar{\mathbf{f}})-\xi^2\bar{\mathbf{f}}$, we have

$$(v_s^2\xi^2-\omega^2)\bar{\mathbf{u}}=\frac{1}{\rho\xi^2}\boldsymbol{\xi}(\boldsymbol{\xi}\cdot\bar{\mathbf{f}})-\frac{1}{\rho\xi^2}\boldsymbol{\xi}\times\boldsymbol{\xi}\times\bar{\mathbf{f}}-v_p^2\boldsymbol{\xi}(\frac{\rho^{-1}\boldsymbol{\xi}\cdot\bar{\mathbf{f}}}{v_p^2\xi^2-\omega^2})$$
$$+v_s^2(\boldsymbol{\xi}\frac{\rho^{-1}\boldsymbol{\xi}\cdot\bar{\mathbf{f}}}{v_p^2\xi^2-\omega^2}) \tag{9.103}$$

Simplifying Eq. (9.103), we obtain an explicit expression of displacement in the Fourier space,

$$\rho\omega^2\bar{\mathbf{u}}=\frac{\omega^2}{\xi^2(v_p^2\xi^2-\omega^2)}\boldsymbol{\xi}(\boldsymbol{\xi}\cdot\bar{\mathbf{f}})-\frac{\omega^2}{\xi^2(v_s^2\xi^2-\omega^2)}\boldsymbol{\xi}\times(\boldsymbol{\xi}\times\bar{\mathbf{f}}) \tag{9.104}$$

Notice that although ω^2 can be canceled on both sides, the left-hand side can be directly transformed into the displacement field in Cartesian coordinate and time domain. Taking the inverse Fourier transform of both sides about the spatial domain, we can write

$$\rho\bar{\mathbf{u}}(\mathbf{x},\omega)=\nabla(\nabla\cdot\phi_{v_p}^{\omega})-\nabla\times(\nabla\times\phi_{v_s}^{\omega}) \tag{9.105}$$

where $\phi_{v_p}^{\omega}$ and $\phi_{v_s}^{\omega}$ are the fundamental solutions to the Helmholtz equation ($\Delta\phi=-\omega^2\phi=0$) in the frequency domain. Applying once more the integral about the frequency domain, we obtain

$$\rho\mathbf{u}(\mathbf{x},t)=\nabla(\nabla\cdot\phi_{v_p})-\nabla\times(\nabla\times\phi_{v_s}) \tag{9.106}$$

where ϕ_{v_p} and ϕ_{v_s} are the fundamental solutions to the Helmholtz equation in the time domain. The fundamental solutions are listed Table 9.1, including $n=1,2,3$. In Table 9.1, v is the speed defined for the primary and secondary waves, $R=|\mathbf{x}-\mathbf{x}'|$, and $H_0^{(1)}$ is the Hankel function of first kind. Note that we only list the fundamental solution of the branch in the positive direction with $H(\mathbf{R})$ and neglect the branch with $H(-\mathbf{R})$. According to Eqs. (9.105) and (9.106), the fundamental solution can be expressed as

$$G_{ij}(\mathbf{x},\mathbf{x}')=\frac{1}{\rho}\left[\partial_i\partial_j\phi_{v_p}+(\delta_{ij}-\partial_i\partial_j)\phi_{v_s}\right] \tag{9.107}$$

TABLE 9.1 Table of the fundamental solutions to Helmholtz equation in frequency and time domain.

Dimension n	Time domain	Frequency domain	Original equation
1	$\frac{\delta(t-\frac{R}{v})}{2v}$	$\frac{i}{2v^2\sqrt{\frac{\omega}{v}}}e^{i\frac{\omega}{v}R}$	$\frac{1}{\xi^2-\omega^2}$
2	$\frac{i}{2\pi v^2}\frac{H(t-\frac{R}{v})}{\sqrt{v^2t^2-R^2}}$	$\frac{i}{4v^2}H_0^{(1)}(\sqrt{\frac{\omega}{v}}R)$	same
3	$\frac{1}{4\pi Rv^2}\delta(t-\frac{R}{v})$	$\frac{e^{i\sqrt{\frac{\omega}{v}}R}}{4\pi v^2 R}$	same

where for the specific case of time-harmonic load, substitute the fundamental solution ϕ_v into ϕ_v^ω. For the 2D or 1D solution, $\nabla\times$ is not defined, but the same procedure can be conducted as in Sections 9.A.1 and 9.A.2, and the same index form of the Green's function can be obtained as Eq. (9.107) by replacing the scalar potential with the new form in Table 9.1.

Appendix 9.B Green's function for transient heat conduction

The above method can be applicable to the transient heat conduction problem as well. Now we derive the Green's functions in thermal analysis in parallel with Appendix 9.A.

The governing equation in the absence of a heat source can be written as

$$\frac{\partial u}{\partial t} = \alpha \Delta u + \frac{q}{c_p} \tag{9.108}$$

where u and q are the temperature and heat source fields, α is the heat diffusivity, and c_p is the specific heat based on volume as $\alpha = \frac{k}{c_p}$ with k being the thermal conductivity. Notice that although we use the same symbol u and α as in Appendix 9.A but with different meaning. Because the symbols have been extensively used in the literature for elastic and thermal problems, we do not change them although it might cause some misunderstanding.

Compared with the derivation in Section 9.A.3, the displacement vector is replaced by the temperature scalar. The Fourier transform relationship in Eq. (9.99) will keep the same. Therefore the temperature field can be similarly transformed from the Fourier domain as

$$u(\mathbf{x}, t) = \int_{-\infty}^{\infty} \frac{d\omega}{2\pi} \int_{-\infty}^{\infty} \frac{1}{(2\pi)^n} e^{-i(\boldsymbol{\xi}\cdot\mathbf{x}-\omega t)} \overline{u}(\boldsymbol{\xi}, \omega) d\boldsymbol{\xi} \tag{9.109}$$

Eq. (9.108) can be transformed into the Fourier domain in the simplified form

$$-i\omega\overline{u}(\boldsymbol{\xi}, \omega) = -\alpha\xi^2\overline{u}(\boldsymbol{\xi}, \omega) + \frac{\overline{q}}{c_p} \tag{9.110}$$

or

$$c_p \alpha \overline{u}(\boldsymbol{\xi}, \omega) = \frac{\overline{q}}{\xi^2 - i\omega/\alpha} \tag{9.111}$$

Taking the inverse Fourier transform of both sides of Eq. (9.111) with $\overline{q} = 1$ to the spatial domain, we obtain the Green's function for harmonic heat point source as

$$G(\mathbf{x}, \omega) = \int_{-\infty}^{\infty} \frac{1}{(2\pi)^3} e^{-i(\boldsymbol{\xi}\cdot\mathbf{x})} \overline{u}(\boldsymbol{\xi}, \omega) d\boldsymbol{\xi} = \frac{\exp\left[-(1-i)\sqrt{\frac{\omega}{2\alpha}} r\right]}{4\pi c_p \alpha r} \tag{9.112}$$

Notice that the same procedure of derivation in Section 9.A.1 can be used by treating $i\omega$ as β^2 in Eq. (9.86). Here we use $\sqrt{-i\omega} = (1-i)\sqrt{\frac{\omega}{2}}$.

Applying the inverse Fourier transform of Eq. (9.111) with $\overline{q} = 1$ to the frequency domain, we obtain

$$G(\boldsymbol{\xi}, t) = \int_{-\infty}^{\infty} \frac{1}{2\pi} e^{i\omega t} \overline{u}(\boldsymbol{\xi}, \omega) d\omega = \frac{\exp(-\alpha t \xi^2)}{c_p} H(t) \tag{9.113}$$

Applying the inverse Fourier transform of the above equation to the spatial domain, we can derive the Green's function in the actual spatial-time domain following the procedure as in Section 9.A.1:

$$\begin{aligned}
G(\mathbf{x}, t) &= \int_{-\infty}^{\infty} \frac{1}{(2\pi)^3} e^{-i(\boldsymbol{\xi}\cdot\mathbf{x})} G(\boldsymbol{\xi}, t) d\boldsymbol{\xi} \\
&= \frac{H(t)}{c_p (2\pi)^3} \int_{-\infty}^{\infty} e^{-\alpha t \xi^2} e^{-i(\boldsymbol{\xi}\cdot\mathbf{x})} d\boldsymbol{\xi} \\
&= \frac{H(t)}{c_p (2\pi)^3} \int_{0}^{\infty} \int_{S^2} e^{-\alpha t \xi^2} e^{-i(\boldsymbol{\xi}\cdot\mathbf{x})} \xi^2 d\xi dS \\
&= \frac{H(t)}{2c_p (2\pi)^3} \int_{-\infty}^{\infty} \int_{S^2} e^{-\alpha t \xi^2} e^{-i(\xi x \mathbf{s}\cdot\mathbf{n})} \xi^2 d\xi dS \\
&= \frac{H(t)}{2c_p (2\pi)^3} \int_{S^2} \frac{\sqrt{\pi}}{4\sqrt{\alpha t}^3} e^{-\frac{(x\mathbf{s}\cdot\mathbf{n})^2}{4\alpha t}} \left(2 - \frac{(x\mathbf{s}\cdot\mathbf{n})^2}{\alpha t}\right) dS \\
&= \frac{H(t)}{2c_p (2\pi)^3} \int_{-1}^{1} \int_{0}^{2\pi} \frac{\sqrt{\pi}}{4\sqrt{\alpha t}^3} e^{-\frac{(x\mathbf{s}\cdot\mathbf{n})^2}{4\alpha t}} \left(2 - \frac{(x\mathbf{s}\cdot\mathbf{n})^2}{\alpha t}\right) d(\mathbf{s}\cdot\mathbf{n}) d\phi \\
&= \frac{H(t)}{2c_p (2\pi)^3} \frac{2\pi\sqrt{\pi}}{\sqrt{\alpha t}^3} \int_{-1}^{1} e^{-\left(\frac{x\mathbf{s}\cdot\mathbf{n}}{2\sqrt{\alpha t}}\right)^2} \left(\frac{1}{2} - \left(\frac{x\mathbf{s}\cdot\mathbf{n}}{2\sqrt{\alpha t}}\right)^2\right) d(\mathbf{s}\cdot\mathbf{n})
\end{aligned}$$

$$
\begin{aligned}
&= \frac{H(t)}{2c_p(2\pi)^3}\frac{4\pi\sqrt{\pi}}{\sqrt{\alpha t}^3}\frac{2\sqrt{\alpha t}}{x} \\
&\times \int_0^{\frac{x}{2\sqrt{\alpha t}}} e^{-\left(\frac{x\mathbf{s}\cdot\mathbf{n}}{2\sqrt{\alpha t}}\right)^2}\left(\frac{1}{2}-\left(\frac{x\mathbf{s}\cdot\mathbf{n}}{2\sqrt{\alpha t}}\right)^2\right) d\left(\frac{x\mathbf{s}\cdot\mathbf{n}}{2\sqrt{\alpha t}}\right) \\
&= \frac{H(t)}{2c_p(2\pi)^3}\frac{4\pi\sqrt{\pi}}{\sqrt{\alpha t}^3}\frac{1}{2}e^{-\frac{x^2}{4\alpha t}}
\end{aligned}
\tag{9.114}
$$

which can be simplified to

$$
G(\mathbf{x}, t) = \frac{1}{c_p(4\pi\alpha t)^{\frac{3}{2}}}e^{\frac{-x^2}{4\alpha t}} \tag{9.115}
$$

The above derivation can be extended to 2D and 1D transient heat problems in parallel with Appendix 9.A.

Appendix 9.C Integral of Green's function for transient heat conduction for experimental validation

To show the phase change material (PCM) in building envelope application, ambient temperature change interaction with the building skin and the phase change of the PCM, iBEM implements the time integral based on the transient Green's function, which is validated by the experiments as well.

9.C.1 Explicit time-integral form of *G*, *T* and discussion of singularity

The explicit time integrals of the Green's function and its derivative are provided as follows [117]:

$$
\begin{aligned}
\overline{G}(\mathbf{r}, t_\Omega, \Delta t) &= \int_{t_f-1}^{t_f} G(\mathbf{r}, \tau)\, dt \\
&= \frac{1}{4\alpha r\pi^{\frac{3}{2}}}[\Gamma(0.5, \frac{r^2}{4\alpha(t_\Omega+\Delta t)}) - \Gamma(0.5, \frac{r^2}{4\alpha t_\Omega})]
\end{aligned}
\tag{9.116}
$$

$$
\begin{aligned}
\overline{T}(\mathbf{r}, t_\Omega, \Delta t) &= \int_{t_f-1}^{t_f} T(\mathbf{r}, \tau)\, dt \\
&= \frac{\mathbf{r}\cdot\mathbf{n}}{2\alpha r^3\pi^{\frac{3}{2}}}[\Gamma(1.5, \frac{r^2}{4\alpha(t_\Omega+\Delta t)}) - \Gamma(1.5, \frac{r^2}{4\alpha t_\Omega})]
\end{aligned}
\tag{9.117}
$$

where Γ is the upper incomplete gamma function.

Because the Green's function is singular at the source points, the integrals over the spatial domain should be carefully conducted. The singularities can be divided into two kinds, weakly singular and strongly singular. As for weak singularity, the integral of the kernel function poses no problem [21], and here we will not discuss the approach. However, the strong singularity generates problems in numerical implementation, and Beer proposed to use method of rigid-body motion, which creates one uniform temperature field (similar to displacement in elastic problem) with zero gradient. Thus the diagonal elements in $\overline{H}, C$ can be expressed by the negative sum of all other elements in the same row. The merit of the method is not only saving computation resources for strong singularities but also avoid errors in the matrix C of direct BEM method. The method of rigid motion works well with steady-state problems, but modifications are required for its application to transient problems. Dargush and Banerjee proposed the method to separate the kernel integral into two parts, where one is time-dependent, and the other is time-independent [133]. Similarly, the kernel function for boundary node temperature can be separated as

$$\int_{\partial D} \overline{T}(\mathbf{r}, t_\Omega, \Delta t)\, ds = \int_{\partial D} \frac{\mathbf{r} \cdot \mathbf{n}}{2\alpha r^3 \pi^{\frac{3}{2}}}[-\gamma(1.5, \frac{r^2}{4\alpha(t_\Omega + \Delta t)}) + \gamma(1.5, \frac{r^2}{4\alpha t_\Omega})]ds \tag{9.118}$$

where $\Gamma(s)$ and $\gamma(s)$ are complete gamma function and lower incomplete gamma functions, and $\Gamma(s,t) = \Gamma(s) - \gamma(s,t)$ is applied. In most cases that $t_\Omega = t_F - t_f \neq 0$ (not the first coefficient matrix), Eq. (9.118) has no singularity as the entire asymptotic series is $O(r^3)$. When the time steps coincide, $\gamma(1.5, \frac{r^2}{4\alpha t_\Omega}) = \gamma(1.5, \infty) \approx \Gamma(1.5)$, which is a constant in the integral equation. Therefore there exist strong singularities, which can be solved by separating constant and lower incomplete gamma function series as ($F - f = 0$):

$$\int_{\partial D} \overline{T}(\mathbf{r}, 0, \Delta t) = \int_{\partial D} \frac{\mathbf{r} \cdot \mathbf{n}}{4\alpha r^3 \pi}\, ds - \int_{\partial D} \gamma(1.5, \frac{r^2}{4\alpha(t_\Omega + \Delta t)})\, ds \tag{9.119}$$

where the first term is the similar term derived in steady-state heat conduction problems. Similarly, the matrix $\overline{H}$ can be divided into "steady-state" and "transient" parts, and the method of rigid-body motion can only be applied to the "steady-state" part. As a consequence of applying method of rigid-body motion directly on $\overline{H}$, the solved heat field will be steady state.

9.C.2 Comparison of the mesh for the experiment verification

As shown in Fig. 9.4, the iBEM only requires a surface mesh, and the number of element seeds on each edge is 12, and thus 144 elements are generated for each surface with totally 866 nodes. The particles in the domain are treated as just four unknowns of the eigenfield, and therefore no internal mesh is needed. The same surface mesh for iBEM can be used for various cases unless the external geometry changes. However, the FEM requires the internal volume mesh. With just one particle, the volume mesh includes 49,342 elements. Considering more

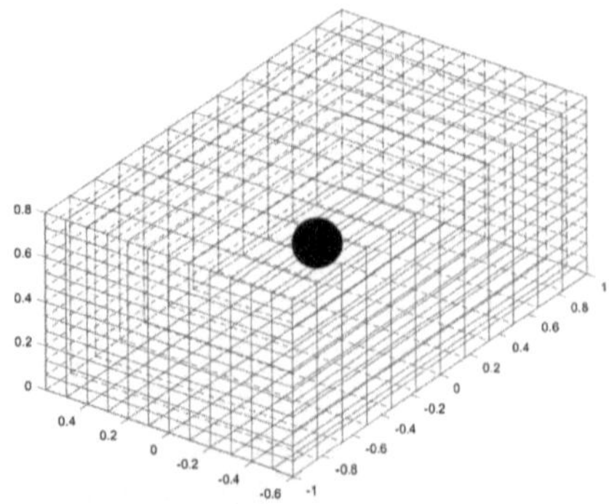

(a) iBEM surface mesh

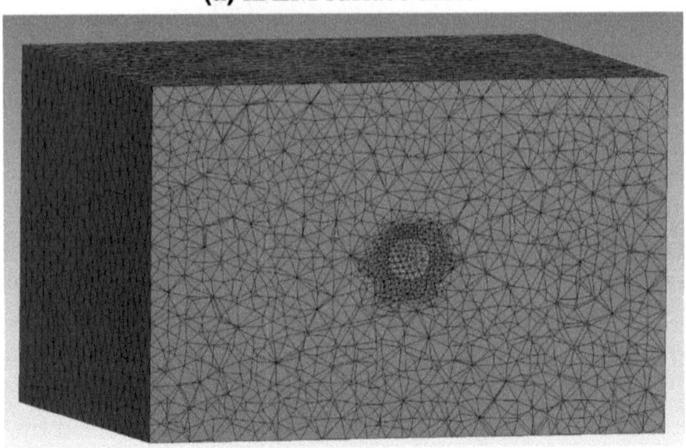

(b) FEM volume mesh

FIGURE 9.4 Schematic illustration of surface/volume mesh for the iBEM and FEM for one PCM particle case.

particles with size gradation, it is quite challenging since the difference between neighbor element sizes is strictly limited for accuracy purposes.

Chapter 10

The iBEM for multiphysical problems

Advanced applications

In this chapter, we demonstrate applications of the iBEM for solving multiphysical problems by integrating different boundary value problems in the same framework on the material modeling and analysis. A case study shows the fabrication and testing of a ferromagnetic composite. During fabrication, ferromagnetic particles are dispersed in a rheological fluid and aligned into chains under an external magnetic field. The magnetic field in the composite can be solved by the potential problem, the magnetic force between particles can be calculated, and the Stokes flow can be simulated step by step. The software can be extended to other multiphysical problems in material manufacturing and processing, which involve different types of forces and phase changes.

10.1 Introduction to multiphysical modeling of composites

With emerging material development through nanotechnologies, additive manufacturing, and computer-aided design, we can precisely control and fabricate materials at multiscale for multifunctional purposes. Understanding the multiphysical behavior of the composites is important for the material design and applications, which typically cover multiple disciplines. Many engineering materials and structures present multiple length/time scale-dependent behaviors, which are coupled with multiphysical phenomena, such as mechanical, kinetic, hydraulic, thermal, electromagnetic, acoustic, and optical aspects. Though multiphysical problems have long been studied in mathematics, physics, and chemistry, the current excitement is driven heavily by the use of the advanced modeling in the applied engineering and science, which advances the fundamental understanding of material behavior and catalyzes the design and discovery of new composite materials.

In addition, multifunctional materials are sought to meet specific requirements through tailored microstructure under changing work environments. For example, smart materials can be considered as multifunctional materials that are able to react upon an external stimulus. The introduction of biomimetics in the material science allows the designing of materials with similar processes as nature does. Properly executed modeling of multiphysics-multiscale-

The Inclusion-Based Boundary Element Method (iBEM). https://doi.org/10.1016/B978-0-12-819384-6.00018-8

multifunctional engineering materials and structures can vastly improve accuracy and efficiency in solutions which were not quite feasible through conventional approaches. In many cases, high-fidelity numerical experiments on computers can reduce the development cost and time for material design and discovery.

Although Eshelby's theory in inclusion and inhomogeneity was developed for elastic problems, it has been extended to multiphysical problems in the past decades. The iBEM is a timely and powerful tool for this need. In this chapter, we use the fabrication and application of a ferromagnetic particulate composite to demonstrate the versatile capability of iBEM on virtual manufacture and experiments of material samples under combined loading conditions with force and magnetic fields.

With further development, iBEM can be a promising tool for coupled (multiphysical) modeling/simulation of engineering materials and structures, materials–structures interaction with environmental effects, and multiphysics coupling of stimuli-responsive smart materials/structures. It can accommodate the multiscale behavior of composites at different time and spatial scales. Some work will be revisited in the next chapter for future work.

10.2 Equivalent inclusion method for magnetostatic problem

Magnetic composites have been used in various engineering applications. During the material fabrication process, using magnetic field can change and control the microstructure of particle distribution for favorable material behavior in the final product. During the service condition, applying magnetic field can yield deformation, called magnetostriction, and change the stiffness of the material in a certain direction, which creates feasibility for design and development of smart materials, such as magnetic active actuators and dampers.

Consider an infinite domain D containing multiple nickel particles Ω_i dispersed in a viscous fluid, as shown in Fig. 10.1. The viscosity of the fluid is μ_0. For convenience of formulation, a solid particle can be treated as a fluid with viscosity $\mu_1 \to \infty$ moving in a Stokes fluid. The magnetic permeability constants of the particles and the fluid are μ_1^m and μ_0^m, respectively. A uniform far-field of magnetic field $\mathbf{H^0}$ is applied to this system. The magnetic field is not high so that both the magnetic and flow behaviors are in the linear range.

Due to the existence of particles, the local magnetic field is disturbed, and interaction forces are induced between particles, which leads to particle motion and alignment. Here we assume that the viscosity of the fluid is fairly high and the motion of particles is governed by the Stokes law. Basically, the whole process includes two coupled parts. First, due to the slow motion of particles, the local magnetic field in the composite can be derived by the magnetostatic theory with the ad hoc microstructure, and then the magnetic force on each particle can be calculated. Second, the magnetic force will drive the particles moving in the fluid, and then the particle distribution or material microstructure will

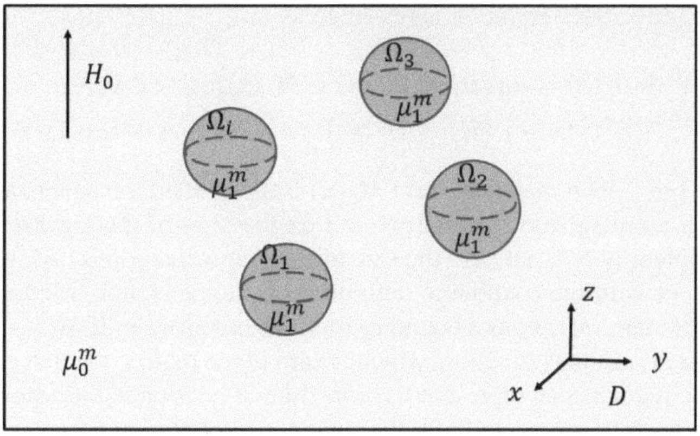

FIGURE 10.1 Configuration of multiple nickel particles dispersed in a viscous fluid.

be changed. Based on the observation in the experiments, the particles will be aligned in the space parallel to the direction of the applied magnetic field, which depends on the volume fraction of particles.

10.2.1 One ellipsoidal inhomogeneity

For an ellipsoidal particle with a magnetic permeability μ_1 embedded in a matrix with a magnetic permeability μ_0, the local magnetic field and magnetic induction satisfies the constitutive relation

$$\begin{cases} B_i(\mathbf{x}) = \mu_1 H_i(\mathbf{x}), & x \in \Omega \\ B_i(\mathbf{x}) = \mu_0 H_i(\mathbf{x}), & x \in D - \Omega \end{cases} \tag{10.1}$$

Here the magnetic field $H_i(\mathbf{x})$ can be written with the magnetic potential U as

$$H_i = -U_{,i} \tag{10.2}$$

When a far-field magnetic load H_i^0 is applied, the disturbed fields H_i' and B_i' will be induced by the material mismatch around inhomogeneity Ω, which satisfies

$$\begin{cases} B_i^0 + B_i'(\mathbf{x}) = \mu_1 \left(H_i^0 + H_i'(\mathbf{x})\right), & x \in \Omega \\ B_i^0 + B_i'(\mathbf{x}) = \mu_0 \left(H_i^0 + H_i'(\mathbf{x})\right), & x \in D - \Omega \end{cases} \tag{10.3}$$

Using the equivalent inclusion method, we can introduce a prescribed magnetic field, similar to eigenstrain in elastic problem, to simulate the material difference. The above equation is rewritten as

$$\begin{cases} B_i^0 + B_i'(\mathbf{x}) = \mu_0 \left(H_i^0 + H_i'(\mathbf{x}) + M_i(\mathbf{x})\right), & x \in \Omega \\ B_i^0 + B_i'(\mathbf{x}) = \mu_0 \left(H_i^0 + H_i'(\mathbf{x})\right), & x \in D - \Omega \end{cases} \tag{10.4}$$

where the prescribed magnetic field $M_i(\mathbf{x})$ is also called a magnetization if the matrix is a nonmagnetic material. Note that the sign of the eigenstrain in the elastic problem is "–", but here the size of the magnetization is "+", which is to be consistent with the traditional definition of magnetization. Mathematically, we can consider $-M_i(\mathbf{x})$ as a counterpart to the eigenstrain. If M_i is considered a piecewise continuous function, which is zero in the matrix, the first equation in the above equations can represent the constitutive relation in the whole domain. For a steady-state magnetic field, the divergence of the magnetic induction is zero, i.e.,

$$B_{i,i} = 0 \tag{10.5}$$

Considering the disturbed magnetic field $H_i' = -U_{,i}$, we can write

$$U_{,ii}(\mathbf{x}) = M_{i,i}(\mathbf{x}) \tag{10.6}$$

Using the Green's function technique, we can write the solution of the magnetic potential as

$$U(\mathbf{x}) = -\int_D G\left(\mathbf{x}, \mathbf{x}'\right) \frac{\partial M_i\left(\mathbf{x}'\right)}{\partial x_i'} \mathrm{d}\mathbf{x}' \tag{10.7}$$

where the Green's function $G\left(\mathbf{x}, \mathbf{x}'\right)$ describes the magnetic potential at point $\mathbf{x}$ due to the magnetization at point $\mathbf{x}'$ in the infinite domain. It reads

$$G\left(\mathbf{x}, \mathbf{x}'\right) = \frac{\phi}{4\pi}, \quad \phi = \frac{1}{|\mathbf{x} - \mathbf{x}'|} \tag{10.8}$$

Because $M(\mathbf{x}) = 0$ as $\mathbf{x} \to \infty$, using the Gauss theorem, we can rewrite this equation as

$$U(\mathbf{x}) = -\int_D \frac{\partial G\left(\mathbf{x}, \mathbf{x}'\right)}{\partial x_i} M_i\left(\mathbf{x}'\right) \mathrm{d}\mathbf{x}', \tag{10.9}$$

where we used $\frac{\partial G(\mathbf{x},\mathbf{x}')}{\partial x_i} = -\frac{\partial G(\mathbf{x},\mathbf{x}')}{\partial x_i'}$. Then the substitution of Eq. (10.9) into $H' = -\nabla U$ yields

$$H_i'(\mathbf{x}) = \int_\Omega \Gamma_{ik}(\mathbf{x}, \mathbf{x}') M_k(\mathbf{x}') \mathrm{d}\mathbf{x}' \tag{10.10}$$

where the modified Green's function is

$$\Gamma_{ik} = \frac{\partial^2 G(\mathbf{x}, \mathbf{x}')}{\partial x_i \partial x_k}. \tag{10.11}$$

For the case of a single ellipsoidal inhomogeneity, the magnetization M_i is constant, written as M_i^0, and we can rewrite the above equation as

$$U(\mathbf{x}) = -\frac{M_i^0}{4\pi}\Phi_{,i} \tag{10.12}$$

Then we can write the disturbed magnetic field as

$$H_i'(\mathbf{x}) = \frac{M_k^0}{4\pi}\Phi_{,ik} \tag{10.13}$$

Using the equivalent inclusion method, we can solve the magnetization M_i^0 from the following equation over the inhomogeneity:

$$\mu_0\left(H_i^0 + \frac{M_k^0}{4\pi}\Phi_{,ik} + M_i^0\right) = \mu_1\left(H_i^0 + \frac{M_k^0}{4\pi}\Phi_{,ik}\right) \tag{10.14}$$

Therefore we obtain

$$M_i^0 = \left(\frac{\mu_0}{\mu_1 - \mu_0}\delta_{ik} - \frac{\Phi_{,ik}}{4\pi}\right)^{-1} H_k^0 \tag{10.15}$$

Then we can get magnetic potential, field, and induction straightforwardly. For more than one inhomogeneity, the magnetization will be in a polynomial form. The same method as in the last section can be used to derive the magnetic fields. The formulation can be extended to multiple particles with multiple material phases for multiphysical problems.

10.2.2 Multiple ellipsoidal inhomogeneities

For the multiparticle system shown in Fig. 10.1, the magnetization $M_i(\mathbf{x})$ is not uniform because of the particle interactions. Considering the Jth particle, expanding the magnetization over the corresponding local coordinate $\mathbf{x}'$ provides

$$M_i^J(\mathbf{x}') = A_i^{J0} + A_{il}^{J1}x_l + A_{ilp}^{J2}x_l x_p \tag{10.16}$$

where A_i^{J0}, A_{il}^{J1}, and A_{ilp}^{J2} are unknown coefficients to be determined through the equivalent condition. Substitute Eq. (10.16) into Eq. (10.10) and let

$$D_{ik}^0 = \int_\Omega \Gamma_{ik}(\mathbf{x},\mathbf{x}')\mathrm{d}\mathbf{x}'$$

$$D_{ikl}^1 = \int_\Omega \Gamma_{ik}(\mathbf{x},\mathbf{x}')x_l'\mathrm{d}\mathbf{x}'$$

$$D_{iklp}^2 = \int_\Omega \Gamma_{ik}(\mathbf{x},\mathbf{x}')x_l'x_p'\mathrm{d}\mathbf{x}'$$

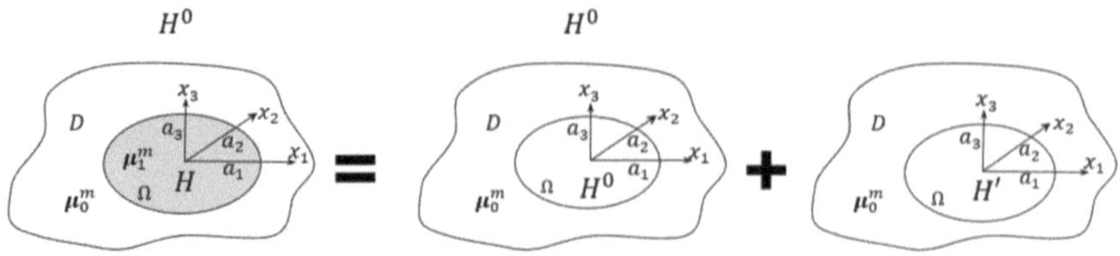

FIGURE 10.2 Equivalent conditions for magnetostatic problem.

where these integrals can be written in an explicit form by Φ, which can be found in Appendix 3.A as follows:

$$D_{ik}^{0} = \frac{1}{4\pi}\Phi_{,ik}, \quad D_{ikl}^{1} = \frac{1}{4\pi}\Phi_{l,ik}, \quad D_{iklp}^{2} = \frac{1}{4\pi}\Phi_{lp,ik} \tag{10.17}$$

Then the perturbed magnetic field caused by the magnetization can be presented by the actual magnetic field from all particles inside the domains as follows:

$$H_i'(\mathbf{x}) = \sum_{J=1}^{N}\left(D_{ik}^{J0}A_k^{J0} + D_{ikl}^{J1}A_{kl}^{J1} + D_{iklp}^{J2}A_{klp}^{J2}\right) \tag{10.18}$$

Consider the equivalent conditions for magnetostatic problem shown in Fig. 10.2, based on our previous work [95]. The magnetic field can be calculated by EIM, in which the mismatch of magnetic properties between the matrix and particles can be simulated by introducing magnetization on the particles, so that the particle is equivalently treated with magnetic permeability μ_0^m with magnetization $\mathbf{M}$. Therefore the solution of the magnetic field can be considered as the superposition of two sources: the first one is the uniform material with μ_0^m under a uniform far-field of $\mathbf{H^0}$, which leads to a uniform magnetic field, and the second one is the perturbed magnetic field induced by $\mathbf{M}$ over the particles. Thus we can write the magnetic field as

$$H_i(\mathbf{x}) = H_i^0 + H_i'(\mathbf{x}) \tag{10.19}$$

In view of Eqs. (10.19) and (10.18), the total magnetic field in one particle subdomain Ω can be expressed by the magnetic field in all the particles and far magnetic field in this system as

$$H_i(\mathbf{x}) = H_i^0 + \sum_{J=1}^{N}\left(D_{ik}^{J0}A_k^{J0} + D_{ikl}^{J1}A_{kl}^{J1} + D_{iklp}^{J2}A_{klp}^{J2}\right) \tag{10.20}$$

Using the equivalent inclusion method, we can get

$$\mu_0^m\left(H_i^0 + H_i'(\mathbf{x}) + M_k^J(\mathbf{x}')\right) = \mu_1^m\left(H_i^0 + H_i'(\mathbf{x})\right) \tag{10.21}$$

Then the magnetization $M_k^J(\mathbf{x}')$ can be solved from the following equation over the particles:

$$\begin{aligned} & A_i^{J0} + A_{ik}^{J1} x_k + A_{ikl}^{J2} x_k x_l \\ & = H_i^0 + \chi_1 \sum_{J=1}^{N} \left(D_{ik}^{J0} A_k^{J0} + D_{ikl}^{J1} A_{kl}^{J1} + D_{iklp}^{J2} A_{klp}^{J2} \right) \quad \text{for } I = 1, 2, \ldots, N \end{aligned} \tag{10.22}$$

where $\chi_1 = \frac{\mu_1^m - \mu_0^m}{\mu_0^m}$. Then expanding the right side of Eq. (10.22) in a Taylor series at the center of each particle and matching the coefficients of the terms of $\mathbf{x}$, we obtain the following equations:

$$A_i^{I0} = H_i^0 + \chi_1 \sum_{J=1}^{N} \left(D_{ik}^{J0} A_k^{J0} + D_{ikl}^{J1} A_{kl}^{J1} + D_{iklp}^{J2} A_{klp}^{J2} \right) \quad \text{for } I = 1, 2, \ldots, N, \tag{10.23}$$

$$A_{ik}^{I1} = \chi_1 \sum_{J=1}^{N} \left(D_{im,k}^{J0} A_m^{J0} + D_{imn,k}^{J1} A_{mn}^{J1} + D_{imnp,k}^{J2} A_{mnp}^{J2} \right) \quad \text{for } I = 1, 2, \ldots, N, \tag{10.24}$$

$$A_{ikl}^{I2} = \chi_1 \sum_{J=1}^{N} \frac{1}{2} \left(D_{im,kl}^{J1} A_m^{J0} + D_{imn,kl}^{J2} A_{mn}^{J1} + D_{imnp,kl}^{J2} A_{mnp}^{J2} \right) \quad \text{for } I = 1, 2, \ldots, N. \tag{10.25}$$

By solving the above linear equation system of A_i^0, A_{ik}^1, and A_{ikl}^2 we can obtain the magnetic field in the whole domain by using these parameters in Eq. (10.19). Notice that the magnetic in particles can also be calculated by Eq. (10.19), which reaches the accuracy up to the quadratic terms in the Taylor expansion of the magnetic field. Generally speaking, we can obtain higher accuracy by using more terms in Eq. (10.19). Based on our previous work [134], very good accuracy can be obtained when the linear form of the magnetic field is used. In this section, we use the quadratic form of the magnetic field, so that the accuracy will be fairly high, especially when the particle spacing is small. However, it also requires much higher computational cost and memory in simulation. When the spacing between particles is not very small, the linear form solution may capture the physics very well.

10.3 Basic theory of a steady-state magnetic problem

10.3.1 Local magnetic field

As illustrated in Fig. 10.1, the far-field magnetic field $\mathbf{H}^0$ keeps constant with the magnitude and direction. Because the particle motion is slow, the Stokes law is applicable, and the inertia force is disregarded. Then, based on the magnetostatic theory, the local magnetic field can be solved in terms of the configuration of particle distribution. Because there is no electric process involving, this specific problem will be degraded to solve the following Maxwell equations:

$$\nabla \cdot \mathbf{B} = 0 \tag{10.26}$$

$$\nabla \times \mathbf{H} = \mathbf{0} \tag{10.27}$$

where $\mathbf{B}$ is the magnetic induction. For diamagnets and paramagnets, the relationships between the magnetic induction $\mathbf{B}$, the magnetic field $\mathbf{H}$, and the magnetization $\mathbf{M}$ are as follows:

$$\mathbf{M} = \chi \mathbf{H}, \ \mathbf{B} = \mu_i^m \mathbf{H} = \mu^m (1 + \chi) \mathbf{H} \tag{10.28}$$

where χ is called the volume magnetic susceptibility, μ_i^m is the magnetic permeability of the material i, and μ^m is the vacuum permeability. By Helmholtz decomposition, let $\mathbf{H} = -\nabla U$, where U is a scalar magnetic potential. Thus Eq. (10.27) can be automatically satisfied. Substituting Eq. (10.28) into Eq. (10.26), we can rewrite the Maxwell equations as

$$\nabla \cdot \mathbf{H} = -\nabla \cdot \mathbf{M} \tag{10.29}$$

Then by using $\mathbf{H} = -\nabla U$ Eq. (10.29) leads to only one scalar [135] equation

$$\nabla^2 U = \nabla \cdot \mathbf{M} \tag{10.30}$$

Since both the particle and matrix phases exist, the interface continuity along the surface of particles should be considered. As the magnetic field $\mathbf{B}$ is continuous in the normal direction, using the linear constitutive relation between $\mathbf{B}$ and $\mathbf{H}$, we can get the following interface conditions:

$$\mu_1^m \mathbf{H}_n^- = \mu_0^m \mathbf{H}_n^+ \tag{10.31}$$

where μ_1^m and μ_0^m are the particle permeability and the matrix permeability, respectively, the subscript n denotes the component of the magnetic field $\mathbf{H}$ in the normal direction of the surface, i.e., $|\mathbf{H}_n| = \mathbf{H} \cdot \mathbf{n}$, and the superscripts $+$ and $-$ denote the outside and inside surface of the particle. In tangential direction, based on the Maxwell equation and the condition that no free surface current exists, we can write

$$\mathbf{H}_t^+ = \mathbf{H}_t^-, \tag{10.32}$$

where the subscript t denotes the component of the magnetic field $\mathbf{H}$ in the tangential direction, i.e., $\mathbf{H}_t = \mathbf{H} - \mathbf{H}_n$.

For linear isotropic magnetic materials, the boundary value problem is reduced to Eq. (10.30), where U is the scalar magnetic potential, and $\mathbf{M}$ is the magnetization with the interface continuity and far-field boundary conditions.

Based on equivalent inclusion method, the magnetic field can be obtained by Eq. (10.19).

10.3.2 Magnetic force

The Lorentz law, one of the fundamental laws in electromagnetic analysis, describes the force that one charge with velocity v and electric quantity Q is subjected to electric field E and magnetic field B. The original Lorentz law [136] is

$$\mathbf{F} = Q(\mathbf{E} + \mathbf{v} \times \mathbf{B}). \tag{10.33}$$

When it is extended to the cases of line current, surface current, and volume current with $\mathbf{E} = 0$, by integrating the force in its domain we can derive

$$\mathbf{F} = \int (\mathbf{v} \times \mathbf{B})\mathrm{d}q \tag{10.34}$$

where the currents $\mathbf{v}$ come from electrons orbiting around nuclei and electrons spinning about their axes on an atomic scale of the magnetic material, and $\mathbf{B}$ is the external magnetic induction. When it is extended to the cases of surface current and volume current, we get

$$\mathbf{F} = \int (\mathbf{v} \times \mathbf{B})\mathrm{d}q = \int_{\partial\Omega} (\mathbf{v} \times \mathbf{B})\sigma \mathrm{d}s = \int_{\partial\Omega} (\mathbf{K} \times \mathbf{B})\mathrm{d}s \tag{10.35}$$

$$\mathbf{F} = \int (\mathbf{v} \times \mathbf{B})\mathrm{d}q = \int_{\Omega} (\mathbf{v} \times \mathbf{B})\rho \mathrm{d}v = \int_{\Omega} (\mathbf{J} \times \mathbf{B})\mathrm{d}v \tag{10.36}$$

where σ and ρ are the surface and volume charge densities, respectively, and $\mathbf{K} = \sigma\mathbf{v}$ and $\mathbf{J} = \rho\mathbf{v}$ are the currents referring to different cases.

Thus, in the magnetized materials, we consider the volume current density $\mathbf{J}_f$ (free), the density $\mathbf{J}_b$ (bound) inside the domain, surface current density $\mathbf{K}_f$ (free), and the density $\mathbf{K}_b$ (bound) on the surface. The bound current is related to the magnetization of the medium, and the free current is related to the current flow through wires embedded in the medium or through the materials themselves. In the present model, we consider only magnetization, and thus $\mathbf{J}_f = 0$ and $\mathbf{K}_f = 0$. Finally, the total force is [137]

$$\mathbf{F} = \int_{\Omega} (\mathbf{J}_f \times \mathbf{B})\mathrm{d}v + \int_{\Omega} (\mathbf{J}_b \times \mathbf{B})\mathrm{d}v + \int_{\partial\Omega} (\mathbf{K}_f \times \mathbf{B})\mathrm{d}s + \int_{\partial\Omega} (\mathbf{K}_b \times \mathbf{B})\mathrm{d}s \tag{10.37}$$

where the induced bound surface and volume current, which is originated from the magnetization, obeys the rules

$$\mathbf{K}_b = \mathbf{M} \times \mathbf{n} \tag{10.38}$$

$$\mathbf{J}_b = \nabla \times \mathbf{M} \tag{10.39}$$

In the present model, assuming that $J_f = 0$ and $K_f = 0$ and substituting Eq. (10.28) into Eq. (10.39), we get

$$\mathbf{J}_b = \nabla \times \mathbf{M} = \nabla \times (\chi_m \mathbf{H}) = \chi_m \mathbf{J}_f \tag{10.40}$$

which means that the bound current inside the domain is proportional to the free current flowing in the domain. Since there is no free current flowing in the material, the bound current does not exist inside the domain. Thus the only term left in the force expression is the force due to the bound surface current induced by magnetization,

$$\mathbf{F} = \int_{\partial\Omega} (\mathbf{K}_b \times \mathbf{B}) ds \tag{10.41}$$

To calculate the force, the bound surface current (10.38) should be evaluated on the surface. As for the conductive ferromagnetic material, if there is no free current prescribed inside the domain **D**, then all the magnetic forces are acting on the surface. The magnetic field **B** can be presented by the simple relation $\mathbf{B} = \mu_1^m \mathbf{H}$. The magnetization **M** can be represented by $\mathbf{M} = \chi_1 \mathbf{H}$, and based on our assumptions, the distribution of local magnetic field **H** follows the polynomial form up to quadratic term as Eq. (10.19). By combining Eqs. (10.19) and (10.41) along with normal vector $\mathbf{n} = \frac{x_m e_m}{a}$ for a spherical domain, we obtain

$$\begin{aligned}
\mathbf{F} &= \int_{\partial\Omega} (\mathbf{M} \times \mathbf{n} \times \mu_1^m \mathbf{H}) ds \\
&= \chi_1 \mu_1^m \int_{\partial\Omega} \left[\begin{array}{c} \left(A_i^0 e_i + A_{ij}^1 x_j e_i + A_{ijk}^2 x_j x_k e_i \right) \times \frac{x_m e_m}{a} \\ \times \left(A_n^0 e_n + A_{np}^1 x_p e_n + A_{npq}^2 x_p x_q e_n \right) \end{array} \right] ds \\
&= \frac{\chi_1 \mu_1^m (\varepsilon_{iml} \varepsilon_{lnr} e_r)}{a} \left[\begin{array}{c} \int_{\partial\Omega} A_i^0 A_n^0 x_m ds + \int_{\partial\Omega} A_i^0 A_{np}^1 x_m x_p ds \\ + \int_{\partial\Omega} A_i^0 A_{npq}^2 x_m x_p x_q ds + \int_{\partial\Omega} A_n^0 A_{ij}^1 x_j ds \\ + \int_{\partial\Omega} A_{ij}^1 A_{np}^1 x_m x_j x_p ds + \int_{\partial\Omega} A_{ij}^1 A_{npq}^2 x_m x_j x_p x_q ds \\ + \int_{\partial\Omega} A_n^0 A_{ijk}^2 x_m x_j x_k ds + \int_{\partial\Omega} A_{np}^1 A_{ijk}^2 x_m x_j x_k x_p ds \\ + \int_{\partial\Omega} A_{ijk}^2 A_{npq}^2 x_m x_j x_k x_p x_q ds \end{array} \right]
\end{aligned} \tag{10.42}$$

where e_i is the base vector in Cartesian coordinate, and ε_{ijk} is the Levi-Civita or permutation symbol. Based on the assumption that our particles are perfect spheres, it is more convenient to evaluate the integral in spherical coordinate. The relationships between Cartesian and spherical coordinates are $x_1 = a \sin\phi \cos\theta$, $x_2 = a \sin\phi \sin\theta$, and $x_3 = a \cos\phi$.

For the spherical surface, the surface integral can be expressed by spherical coordinate as

$$\int_{\Omega} ds = \int_0^{\pi} \sin\phi d\phi \int_0^{2\pi} a^2 d\theta \tag{10.43}$$

Then we can analytically evaluate the force. With the help of the identities of the integrals shown in Eq. (10.45), Eq. (10.42) can be finally simplified as follows:

$$F_r = \frac{\chi \mu \epsilon_{iml} \epsilon_{lnr}}{a}[\int_{\partial\Omega} A_i^0 A_{np}^1 x_m x_p \mathrm{d}s + \int_{\partial\Omega} A_n^0 A_{ij}^1 x_m x_j \mathrm{d}s$$
$$+ \int_{\partial\Omega} A_{ij}^1 A_{npq}^2 x_m x_j x_p x_q \mathrm{d}s + \int_{\partial\Omega} A_{np}^1 A_{ijk}^2 x_m x_j x_k x_p \mathrm{d}s] \quad (10.44)$$

where ε_{ijk} is the three-dimensional Levi-Civita symbol, and the following identities are applied:

$$\int_{\partial\Omega} x_m \mathrm{d}s = 0$$

$$\int_{\partial\Omega} x_m x_p \mathrm{d}s = \frac{4\pi a^4}{3}\delta_{mp}$$

$$\int_{\partial\Omega} x_m x_p x_q \mathrm{d}s = 0$$

$$\int_{\partial\Omega} x_m x_n x_p x_q \mathrm{d}s = \begin{cases} \frac{4\pi a^6}{5}, & \text{four indices are equal} \\ 0, & \text{two indices are equal, the other two indices are different} \\ \frac{4\pi a^6}{15}, & \text{two indices are equal, the other two indices are equal with different value} \\ 0, & \text{three indices are equal} \end{cases}$$

$$\int_{\partial\Omega} x_m x_n x_p x_q x_j \mathrm{d}s = 0. \quad (10.45)$$

10.4 Particle motion in a rheological fluid

Consider a nickel particle subdomain D with viscosity μ_1 moving in a homogeneous fluid domain Ω [138]. The viscosity μ_0 is constant in the fluid. In such a high viscous fluid with small Reynolds number, the inertia forces of the fluid may be disregarded in the Navier–Stokes equation, which reduces to the Stokes equation. Since the radius of the particle is small in our problem, we can assume that the driving magnetic force from the far-field magnetic loads is uniformly distributed on each nickel particle. The body force in this system can be simply described as follows:

$$f_i(x) = \begin{cases} f_i^0, & x \in D_i \\ 0, & x \in \Omega - \{D_1, D_2, \ldots, D_i\} \end{cases} \quad (10.46)$$

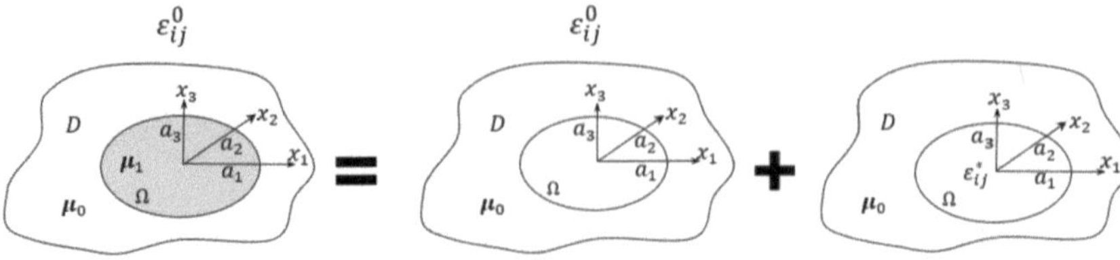

FIGURE 10.3 Equivalent condition for Stokes flow.

Using the equivalent inclusion method, the material mismatch of different viscosity in fluid and particles can be represented by eigenstrain rate $e^*_{ij}(x)$. Then the material property of the whole domain is the same as that of the fluid. Then the Stokes equation tells us that

$$2\mu_0 e_{ij,i} - p_{,j} = 0, \quad v_{i,i} = 0, \quad x \in \Omega - D \tag{10.47}$$

$$2\mu_0 \left(e_{ij,i} - e^*_{ij,i}\right) - p_{,j} = -f_i, \quad v_{i,i} = 0, \quad x \in D \tag{10.48}$$

Based on the equivalent inclusion method [10], e^*_{ij} satisfies the stress equivalent condition

$$\mu_0(e_{ij} - e^*_{ij}) = \mu_1 e_{ij}, \quad x \in D \tag{10.49}$$

Here the eigenstrain rate is a fictitious nonmechanical field, which makes the stress field in two problems equivalent as shown in Fig. 10.3. Then the original inhomogeneity problem can be treated as an inclusion problem with the eigenstrain rate satisfying Eq. (10.49). We assume that the eigenstrain rate follows the polynomials form up to quadratic term:

$$e^*_{ij}(\mathbf{x}) = e^0_{ij} + e^1_{ijk}x_k + e^2_{ijkl}x_k x_l \tag{10.50}$$

With the aid of the Green's function techniques, the velocity field in the fluid can be expressed by the integral of body force f and eigenstrain rate e^* in all particles as follows:

$$v_i = \int_{\Omega} \left(G_{ik}(\mathbf{x}, \mathbf{x}') f_k - 2\mu_0 G_{im,n}(\mathbf{x}, \mathbf{x}') e^*_{mn}(\mathbf{x}')\right) \mathrm{d}\mathbf{x}' \tag{10.51}$$

where the Green's function is the following:

$$G_{ij}(\mathbf{x}, \mathbf{x}') = \frac{1}{4\pi\mu_0}\left(\delta_{ij}\phi - \frac{1}{2}\frac{\partial^2 \psi}{\partial x_i \partial x_j}\right), \quad \psi = |\mathbf{x} - \mathbf{x}'| \tag{10.52}$$

This integral can be analytically carried out in a spherical domain. Finally, we can obtain the velocity

$$v_i = \sum_{J=1}^{N} \left(L_{ik} f_k^J + M_{imn} e_{mn}^{0J} + N_{imnp} e_{mnp}^{1J} + Q_{imnpq} e_{mnpq}^{2J} \right) \tag{10.53}$$

with

$$L_{ik} = \frac{1}{4\pi\mu_0} \left[\delta_{ik}\Phi - \frac{1}{2}\Psi_{,ik} \right]$$

$$M_{imn} = -\frac{1}{4\pi} \left[\delta_{im}\Phi_{,n} + \delta_{in}\Phi_{,m} - \Psi_{,imn} \right]$$

$$N_{imnp} = -\frac{1}{4\pi} \left[\delta_{im}\Phi_{p,n} + \delta_{in}\Phi_{p,m} - \Psi_{p,imn} \right]$$

$$Q_{imnpq} = -\frac{1}{4\pi} \left[\delta_{im}\Phi_{pq,n} + \delta_{in}\Phi_{pq,m} - \Psi_{pq,imn} \right]$$

where the detail expression for Φ and Ψ can be found in Appendix 3.A.

Based on the strain rate and velocity relation, we can also easily derive the strain rate, which is caused by the body force and eigenstrain rate:

$$e_{ij} = \sum_{J=1}^{N} \left(F_{ijk} f_k^J + D_{ijmn}^0 e_{mn}^{J0} + D_{ijmnp}^1 e_{mnp}^{J1} + D_{ijmnpq}^2 e_{mnpq}^{J2} \right) \tag{10.54}$$

with

$$F_{ijk} = \frac{1}{8\pi\mu_0} \left[\delta_{ik}\Phi_{,j} + \delta_{jk}\Phi_{,i} - \Psi_{,ijk} \right]$$

$$D_{ijmn}^0 = -\frac{1}{8\pi} \left[\delta_{im}\Phi_{,jn} + \delta_{in}\Phi_{,jm} + \delta_{jm}\Phi_{,in} + \delta_{jn}\Phi_{,im} - 2\Psi_{,ijmn} \right]$$

$$D_{ijmnp}^1 = -\frac{1}{8\pi} \left[\delta_{im}\Phi_{p,jn} + \delta_{in}\Phi_{p,jm} + \delta_{jm}\Phi_{p,in} + \delta_{jn}\Phi_{p,im} - 2\Psi_{p,ijmn} \right]$$

$$D_{ijmnpq}^2 = -\frac{1}{8\pi} \left[\delta_{im}\Phi_{pq,jn} + \delta_{in}\Phi_{pq,jm} + \delta_{jm}\Phi_{pq,in} + \delta_{jn}\Phi_{pq,im} - 2\Psi_{pq,ijmn} \right]$$

Substituting Eq. (10.53) into Eq. (10.49), we get equivalent conditions for all the particles:

$$
\begin{aligned}
&e_{ij}^{I0} + e_{ijr}^{I1} x_r + e_{ijrz}^{I2} x_r x_z \\
&= \frac{\mu_0 - \mu_1}{\mu_0} \sum_{J=1}^{N} \left(F_{ijk} f_k^J + D_{ijmn}^{J0} e_{mn}^{J0} + D_{ijmnp}^{J1} e_{mnp}^{J1} + D_{ijmnpq}^{J2} e_{mnpq}^{J2} \right) \\
&\text{for } I = 1, 2, \ldots, N.
\end{aligned}
\tag{10.55}
$$

By expanding the right side of this equation into a Taylor series of x and by comparing the coefficients of the terms of r up to quadratic terms the above equation can be written as

$$
\begin{aligned}
-\sum_{J=1}^{N} F_{ijk} f_k^J &= -\frac{\mu_0}{\mu_0 - \mu_1} e_{ij}^{I0} \\
&\quad + \sum_{J=1}^{N} \left(D_{ijmn}^{J0} e_{mn}^{J0} + D_{ijmnp}^{J1} e_{mnp}^{J1} + D_{ijmnpq}^{J2} e_{mnpq}^{J2} \right) \\
&\quad \text{for } I = 1, 2, \ldots, N
\end{aligned}
\tag{10.56}
$$

$$
\begin{aligned}
-\sum_{J=1}^{N} F_{ijk,r} f_k^J &= -\frac{\mu_0}{\mu_0 - \mu_1} e_{ijr}^{I1} \\
&\quad + \sum_{J=1}^{N} \left(D_{ijmn,r}^{J0} e_{mn}^{J0} + D_{ijmnp,r}^{J1} e_{mnp}^{J1} + D_{ijmnpq,r}^{J2} e_{mnpq}^{J2} \right) \\
&\quad \text{for } I = 1, 2, \ldots, N
\end{aligned}
\tag{10.57}
$$

$$
\begin{aligned}
-\sum_{J=1}^{N} F_{ijk,rz} f_k^J &= -\frac{\mu_0}{\mu_0 - \mu_1} e_{ijrz}^{I2} \\
&\quad + \sum_{J=1}^{N} \left(D_{ijmn,rz}^{J0} e_{mn}^{J0} + D_{ijmnp,rz}^{J1} e_{mnp}^{J1} + D_{ijmnpq,rz}^{J2} e_{mnpq}^{J2} \right) \\
&\quad \text{for } I = 1, 2, \ldots, N
\end{aligned}
\tag{10.58}
$$

By solving the linear system of equations we can get all the unknown eigenstrain rates in all particles and express the local velocity field by the body force and eigenstrain rate term based on Eqs. (10.53) and (10.54).

*10.5 Numerical simulation and case studies

The implementation of our algorithms is very simple and straightforward. At each time step, Eqs. (10.23), (10.24), and (10.25) are first solved to get the terms of $\mathbf{A^0}$, $\mathbf{A^1}$, and $\mathbf{A^2}$ on all particles based on current configurations. After that, Eq. (10.44) is deployed to calculate the driving forces among particles from $\mathbf{A^0}$, $\mathbf{A^1}$, and $\mathbf{A^2}$ previously obtained. Then equivalent conditions for a Stokes flow are applied by Eqs. (10.56), (10.57), and (10.58) to obtain the eigenstrain rates $\mathbf{e}^0$, $\mathbf{e}^1$, and $\mathbf{e}^2$ for each particle. So, finally, the velocity field is derived by Eq. (10.53), and the positions of particles will be updated by numerical integration of velocity with respect to the predefined time interval. In such a way the problem will be solved iteratively in time and space.

*10.5.1 Case study for solving the magnetostatic problem by EIM

Consider two particles are located in the infinite domain under magnetic field H_0 in the z-direction. The conventional dipole model assumes that the particles are so small that they can be treated as two dipoles with magnetization $M_i = 3\left(\mu_1^m - \mu_0^m\right)/\left(\mu_1^m + 2\mu_0^m\right) H_i^0$, which is the solution that one particle is in an infinite domain. This approach definitely ignores the interactions between particles. The local distribution of magnetization of each particle is actually redistributed by the interactions between particles, which cannot be captured by the traditional dipole model. Theoretically speaking, this problem can also be solved by the finite element method (FEM). However, as this problem is extended to three-dimensional (3D) infinite domain and more particles are considered, the scale of the problem increases dramatically. On the contrary, our method is just to solve linear equations with $39 \times N$ unknowns (N is the number of particles), which is a much smaller system. Let us consider a two-particle case, in which the magnetic permeabilities of the particles and matrix are $\mu_1^m = 10^{-7}$ and $\mu_0^m = 7 \times 10^{-4}$ Ns^2/C^2, respectively. The far magnetic field H_0 is 1 A/m. The distribution of the magnetic field H is shown in Figs. 10.4 and 10.5 by solving Eqs. (10.23), (10.24), and (10.25).

From Fig. 10.4 we can see that the distribution of the magnetic field is not uniform anymore in the two-particle case. The magnetic field is mostly concentrated at the bottom area of the up particle and top area of the down particle, which may be considered as a tip effect.

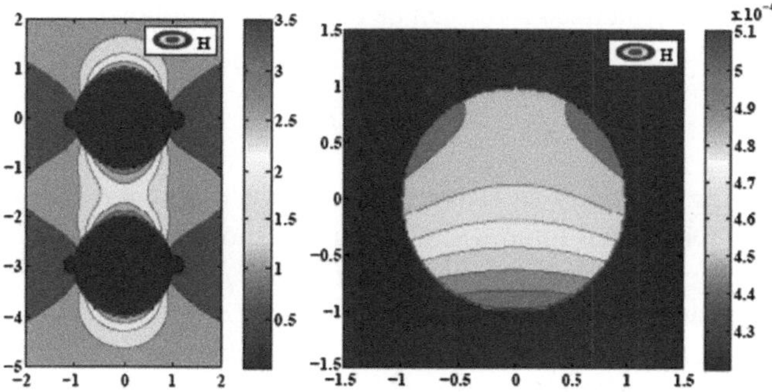

FIGURE 10.4 Two particles are located up and down.

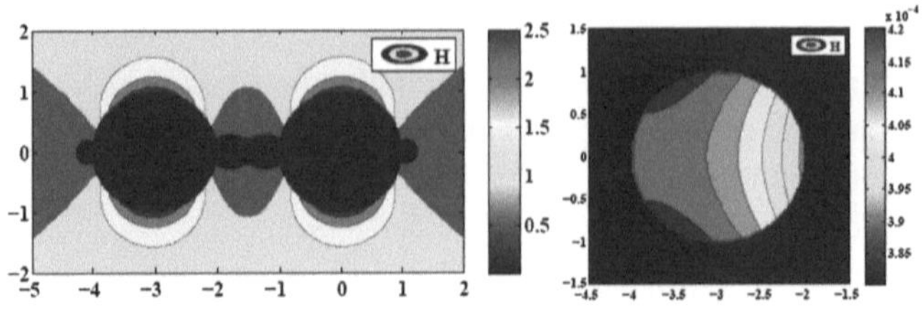

FIGURE 10.5 Two particles are located side by side.

*10.5.2 Case study for calculating magnetic force between two particles

The forces exerted on the particles are the result of magnetization. If two particles are far from each other, then it is acceptable to assume that the distribution of magnetization is uniform in each particle. Then the classical dipole model can be applied that all the magnetization can be considered to be concentrated at the center of each particle. However, if the two particles are close to each other, then the particle cannot be treated as a point anymore. The shape effect significantly affects the direction and magnitude of the magnetic force. The distribution of the magnetization is not uniform anymore. As we can see from the up and down case, shown in Fig. 10.4, the magnetization is more concentrated at the edge of each particle in the nearest distance between two particles. It can be similarly considered that the opposite charges move to the nearest edges when two objectives are put together in electrostatics. Then definitely the actual force should be much higher than the force evaluated by the dipole model.

To make a comparison between the classical dipole model with the proposed formula, two particles are in the infinite domain in up and down configuration. Here we collected the effect of particle distance ratio b/a and the mismatch ratio μ_1^m/μ_0^m of the material properties to the force variation, where a is the radius of each particle, and b is the center to center distance of two particles, respectively. First, for the effect of particle distance ratio, we set the material mismatch ratio as constant, say $\mu_1^m/\mu_0^m = 10$. The uniform far-field magnetic field is H_0 in the z-direction. The normalized force can be calculated as follows:

$$f = \frac{F}{\mu_0^m a^2 (H_0)^2} \tag{10.59}$$

By changing the ratio b/a to correspond to the different distances between particles, the normalized force will also vary as shown in Fig. 10.6 (left). Second, for the effect of the material mismatch ratio μ_1^m/μ_0^m on the force at $b/a = 3$, the result is shown in Fig. 10.6 (right).

Fig. 10.6 (left) shows that the present model delivers a much higher estimate of force than the dipole force model. The magnetic force rapidly decreases with the increase of the center-to-center distance between two particles when

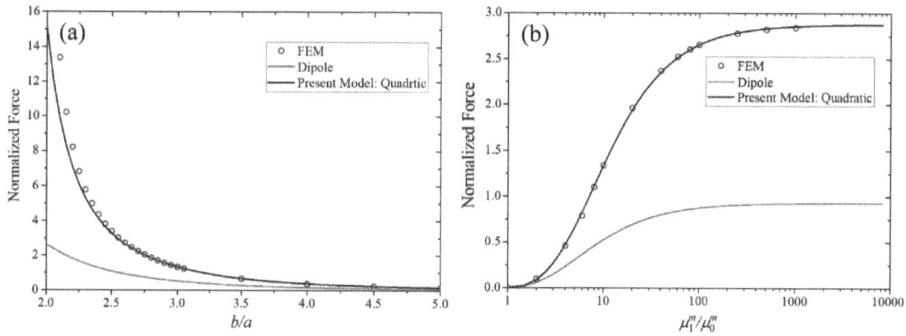

FIGURE 10.6 The normalized force variation with respect to the distance ratio b/a (left) and the material mismatch ratio μ_1^m/μ_0^m (right).

$\mu_1^m/\mu_0^m = 10$, and the difference between the proposed model and dipole model is also decreased. The result from the present model gradually converges to the dipole model. Thus it can be demonstrated that for a large center-to-center distance, the dipole model will give approximately the same results as the proposed model. Fig. 10.6 (right) shows that the interaction force will converge when $\mu_1^m/\mu_0^m > 100$. The trend from our model is the same as that from the dipole model. However, the converged value is much larger than that in the dipole model. It is worth noting that, the present model has great agreement with FEM results.

*10.5.3 Particle chain rotating the magnetic field

From the experiments, if the particle chain is initiated at a different direction, it will definitely rotate to the direction corresponding to the external magnetic field. In such a process the chain will break and reform periodically [140]. To exclude the overlap between particles, in each location update procedure, we keep the zero relative velocity in the center-to-center direction of the two particles. So in this case, if two particles will contact each other in the next step, then their modified velocity will just lead them to rotate with respect to each other tangentially to the center-to-center direction. We simulate four and five particles, respectively. The radius of the particle $a = 1$. The magnetic permeabilities of the matrix and particles are $\mu_0^m = 1$ and $\mu_1^m = 10{,}000$. The magnitude of the external field H is 1 in the z-direction. The viscosity of the Newtonian fluid is $\eta = 1$. In EIM the viscosity of the particle should be much larger than that of fluid to simulate a rigid drop. Here we give it as $\eta_p = 10{,}000$. The time step $dt = 0.02$. The initial configuration is shown in Fig. 10.7. The positions of the five particles at different times are recorded and shown in Fig. 10.8.

In the beginning the upper two particles and lower two particles try to rotate to the external magnetic direction. At $n = 182$ steps the chain will be broken up and then gradually reformed later. Since the tangential velocity, which leads the particles to rotate, will decrease when the chain is approaching in the direction

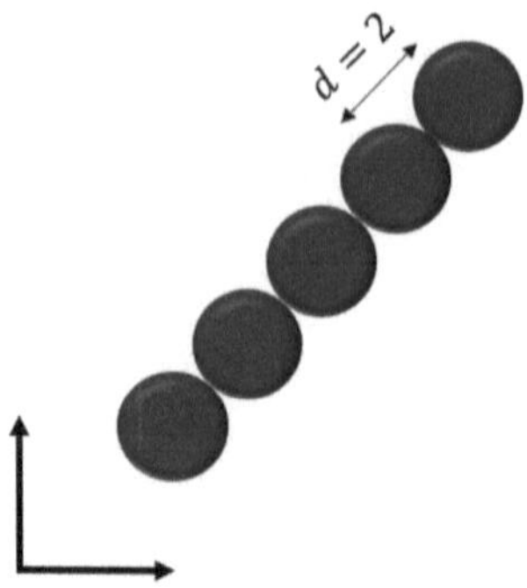

FIGURE 10.7 Initial configuration of the chain.

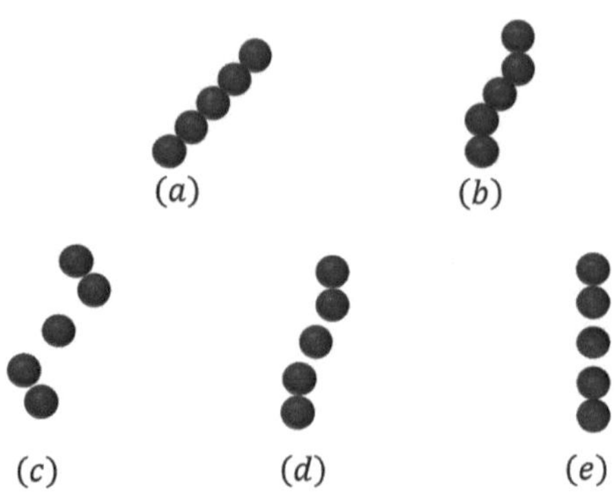

FIGURE 10.8 The configuration of the particles at (a) $n = 0$, (b) $n = 182$, (c) $n = 220$, (d) $n = 354$, and (e) $n = 1100$.

of the external magnetic field. Then, after a relatively long time, the direction of the chain will finally be consistent with the external magnetic field.

*10.6 Validation with laboratory tests

An experiment of observing nickel particles motion in polydimethylsiloxane (PDMS) in far-field magnetic field under a microscope is carried out as shown in Fig. 10.9. PDMS composites with chain-structured nickel particles were fabricated by using a solution casting method and an external magnetic field. Two magnets are placed close to each other to produce a constant far-field magnetic strength. We randomly disperse the nickel particles in the PDMS. Then a microscopic slide with that solution is placed between the two magnets. After releasing the magnetic field, a camera is used to record the motion of nickel particles. To obtain clear pictures for further comparison, the mixer is dilute. So the volume fraction of the particle is quite low. The viscosity of the PDMS is 5 mPa, and the magnetic permeability is 1.26 μNA^{-2}. The magnetic permeability of nickel particle with a diameter 25 μm is 126 μNA^{-2}, which is much

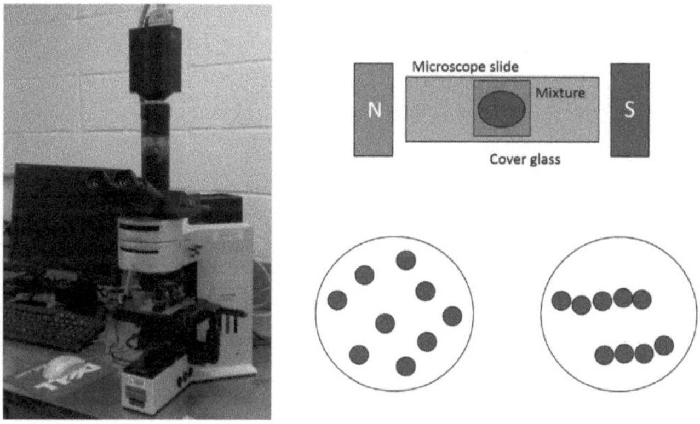

FIGURE 10.9 Experimental setup.

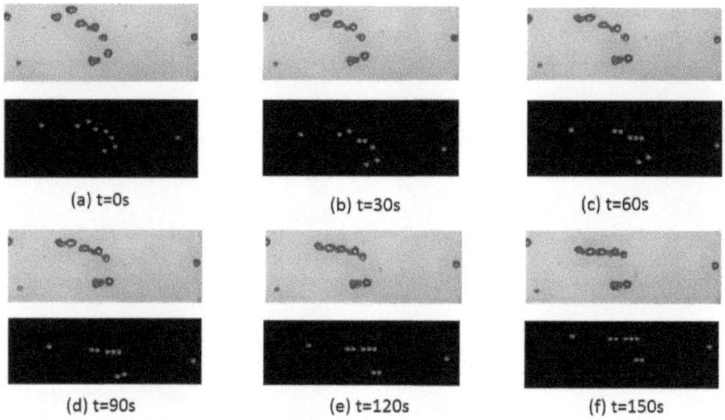

FIGURE 10.10 Experimental validation of our proposed model.

larger than that of PDMS. A magnetic field $B = 3\,\text{mT}$ is applied by two magnets. The strength of the magnetic field is measured by magnetometers. An area with nine particles is selected for the observation and simulation. The comparison is shown in Fig. 10.10 at different time steps ($t = 0, 30, 60, 90, 120, 150\,\text{s}$), which demonstrates that the proposed model can successfully catch the physics of this complex magnetostatic and stokes coupling problem.

*10.7 Virtual experiments

As the number of particles increases and the domain becomes large, the traditional FEM [139] may not be applicable for modeling of the particle motion because of the formidable computational cost. On the other hand, even when a small number of particles are modeled, the size discrepancy between particles will also cause problems in the meshing process. However, the present method can be applied by solving a linear system of equations with the $39N$ unknowns

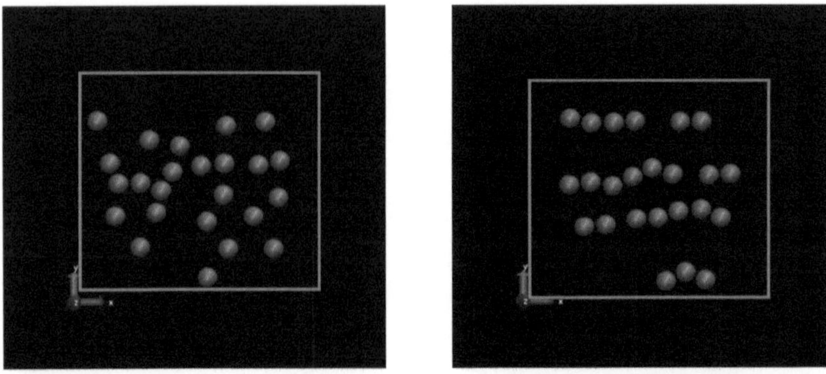

FIGURE 10.11 The initial configuration and final configuration of the simulation of particles with volume fraction 15%.

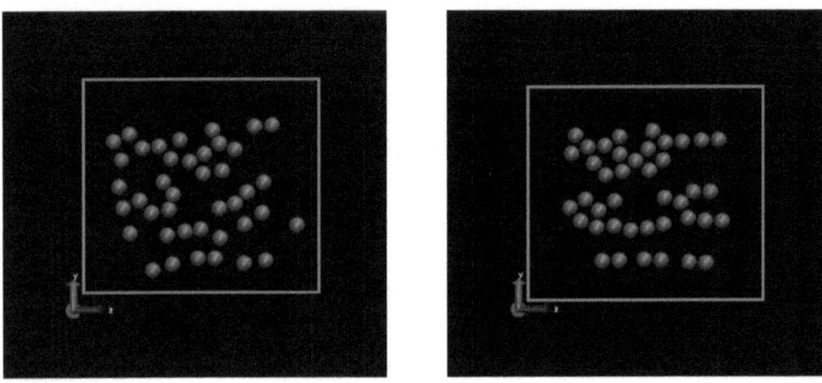

FIGURE 10.12 The initial configuration and final configuration of the simulation of particles with volume fraction 20%.

for the magnetic problem and $60N$ for the Stokes flow problem, where N is the number of particles. Here we show a case for simulating multiple magnetic particle motion in the external magnetic field. The parameters are the same as in Section 10.5.1, except for the time steps $dt = 1$. Some papers have reported that the optimal performance of the magnetic induced chain structures appears when the volume fraction of the particles is about 15% [141].

A significant reduction of the performance is observed under the volume fraction of 20%. To simulate cases with different volume fractions, a boundary is added in the simulation box. The boundary effect is handled by the boundary element method. The simulation results for two cases are shown in Figs. 10.11 and 10.12.

We can conclude that for the case of volume fraction 15%, the particles are initially separated far enough. The quality of the chain is ensured, and four chains are observed clearly. However, for the case of volume fraction 20%, the randomness of the particles distribution under the magnetic field is almost the same as that without a magnetic field. The particles are clustering and can not move freely. The anisotropy in both thermal conductivity and diffusivity is relatively weaker compared with the low volume fraction case. Therefore more attentions should be paid to control the volume fraction of the particle reinforcements to achieve optimal performance in the industry.

Chapter 11

Recent development toward future evolution

A powerful tool for virtual experiments

This chapter concludes the book with some active research topics and visions for future research directions. Based on the teaching notes and PhD dissertations, the book provides a practical and comprehensive approach to solve three types of problems (elastic, thermal, and Stokes flows) for general heterogeneous material systems and composite structures. The iBEM software package is available for research purposes. The ongoing and future research topics include adaptively forming inclusion for multiscale/multiphysical modeling and simulation, parallel computing, and optimization for high-speed large-scale calculation, extending the concept of eigenstrain for nonlinear material behavior with damaging and degradation, and implementation of iBEM for emerging engineering technologies. The applications of iBEM to fracture analysis, elastoplastic analysis, and the size effects of nanocomposites are discussed.

11.1 Recent development of iBEM

The recent work [42,43] has prepared the foundation to extensively use iBEM in modeling and simulation of composites with arbitrary shapes of particles, which extended the traditional Eshelby's theory from ellipsoidal particles to general heterogeneous material systems. Although the analytical formulation is long and time consuming for rederivation, we have encapsulated the equations in the standardized application program interface (API) functions, which is summarized in Appendix, in hope that the readers can exercise the method and directly use it for customized problems or extend it through second development.

Meanwhile, we are developing parallel computing algorithms to accelerate the computational speed for large-scale virtual experiments and virtual material manufacturing/processing. Using an adaptive cut-off with the influence zone of the Green's function based on the required accuracy and speed, we can automatically adjust the model scale. The research team is also enriching the program with new problems and case studies. Although the iBEM is currently mainly applicable for linear problems, we are extending it for nonlinear material behaviors by using eigenstrain in adaptive area/volume, which paves a clear way to simu-

The Inclusion-Based Boundary Element Method (iBEM). https://doi.org/10.1016/B978-0-12-819384-6.00019-X

late plastic behavior, fracture and damage mechanics, and material degradation and phase change.

Specifically, we list some ongoing research activities the team is working, which can be found in journal or conference publications in near future:

Using the fundamental solutions for the bimaterial infinite domain, we can extend BEM from one material to a bimaterial domain. In general, the conventional BEM can solve the bimaterial problem by the multiregion method, which requires the interfacial continuity equations. Because of the singularity effect at the ends of the interface, the results are sensitive to the interface discretization [142]. Using the bimaterial fundamental solution can analytically show the interface discontinuity and singularity without mesh along the interface. This is particularly important for mechanical analysis of layered materials, such as building-integrated photovoltaic panels, laminated composites, and microelectromechanical systems. When voids or bubbles occur in the manufacture process or service life, they can be treated by equivalent inclusions, and the stress concentration in their neighborhood can be calculated by the iBEM.

Particularly, full-spectrum functionally graded materials exhibit material phase change from one to another gradually with a continuous variation of the particle concentration. It typically exhibits three material zones [143], in which the central zone exhibits a fuzzy microstructure with unclear particle and matrix phases, but the other two zones exhibit opposite particle and matrix phases. They can be simplified into a bimaterial containing inhomogeneities, in which the bimaterial fundamental solution can be directly used in the EIM for calculating local fields and modeling the effective material behavior.

The bimaterial solution can be reduced to a semi-infinite domain by assuming the material constants with values at zero or infinity. Particularly, geothermal energy has attracted intensive applications. The heat and stress transfer from the surface to the ground through a pipe-storage system can be simulated by the solution and implemented into the iBEM. The static, harmonic, and transient heat conduction, thermoelastic analysis, and mechanical analysis can be conducted for future geothermal system design and analysis. Depending on the geometry of the applications, 2D or 3D problems can be formulated.

Notice that in the literature, a single-region boundary element method (BEM) [144] has been used to conduct the elastic analysis of multilayered materials using the generalized Kelvin solution [145]. However, the fundamental solution is given in the integral form. It is formidable to extend it to the EIM. Although the bimaterial solution is not as general as the above mentioned method for multilayered materials, the closed-form fundamental solution in Chapter 2 for a bimaterial infinite domain is particularly suitable for iBEM, which can disclose the physics and mechanics of bilayered material in an elegant way. Moreover, using the periodic boundary condition, this method can be extended to multilayered composites with a repetitive pattern.

Many composites were used in vibrational environments with time-dependent material performance. During various practical operation or material fabrication conditions, the particle–matrix systems are inevitably subject to intensive oscillations [146,147]. Due to the considerable hysteresis loss and relatively low thermal conductivity of the viscoelastic polymers, self-heating phenomena are often observed as hot spots and/or melting/softening when the polymer is subject to external excitations [148]. During a particle mixing process [149,150], the temperature increases with the energy dissipation as well, which significantly changes the mixing behavior. Because of a high sensitivity of the mechanical properties of the polymer to temperature variations, unanticipated fatigues and even catastrophic damages were often reported [150].

To prevent any unexpected damage or failure, better understanding of the thermomechanical responses of polymeric particulate composite under external excitations is therefore of great importance to materials engineers and designers. It is challenging to simulate the time-dependent material behavior and failure with consideration of dynamic loading as well. In Chapter 9, we presented some novel applications, which can be extended to more challenging problems, such as microstructural evolution and particle motion in the matrix, self-heating caused by hysteresis behavior, and thermal-mechanical coupling behavior of composites [151]. The iBEM can be an excellent tool to calculate the stress distribution in the composite and simulate the microstructure evolution with the vibration process.

When the inhomogeneities are reduced to planes or lines, they can represent cracking, sliding, debonding, or dislocation. In the literature the planes or lines are straight in general. With the domain discretization, we can use arbitrary shapes of inhomogeneities to simulate curved planes or lines, so that a mixed mode cracking can be simulated with adaptive domain discretization. Some unique crack patterns can be simulated or predicted by virtual experiments. For example, spiral cracking has been a fascinating behavior of confined layered materials [152,153]. Applications of iBEM to mixed-mode fracture problems are underway.

When an elastoplastic material or structure is subjected to an increasing load, the plastic deformation changes in magnitude and spatial domain with the growth of yielding surface. The plastic strain within the yielding surface can be considered as a varying eigenstrain over an inclusion. With domain discretization of an arbitrary inclusion, it is practical to use eigenstrain distribution in a varying plastic zone for efficient simulating the elastoplastic material behavior. The return-mapping algorithm can be used to update the plastic eigenstrain, and the plastic zone and the iBEM can be applied to the calculation of the local stress and displacement field [154]. Compared to the classical plasticity with the finite element method or BEM [154,155], the iBEM can handle the plastic strain with an eigenstrain directly with an adaptive mesh on the plastic zone using the arbitrary shape of inclusions for the domain integral. Particularly, when the plastic zone is small, the iBEM can use a small-scale computational model to solve the problem. Compared with the micromechanics-based elastoplastic model [113,156,157], the iBEM can provide the local fields and accurately investigate the material yielding behavior with more realistic and complex elastoplastic constitutive models.

When particles in a heterogeneous material system exhibit a large range of particle size, aspect ratio, and geometric shapes, such as fiber-reinforced concrete or composites, there is no good modeling method for the challenging features such as the particle fiber interaction, void/crack initiation and propagation, and computational instability and efficiency. We will use iBEM for these challenging applications.

Size effects of composites on the effective stiffness and strength have been a challenging problem in the past decades. The singularity of Eshelby's tensor can be a good tool for material behavior modeling. We can predict the material behavior from nano- and micro- to macroscales and covers the mechanical, thermal, electrical, optical, acoustic, and magnetic properties.

11.2 Future research directions

Although the boundary element method (BEM) has been in the literature for decades with elegant mathematics and a unique capability to simplify problems, it is not yet a mainstream software like the finite element method (FEM). The main drawbacks of BEM are the inconvenience in handling with multiregion domains or inelastic behaviors, as the BEM relies on the fundamental solution of the Green's function. Eshelby's equivalent inclusion method uses the eigenstrain concept to release these two constraints to BEM by using inelastic strains, either material mismatch eigenstrain or plastic strain in the specific subdomains. Particularly, the breakthrough of the domain integral of polynomial eigenstrain in arbitrary subdomain enables to solve general problems for 2D and 3D, static and dynamic, and linear and nonlinear problems and creates the opportunity to make iBEM a mainstream method in computational mechanics. To make a new era of iBEM, the following future works are envisioned:

1. The closed-form domain integral of the Green's function is exact but complicated and cumbersome to be applied for the community. It takes time to establish a linear equation system. It is of significance to simplify those formulae and standardize the functions, especially for the polyhedral problem, so that the algorithm can be straightforwardly generalized to actual applications.
2. The linear equation system for the particle domain discretization method is established on the specific points by the collocation method, which is straightforward but with singularity issues. Although good results can be obtained with a low number of elements, because the global matrix is not symmetric, the numerical solver of the linear equation system can be slow. The Galerkin approximation [158] may provide a weak form solution with a symmetric matrix, in which the singularity issue can be addressed in the volume/area integral. However, the gain in accuracy/stability and the cost in computation should be investigated. It may not be better than the current one due to the time-consuming volume integral.

3. The domain discretization only considers the element shapes of triangle or tetrahedron. It can be straightforwardly extended to multiple shapes and types of elements, such as cubic, cylinder, cone, or sector of them. Particularly, the eigenstrain variation exhibits some patterns, specific elements in certain parts of inhomogeneities may provide higher efficiency and accuracy.
4. When the accuracy is allowed to be compromised, alternative numerical integral methods can be explored to expedite the equation construction. Particularly, the influence zone of eigenstrain is small, and a certain approximation of eigenstrain may still provide similar accuracy.
5. As for the BEM part, the isoparametric elements and their related numerical integration scheme have been proposed in the literature [24,68,76] and can be used in the iBEM as well.
6. The fundamental solution for the bimaterial infinite domain can be introduced to the iBEM so that we can solve inhomogeneity problems in layered materials.
7. We can generalize the eigenstrain concept to plastic strain, displacement discontinuity, dislocation, among other physical phenomena, and solve those problems by iBEM.
8. To accelerate the modeling and simulation, the concept of the fast multipole method can be used, so that a cut-off of particle interaction distance can be introduced and parallel computing schemes can be explored.

Overall, with the development of new materials and structures, iBEM can be a powerful tool for virtual experiments and computer-aided design of the material systems. Some implementations and applications are envisioned as follows:

1. *Virtual material testing and design*. The conventional material design in civil engineering requires a number of time-consuming experiments to fabricate material samples for a series of characterization and performance tests to determine the job mix formula (JMF). During the construction process, cores and beams were sampled from the field and tested in the lab for quality control and quality assurance tests. During the service life, samples may be taken for forensic studies of material failures or damages and for remaining lifetime prediction. Using iBEM, we can make virtual material samples for virtual experiments with very few experimental validations. Using imaging methods, we can obtain the microstructure of material after the construction and predict the material performance with iBEM. Therefore iBEM can be a core algorithm for such applications to computer-aided design and analysis of various composite materials for the optimization of JMF.

2. *Image-aided design and testing*. Imaging process has become a powerful tool for microstructural characterization and material diagnosis. Various microscopes, such as transmission (TEM) and scanning (SEM) electron microscopes, high-resolution optical microscopes, atomic force microscopes, and micro-CT, can render the microstructural images from nano- to macroscale and from 1D to 3D cases. The images can be transferred to a finite element mesh for material design and testing. However, FEM requires the domain discretization for

both particles and matrix and thus is expensive in data processing and computation. As Eshelby's equivalent inclusion method (EIM) has been a foundation for micromechanics, we can extend iBEM to micromechanics-based modeling and rapidly process the image data for performance prediction. In addition, the image data can be used for further material handling and processing by robots. For example, when a tumor in an animal tissue should be taken out, the image can show the size and location in the in situ condition. When a tool approaches it, the size and location may change. The iBEM can focus on the region of interest (ROI) with moving boundary conditions and reproduce the operation on computer and assist the actual robot operation.

3. *Nondestructive testing and evaluation (NDT & NDE).* NDT and NDE are commonly used in the science and technology industry to evaluate the properties of a material, component, or system without causing damage. Typically, by providing excitation to an object, from the response on the surface or boundary, we can inversely calculate the microstructure or evaluate the material status. We can extend the iBEM to inverse problems as a core algorithm for NDT & NDE.

In addition, as the iBEM exhibits very good flexibility and capability to handle multiregion domains and plastic or other eigenstrains that are not easy for conventional BEM to accommodate, the iBEM can be a good alternative computational method to the finite element method for general engineering design and analysis.

Appendix A

Introduction and documentation of the iBEM software package

Code structure and a case study

In this appendix, we provide a brief introduction and documentation of the current version of the iBEM software package in hope that it can facilitate the further research and development of iBEM among the research community. All the integration of Green's functions for elasticity, heat conduction, Stokes flow have been coded up with explicit expressions of ψ and ϕ, which can be used as standard Application Programming Interface (API) functions for redevelopment. It will be a very helpful tool for our research group to communicate with other teams and provide a foundation for the future development by other teams. The software uses the object-oriented programming (OOP) and can be easily scaled up or extended to other applications.

A.1 Overview of iBEM

The iBEM software package has been rooted at our research on the equivalent inclusion method since 2000 for Dr. Yin's PhD studies. Although some code can be traced back to the early age with the EIM, the iBEM was not used to represent the inclusion-based boundary element method until 2015 in Dr. Song's PhD studies, and the code was assembled with the introduction of the boundary element method to EIM. Since then, we have developed and expanded the C++-based iBEM software package with our research progress, and more advanced aspects of iBEM are still under development. Our research team is still implementing new algorithms for solving various composite material BVPs. However, the structure of the program and the standard integral functions have been well established. The source code and relevant documentation can be downloaded from http://www.columbia.edu/cu/civileng/yin/iBEM.html. We provide the code structure in Fig. A.1, so that a future developer with a specific problem could see which part is relevant to the problem for redevelopment of the code. We also provide a case study, so that the users can do exercise with an established problem and use it as a template for other specific problems.

Basically, the iBEM software package uses the continuum method to solve the local fields for multiphysical problems. It can homogenize the material volume and provide the effective material behavior with boundary information. Therefore it is an ideal tool for virtual experiments.

The code can be compiled for laptop, desktop, or high-performance clusters with parallel computing depending on the scale of the problem.

Currently, iBEM is an open-source code, available free of charge, and distributed under the terms of the GNU Public License Version 2 (GPLv2) (https://www.gnu.org/), which means that you can use or modify the code for your own purposes but have to adhere to certain rules when redistributing it – specifically in binary form – or distributing software derived from (parts of) it. There is no warranty of any kind. As each source file states in its header, it is a copyrighted code and thus not in the public domain. For more information about open-source software and open-source distribution, see www.gnu.org or www.opensource.org. The legal text of the GPL is applied to iBEM.

A.2 Structure of iBEM software

The backend of iBEM C++ code is a directed acyclic graph (DAG), which describes the dependencies of service. The iBEMRunner is able to run any main services, and services will figure out its dependencies by applying topological sorting and complete the dependent service in series. Each main service will call the method of the key auxiliary classes to complete their tasks. All the relations between services and between services and key classes are shown in Fig. A.1.

Currently, there are five main services developed: 1) preprocessService, 2) iBEMSystemService, 3) systemSolverService, and 4) postProcessService. One thing is to keep in mind that iBEMSystem is the master service where the system of the iBEM equations is built. As we described in the previous chapters, the system equations consisting of conventional BEM equations (boundary-to-boundary interactions), conventional EIM equations (particle-to-particle interactions), and coupled terms (particle-to-boundary and boundary-to-particle interactions). For the remaining three services, preprocessService allocates memory resources for the run, the systemSolverService solves the linear equations after applying boundary conditions on our iBEM system equations, and, finally, the postProcessService generates the field or flux on the points we are interested in after the unknown boundary response and eigenstrain on the particles are resolved.

We provide the more detailed flowchart, which shows how we build up iBEM system of the equations step by step for an iBEM case study:

As shown in Fig. A.2, the main purpose of iBEM is establishing the linear equation system with boundary displacements and tractions as well as eigenstrains, which includes the *coupled terms* introduced by boundary and multiple

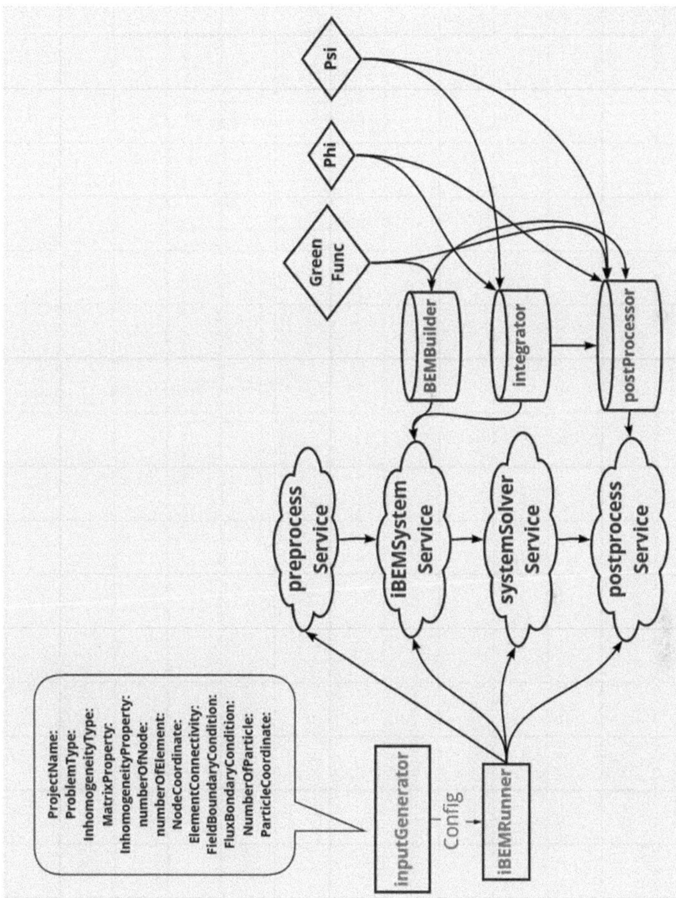

FIGURE A.1 The programming structure and service flow of the iBEM software package.

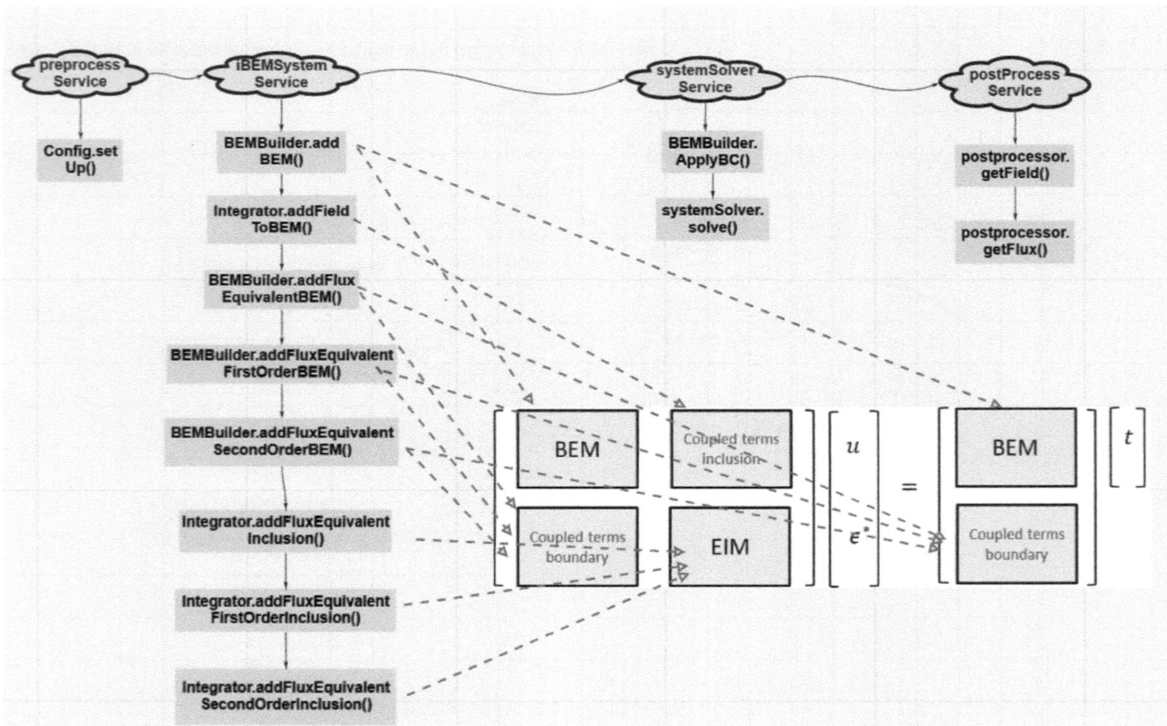

FIGURE A.2 The flowchart of the main steps of the iBEM programming and operation.

inhomogeneities. As we have briefly presented the architecture of iBEM in Section 5.4, the algorithm of iBEM mainly includes four master services:

(i) **Preprocess service**. This service requires users to provide input files for iBEM modeling. A detailed example is provided in the case study for 3D elasticity problem. The input file provides the information for the dimension of the problem including 2D or 3D cases, the type of the problem including elastic, heat conduction, and Stokes flows, the boundary element mesh, and the types of inhomogeneities, such as polygon, polyhedron, ellipse, or ellipsoid. When domain discretization is used, the mesh information will be uploaded.

(ii) **iBEMSystem Service**. The specific master service executes based on the input files to generate the global matrix of iBEM, including convectional BEM part, coupled-term part, and EIM part. Accordingly, because users have defined the dimension, types, and boundary mesh, the coefficient matrix for BEM can be built. Notice that besides the BEM coefficient matrix, the coupled terms by boundary effects are also calculated in this service, and the typical functions are listed in Section A.3.2. Subsequently, the coupled terms by the inhomogeneity effects simulated by an eigenfield to boundary are calculated, which generally requires the third-order partial derivatives of the domain integrals. Finally, the EIM part, including higher-order derivatives of domain integrals, is calculated according to user's needs, such as uniform, linear, and quadratic polynomial form of eigenfields. Therefore the linear equation system for displacements or tractions on the boundary and eigenstrain on inclusions can be established.

(iii) **SystemSolver Service**. Firstly, this service rearranges the global iBEM matrix based on the boundary conditions. Secondly, this service provides several tools (numerical algorithms) to solve the global iBEM matrix. Developed for virtual experiments, the iBEM package may face various situations defined by the distribution, size gradation, and dimension of the particles. Moreover, considering the computational capability and admissible errors based on different applications, the users can choose whatever they like according to the EIGEN library, such as the *HouseHolder*, *FullLU*, or even *PartialLU*, where the detailed descriptions are available at their websites. Based on our recommendation, for system with smaller DOFs, *FullLU* would provide the best accuracy among others; for system with large number of particles or DOFs, *PartialLU* with multiple processor solving is preferred though the accuracy may be compromised to some extent. Oncc the unknowns are calculated from the linear equation system, the local field and effective material behavior can be analyzed.

(iv) **PostProcess Service**. In this service, using the boundary responses and eigenfields by the **SystemSolver Service**, the users can set up observing points of their interests, such as axis line of displacement/stress and uniformly distributed points for contour plots. Moreover, with specific boundary conditions, i.e., uniload test, periodic boundary conditions, the effective behaviors of composites can be obtained.

Overall, iBEM software package executes each master service from (i) to (iv) in a sequence based on the dependent relation. In those master services the operation is divided into several steps (essential functions). Among the four services, the second one of **iBEMSystem** is of most importance as it generates each building block for the global matrix of the linear equation system, and the fourth service can be replaced based on customized needs or demands. Each part is produced by dedicated step from auxiliary class (BEMBuilder or Integrator) as shown in Fig. A.2.

A.3 Key classes

The code is structured so that each type of problems will be associated with several important classes. There are import classes required to run the iBEM algorithms for each type of problems. In the following, four master services are explained in detail.

A.3.1 Main class *Config* – Preprocess service

The main class *Config*, serving as the input file reader and problem classification of the iBEM, includes a series of essential functions. For simplicity, the input reading functions are defined as "Read + keywords", and the types of problems are defined as "get + keywords". We provide some examples:

(1) keywords = BEM_num, the function $ReadBEM\_num$ reads the number of nodes and elements;
(2) keywords = BEM_info, the function $ReadBEM\_info$ reads the coordinates and connection of nodes and elements;
(3) keywords = $BEMbuilder$, the function $getBEMbuilder$ returns the types of problems that the users want to implement, thus linking to the specific Green's function and returning the corresponding BEMBuilder class name.

Note that the names of the input files can be modified by the reader's needs. Currently, the default names are "BEM.txt", "calculationtype.txt", and "postprocess.txt" In addition, the class is operated by a void function **setup**(). When a specific iBEM model is redeveloped, the base class of Config needs to be inherited, and those methods can be overwritten. For example, when we are finishing the iBEM for 3D elasticity, a child class *configElastic3D* is generated, and the configurations are changed accordingly.

A.3.2 Class *BEMBuilder* – iBEMSystem Service

Based on the input files and class *Config*, the Green's function of the BEM is selected accordingly. This class is, however, responsible for creating the BEM coefficient matrix and *coupled terms* by boundary effects. For simplicity, the functions are named as "add + keywords". We show some examples.

(1) keywords = BEM, the function $addBEM$ generates the coefficient matrix, which is the same as the indirect conventional BEM;
(2) keywords = $FluxEquivalentFirstOrderBEM$,
the function $addFluxEquivalentFirstOrderBEM$ generates the coefficients for coupled terms by boundary effects for the equivalent flux conditions up to linear eigenstrain terms.

In the same class, the coefficient matrix is reorganized by various types of boundary conditions, such as Dirichlet, Neumann, and PBC. The details can be found in [21] or reading the function $voidApplyBC$ in the corresponding files. Similarly, when the users develop their own iBEM for specific type, the base class *BEMBuilder* needs to be inherited, and those methods should be overwritten. For example, when we finish the iBEM for 3D elasticity, a child class *BEMBuilderElastic3D* is generated and defines all those boundary-integral-related equations.

A.3.3 Class *integrator* – iBEMSystem Service

Continuing from the class *BEMBuilder*, this class is responsible for calculating the rest of the global iBEM matrix, the coupled terms by inhomogeneity effects and EIM stress equivalent conditions. For simplicity, the functions are named as "add + keywords". We show some examples.

(1) keywords = $FieldToBEM$, the function $addFieldToBEM$ generates the coefficients matrix for coupled terms by eigenfields, which can also be interpreted as particle boundary interactions.
(2) keywords = $FluxEquivalentFirstOrderInclusion$,
the function $addFluxEquivalentFirstOrderInclusion$ generates the coefficients matrix with the stress equivalent conditions up to linear terms of the eigenfields.

When a user develops an iBEM model for a specific problem, the base class of *integrator* needs to be inherited, and those methods can be overwritten. For example, when we finish the iBEM for 3D elasticity, a child class *integratorElastic3D* is generated and defines all those inclusion-related domain integrals.

A.3.4 Class *postprocessor* – SystemSolver and PostProcessor Service

When the global iBEM matrix is obtained in previous classes, this class solves the linear equation system and generates postprocess results, such as stress and strain fields. The function *Solve* solves the global matrix and stores the solution to *Ufile.txt* (field), *Tfile.txt*(traction), and *Eigen.txt* (eigen–field), respectively. For simplicity, the functions are named as "*get + keywords*", i.e., keywords = Field, the function $getField$ returns the field value for specific interested points, such as temperature and displacements. Another important method, keywords = Flux, the function $getFlux$ returns the flux value for specific interested points, such as temperature and displacements.

Notice that the observing points have been read in the class *Config* by the function $ReadPostprocess\_info()$. However, for some specific geometry, such as points along axis line, the users can choose to provide coordinates of points directly or simply revise the functions in this class. When you develop your iBEM for specific type, the base postprocessor needs to be inherited, and those methods should be overwritten. For example, when we finish the iBEM for 3D elasticity, a child class *postprocessorElastic3D* is created and defines all those methods to resolve the field and flux values at the points of interest.

A.4 Installation of iBEM

Currently, the software package has been developed with Visual Studio, and we have compressed the header, source files, and recent version of EIGEN library for the users. For the users using Visual Studio, the iBEM can be installed by the following steps:

(1) Download and release the folder.
(2) Import all the files following the sequence of folder (to avoid chaos) to IDE and make sure the compiler version is set up consistently with the IDE (i.e., VS120 or VS140).
(3) To enable your program for large number of particles, change the platform from $win32$ to $x64$, which releases the limit of 2 Gb RAM.
(4) Check the configuration of language and enable the OMP selection; otherwise, the program can only be operated in the single core mode.

Note that for users using other IDEs, such as Xcode, in step (2), make sure the EIGEN library is imported or linked properly to your current working directories. The C++ source code can be compiled in other computer operation systems as well.

A.5 Run iBEM

Currently, the iBEM runs in the console mode. Graphic user interface (GUI) can be developed in future for convenience of model construction and revision. As shown in the case study in the next section, the input file follows the predefined format. After downloading the software packages and example inputs txt files for each type of the problems, the users can modify the example input of the problem in accordance with the specific model parameters and configurations.

Once your input is prepared, compile and execute the C++ code. The code will output the solutions of the iBEM system equations and the field and flux values from the *postprocess*. As an alternative way, we can obtain the output files and illustrate the data with other software packages.

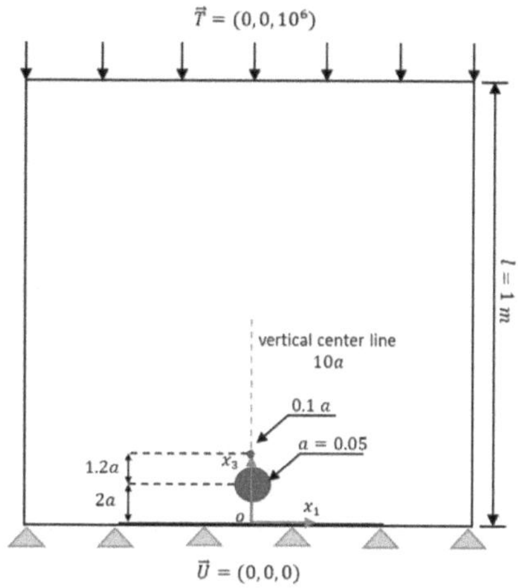

FIGURE A.3 Problem description of case study with two spherical inhomogeneities (one with radius a and the other with $0.1a$).

A.6 A case study for tutorial

As a verification case study, consider the example shown in Fig. 5.6, which contains two spherical inhomogeneities (one with radius a and the other with $0.1a$) under intensive boundary effects. In the following, the problem is described with geometric dimensions and boundary element mesh, and the input/output files of iBEM are explained in detail.

As shown in Fig. A.3, a box of matrix material (Young's modulus $E^0 = 1\,\text{MPa}$ and Poisson's ratio $\nu^0 = 0.25$) with length 1 m is subjected to uniaxial compression at its top surface and displacement constraint at its bottom, whereas the side surfaces are free of traction. Two spherical inhomogeneities (Young's modulus $E^0 = 2\,\text{MPa}$ and Poisson's ratio $\nu^0 = 0.25$) with $a = 0.05$ and $0.1a$ radius vertically with distance $0.1a$, whereas its (large inhomogeneity) distance to boundary is a. Because the ratio $\frac{d}{r}$ is small, the interaction is comparatively intensive. Due to the small distance of two inhomogeneities, the accuracy of the solution depends on the smooth transition and overall quality of the mesh, which requires several adjustments (test the mesh sensitivity) and occupies high computational resources (RAM). As for the multiregion BEM, it would experience similar issues. As shown in the input file BEM.txt, the boundary mesh includes 1176 4-node quadrilateral elements, and two inhomogeneities require quadratic eigenstrain with 108 unique terms. As for numerical verification, FEM employs the 4-node tetrahedral element, and the number of elements is 1,263,575.

For detailed stress comparison, the readers can also refer to Section 5.6.3. In the following, the input files are explained line by line, which is available in the folder of *case study* in the compressed folder.

(i) **BEM.txt**

```
# comment line, add # if you want to skip the line
# name of your input file
Input_file_of_test_case

# Young's moduli of the matrix
1000000
# Poisson's ratio of the matrix
0.25

# number of boundary nodes
1178
# number of boundary elements
1176

# coordinates of boundary nodes
0 1 0
0 0 0
0 0.928571 0
0 0.857143 0
... ...
1 0.714286 0.0714286
1 0.785714 0.0714286
1 0.857143 0.0714286
1 0.928571 0.0714286

# elements connection
2 15 57 44
15 14 70 57
14 13 83 70
13 12 96 83
... ...
1139 1152 648 649
1152 1165 647 648
1165 1178 646 647
1178 43 16 646
```

```
# specification of boundary conditions in displacement of elasticity
# 0 and 1 in first column represent known or unknown for displacements;
# second column is the specific value of the displacement;
# 3D problem has three DOFs of each node, order divided by 3; 1 represents u_x, 2 represents u_y, and 0 represents u_z.
0 0
0 0
... ...
1 0
1 0
# specification of boundary conditions in surface traction of elasticity in elements
# 0 and 1 in first 12 columns represent unknown or known for surface tractions;
# i.e., quadrilateral elements has 4 nodes with 3 DOFs
# i.e, the 5th column means t_y of second node in that element and (5 + 12)th column is the specific value of the traction
# example of all displacement specified BCs
0 0 0 0 0 0 0 0 0 0 0 0 0 0 0 0 0 0 0 0 0 0 0 0
0 0 0 0 0 0 0 0 0 0 0 0 0 0 0 0 0 0 0 0 0 0 0 0
... ...
# example of given uniaxial compression σ = 1 MPa in the z direction
1 1 1 1 1 1 1 1 1 1 1 1 0 0 -1000000 0 0 -1000000 0 0 -1000000 0 0 -1000000
1 1 1 1 1 1 1 1 1 1 1 1 0 0 -1000000 0 0 -1000000 0 0 -1000000 0 0 -1000000
... ...
# example of no tractions
1 1 1 1 1 1 1 1 1 1 1 1 0 0 0 0 0 0 0 0 0 0 0 0
1 1 1 1 1 1 1 1 1 1 1 1 0 0 0 0 0 0 0 0 0 0 0 0
```

(2) **position.txt**

```
# number of spherical particles
2

# Young's moduli of the inhomogeneity 1000000 # Poisson's ratio of the inhomogeneity 0.25
# coordinates and radius of the particles; # first 3 columns are coordinates (3D) and then radius
0.5 0.5 0.1 0.05
0.5 0.5 0.16 0.005
```

(3) **calculationtype.txt**

```
# types of problems, either integers or abbrevs
# 1 or 3D_elasticity (enum function)
1
```

(4) **postprocess.txt**

```
# number of postprocess points; if there are more points than coordinates, then
they will be set as the origin.
103
# coordinates of points in 3D
5.000000e-01 5.000000e-01 1.000000e-02
5.000000e-01 5.000000e-01 1.500000e-02
5.000000e-01 5.000000e-01 2.000000e-02
5.000000e-01 5.000000e-01 2.500000e-02
5.000000e-01 5.000000e-01 3.000000e-02
... ...
5.000000e-01 5.000000e-01 4.900000e-01
5.000000e-01 5.000000e-01 4.950000e-01
5.000000e-01 5.000000e-01 5.000000e-01
```

After the iBEM software executes the four main services, it returns the output files based on user's definition. By default the output files include stresses (stress.txt) of observing points defined by input file (postprocess.txt), rearranged boundary responses (Ufile.txt, Tfile.txt), and eigenstrains (Eigen.txt) associated with the inhomogeneities. Ufile.txt and Tfile.txt record the boundary conditions and responses of displacements and surface tractions, respectively.

(i) **Ufile.txt**

```
# displacements of boundary nodes, DOF = 3
# total number of results = 3 * DOF
# i.e., row 16 means u_x of the 6th node
0.000000e+00
0.000000e+00
0.000000e+00
... ...
-9.278741e-04
-9.821535e-03
9.269714e-04
... ...
```

(ii) **Tfile.txt**

```
# surface traction of boundary nodes, DOF = 3
# 1 element containing 4 nodes, 4 * 3 = 12
# total number of results = 12 * DOF
# i.e., row 26 represents surface traction t_y of the 1st
# node belonging to the 3rd element
6.237096e+03
6.237096e+03
2.321389e+04
... ...
6.908847e+02
8.759218e+03
1.214350e+03
... ...
```

(iii) **Eigen.txt**

```
# Eigenfield associated with the inhomogeneity problem
# i.e., quadratic order expansion, 78 eigenstrains (54 unique eigenstrains if us-
ing the minor symmetry)
# total number of results = 78 * number of ellipsoidal inhomogeneities
-6.194188e-04
9.150108e-19
6.816731e-11
... ...
-1.337508e-03
-2.237315e-11
-3.438119e-10
... ...
```

(iv) **stress.txt**

```
# stress field defined by users, 3D problem has 6 unique terms
# sequence σ11, σ12, σ13, σ22, σ23, σ33
# number of rows = number of observing points, i.e., 103
-2.982e+03 -8.579e-12 -7.532e-05 -2.982e+03 -7.532e-05 -1.064e+04
-2.879e+03 -8.690e-12 -7.594e-05 -2.879e+03 -7.594e-05 -1.081e+04
-2.801e+03 -8.865e-12 -7.665e-05 -2.804e+03 -7.665e-05 -1.099e+04
... ...
```

Although FEM constructs a sparse matrix, the accuracy of solution heavily depends on the transition of the mesh between two inhomogeneities even without the stress singularities. The numerical comparisons are enclosed in the same folder, as it is shown in Fig. A.4.

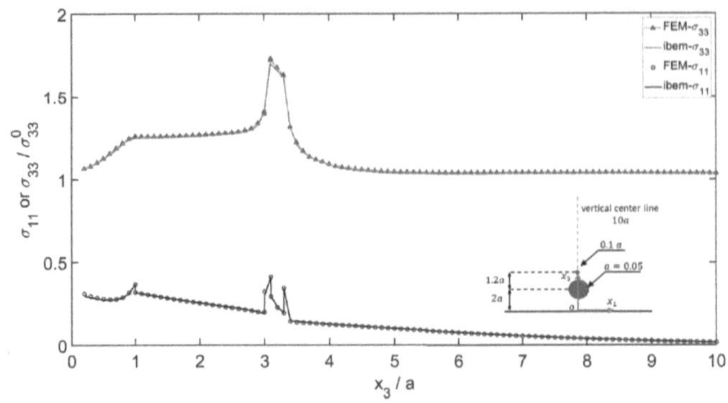

FIGURE A.4 Numerical comparison of normal stresses with FEM of case study along the center axis line.

A.7 Notes for redevelopment of iBEM

It is straightforward to extend iBEM based on users' needs, such as other multiphysical problems and time-dependent problems. Compared with the conventional boundary element method (BEM), iBEM can handle material inhomogeneity and inelastic strain by eigenstrain concept. The current iBEM code is at the early stage. The recent development and advancement of BEM and EIM can be integrated into iBEM for better performance. For some emerging applications and ongoing research introduced in Chapter 11, the current iBEM code may not be able to solve directly, and redevelopment of iBEM will be needed. As demonstrated in Fig. A.1, the iBEM code can be expanded in three directions.

(i) *Extension to other multiphysical problems*. In Chapters 2 and 9, we introduced the fundamental solutions for different physical problems and boundary conditions such as bimaterial domains, and can extend the iBEM code to dynamic and time-dependent problems. As an algorithm based on the Green's function, users need to change the Green's function (potential, traction, etc.) to the corresponding problems accordingly. To do this, we can follow the class *Config* and mimic the way we created for specific problems (3D elasticity). However, the most important part, the domain integrals of Green's function may result in challenges to pursue closed-form formulae. Alternatively, numerical integration is worth implementing in the class *Integrator* if the polynomial-form assumption of eigenfield remains the same.
(ii) *Extension to other boundary element methods*. Currently, the iBEM package only uses the conventional strong-form collocation method. Although the strong singularities are well handled by the rigid-body motion, the weak-form Galerkin scheme may provide better numerical stability and accuracy. Moreover, the BEM can also be integrated with cubic splines as well for isogeometric analysis (IGA). Notice that the redevelopment of BEM part will only influence

the class *BEMbuilder*, and the users can keep coupled terms and EIM parts the same as we did.
(iii) *Paralleling the computation and scaling up the modeling*. Although the analytical integrals show the rigorousness and elegance in mathematics, the speed of handling the special functions can be slow. The fast multipole method can be used to speed up the calculation. The large linear equation system with sparse matrix can be saved or solved through parallel computing.

Bibliography

[1] Toshio Mura, Micromechanics of Defects in Solids, Springer Netherlands, 1987.

[2] J.D. Eshelby, The determination of the elastic field of an ellipsoidal inclusion, and related problems, Proceedings of the Royal Society of London. Series A, Mathematical and Physical Sciences 241 (1226) (aug 1957) 376–396.

[3] J.D. Eshelby, The elastic field outside an ellipsoidal inclusion, Proceedings of the Royal Society of London A: Mathematical, Physical and Engineering Sciences 252 (1959) 561–569, The Royal Society.

[4] Z.A. Moschovidis, T. Mura, Two-ellipsoidal inhomogeneities by the equivalent inclusion method, Journal of Applied Mechanics 42 (4) (dec 1975) 847–852.

[5] H.M. Shodja, I.Z. Rad, R. Soheilifard, Interacting cracks and ellipsoidal inhomogeneities by the equivalent inclusion method, Journal of the Mechanics and Physics of Solids 51 (5) (may 2003) 945–960.

[6] H.M. Shodja, A.S. Sarvestani, Elastic fields in double inhomogeneity by the equivalent inclusion method, Journal of Applied Mechanics 68 (1) (2001) 3–10.

[7] Jin H. Huang, Equivalent inclusion method for the work-hardening behavior of piezoelectric composites, International Journal of Solids and Structures 33 (10) (1996) 1439–1451.

[8] Yuji Nakasone, Hirotada Nishiyama, Tetsuharu Nojiri, Numerical equivalent inclusion method: a new computational method for analyzing stress fields in and around inclusions of various shapes, Materials Science and Engineering: A 285 (1–2) (jun 2000) 229–238.

[9] Hatta Hiroshi, Taya Minoru, Equivalent inclusion method for steady state heat conduction in composites, International Journal of Engineering Science 24 (7) (jan 1986) 1159–1172.

[10] H.M. Yin, P.-H. Lee, Y.J. Liu, Equivalent inclusion method for the stokes flow of drops moving in a viscous fluid, Journal of Applied Mechanics 81 (7) (apr 2014).

[11] H.M. Yin, L.Z. Sun, Magnetoelasticity of chain-structured ferromagnetic composites, Applied Physics Letters 86 (26) (2005) 261901.

[12] L.S. Fu, T. Mura, The determination of the elastodynamic fields of an ellipsoidal inhomogeneity, Journal of Applied Mechanics 50 (2) (1983) 390–396.

[13] Y.J. Liu, H.M. Yin, Stress concentration of a microvoid embedded in an adhesive layer during stress transfer, Journal of Engineering Mechanics 140 (10) (2014) 04014075.

[14] Frank J. Rizzo, An integral equation approach to boundary value problems of classical elastostatics, Quarterly of Applied Mathematics (1967) 83–95.

[15] Frank Joseph Rizzo, Some integral equation methods for plane problems of classical elastostatics, PhD thesis, University of Illinois at Urbana-Champaign, 1964.

[16] R. Lanzafame, M. Messina, Fluid dynamics wind turbine design: critical analysis, optimization and application of BEM theory, Renewable Energy 32 (14) (2007) 2291–2305.

[17] Otto von Estorff, Boundary Elements in Acoustics: Advances and Applications, vol. 9, Wit Pr/Computational Mechanics, 2000.

[18] J. Shen, Computational Electromagnetics Using Boundary Elements: Advances in Modelling Eddy Currents, Computational Mechanics Publications, Southampton, UK Boston, USA, 1995.

[19] M.H. Aliabadi, Boundary element formulations in fracture mechanics, Applied Mechanics Reviews 50 (2) (1997) 83–96.
[20] Alexander H.-D. Cheng, Daisy T. Cheng, Heritage and early history of the boundary element method, Engineering Analysis with Boundary Elements 29 (3) (2005) 268–302.
[21] Gernot Beer, Ian Smith, Christian Duenser, The Boundary Element Method With Programming: For Engineers and Scientists, Springer Science & Business Media, 2008.
[22] Vladimír Sládek, Ján Sladek, Singular Integrals in Boundary Element Methods, vol. 3, Computational Mechanics, 1998.
[23] Quan-Zhang Huang, Zhi-Gao Xu, Hong-Fu Qiang, Guang Wang, Xiao-Ping Zheng, Boundary element method for solid materials with multiple types of inclusions, Acta Mechanica 226 (2) (jul 2014) 547–570.
[24] Gernot Beer, Christian Dünser, Eugenio Ruocco, Vincenzo Mallardo, Efficient simulation of inclusions and reinforcement bars with the isogeometric boundary element method, Computer Methods in Applied Mechanics and Engineering 372 (dec 2020) 113409.
[25] Zhipeng Wu, Qinye Lin, Yongqiang Chen, Application of fast multipole boundary element method for two-dimensional nonlinear interface debonding of particulate composites, Engineering Analysis with Boundary Elements 113 (apr 2020) 72–81.
[26] C.Y. Dong, S.H. Lo, Y.K. Cheung, Application of the boundary-domain integral equation in elastic inclusion problems, Engineering Analysis with Boundary Elements 26 (6) (jun 2002) 471–477.
[27] H.M. Ma, X.L. Gao, Eshelby's tensors for plane strain and cylindrical inclusions based on a simplified strain gradient elasticity theory, Acta Mechanica 211 (1–2) (sep 2009) 115–129.
[28] Hang Ma, Zhao Guo, Manicka Dhanasekar, Cheng Yan, Yijun Liu, Efficient solution of multiple cracks in great number using eigen COD boundary integral equations with iteration procedure, Engineering Analysis with Boundary Elements 37 (3) (mar 2013) 487–500.
[29] Hang Ma, Cheng Yan, Qing hua Qin, Eigenstrain boundary integral equations with local Eshelby matrix for stress analysis of ellipsoidal particles, Mathematical Problems in Engineering 2014 (2014) 1–10.
[30] D. Nardini, C.A. Brebbia, A new approach to free vibration analysis using boundary elements, Applied Mathematical Modelling 7 (3) (jun 1983) 157–162.
[31] J.P. Agnantiaris, D. Polyzos, D.E. Beskos, Some studies on dual reciprocity BEM for elastodynamic analysis, Computational Mechanics 17 (4) (jan 1996) 270–277.
[32] Fenglin Zhou, Jianming Zhang, Xiaomin Sheng, Guangyao Li, A dual reciprocity boundary face method for 3D non-homogeneous elasticity problems, Engineering Analysis with Boundary Elements 36 (9) (sep 2012) 1301–1310.
[33] S.-C. Hsiao, A.A. Mammoli, M.S. Ingber, The evaluation of domain integrals in complex multiply-connected three-dimensional geometries for boundary element methods, Computational Mechanics 32 (4–6) (dec 2003) 226–233.
[34] Marc S. Ingber, Andrea A. Mammoli, Mary J. Brown, A comparison of domain integral evaluation techniques for boundary element methods, International Journal for Numerical Methods in Engineering 52 (4) (oct 2001) 417–432.
[35] G. Song, L. Wang, L. Deng, H.M. Yin, Mechanical characterization and inclusion based boundary element modeling of lightweight concrete containing foam particles, Mechanics of Materials 91 (dec 2015) 208–225.
[36] Y.J. Liu, G. Song, H.M. Yin, Boundary effect on the elastic field of a semi-infinite solid containing inhomogeneities, Proceedings of the Royal Society A. Mathematical, Physical and Engineering Sciences 471 (2179) (jul 2015) 20150174.
[37] Y.J. Liu, H.M. Yin, Equivalent inclusion method-based simulation of particle sedimentation toward functionally graded material manufacturing, Acta Mechanica 225 (4–5) (2014) 1429–1445.
[38] Gan Song, Inclusion Based Boundary Element Method for Modeling and Simulation of Composite Materials, PhD thesis, Columbia University, 2017.

[39] G. Song, H.M. Yin, Stress concentration of one microvoid embedded in an adhesive layer under harmonic load, Journal of Engineering Mechanics 144 (3) (mar 2018) 04018002.
[40] Chunlin Wu, H.M. Yin, The inclusion-based boundary element method (iBEM) for virtual experiments of elastic composites, Engineering Analysis with Boundary Elements 124 (mar 2021) 245–258.
[41] Chunlin Wu, Zhenhua Wei, Huiming Yin, Virtual and physical experiments of encapsulated phase change material embedded in building envelopes, International Journal of Heat and Mass Transfer 172 (jun 2021) 121083.
[42] Chunlin Wu, H.M. Yin, Elastic solution of a polygon-shaped inclusion with a polynomial eigenstrain, Journal of Applied Mechanics (feb 2021) 1–25.
[43] Chunlin Wu, Liangliang Zhang, Huiming Yin, Elastic solution of a polyhedral particle with a polynomial eigenstrain and particle discretization, Journal of Applied Mechanics 88 (12) (Dec 2021) 121001.
[44] Carlos Alberto Brebbia, José Claudio Faria Telles, Luiz C. Wrobel, Boundary Element Techniques: Theory and Applications in Engineering, Springer Science & Business Media, 2012.
[45] H. Okada, Y. Fukui, N. Kumazawa, A boundary element based meso-analysis on the evolution of material damage, in: K. Ravi-Chandar, B.L. Karihaloo, T. Kishi, R.O. Ritchie, A.T. Yokobori Jr., T. Yokobori (Eds.), Advances in Fracture Research (Proceedings of ICF10 International Conference on Fracture), Pergamon (an imprint of Elsevier Science), CD-ROM, Citeseer, 2001.
[46] Hiroshi Okada, Yasuyoshi Fukui, Noriyoshi Kumazawa, Homogenization analysis for particulate composite materials using the boundary element method, Nippon Kikai Gakkai Ronbunshu, A Hen/Transactions of the Japan Society of Mechanical Engineers, Part A 68 (6) (2002) 893–900.
[47] Hiroshi Okada, Yasuyoshi Fukui, Noriyoshi Kumazawa, Homogenization analysis for particulate composite materials using the boundary element method, CMES-Computer Modeling in Engineering and Sciences 5 (2) (2004) 135–149.
[48] C.Q. Ru, Analytic solution for Eshelby's problem of an inclusion of arbitrary shape in a plane or half-plane, Journal of Applied Mechanics 66 (2) (jun 1999) 315–523.
[49] Gregory J. Rodin, Eshelby's inclusion problem for polygons and polyhedra, Journal of the Mechanics and Physics of Solids 44 (12) (dec 1996) 1977–1995.
[50] H. Nozaki, M. Taya, Elastic fields in a polygon-shaped inclusion with uniform eigenstrains, Journal of Applied Mechanics 64 (3) (sep 1997) 495–502.
[51] H. Nozaki, M. Taya, Elastic fields in a polyhedral inclusion with uniform eigenstrains and related problems, Journal of Applied Mechanics 68 (3) (apr 2000) 441–452.
[52] S. Trotta, F. Marmo, L. Rosati, Evaluation of the Eshelby tensor for polygonal inclusions, Composites. Part B, Engineering 115 (apr 2017) 170–181.
[53] Wennan Zou, Qichang He, Mojia Huang, Quanshui Zheng, Eshelby's problem of non-elliptical inclusions, Journal of the Mechanics and Physics of Solids 58 (3) (mar 2010) 346–372.
[54] M.Q. Liu, X.-L. Gao, Strain gradient solution for the Eshelby-type polygonal inclusion problem, International Journal of Solids and Structures 50 (2) (jan 2013) 328–338.
[55] Pu Li, Xiangning Zhang, Yuhuan An, Rui Zhang, Xiaoqing Jin, Ning Hu, Leon M. Keer, Analytical solution for the displacement of a polygonal inclusion with a special application to the case of linear eigenstrain, European Journal of Mechanics. A, Solids 84 (nov 2020) 104049.
[56] Qinghua Zhou, Xiaoqing Jin, Zhanjiang Wang, Jiaxu Wang, Leon M. Keer, Qian Wang, Numerical implementation of the equivalent inclusion method for 2D arbitrarily shaped inhomogeneities, Journal of Elasticity 118 (1) (apr 2014) 39–61.
[57] Hou Jiahui, Li Pu, Li Jianglin, Jin Xiaoqing, A triangular element discretization for computing displacement of an arbitrarily shaped thermal inclusion, Chinese Journal of Theoretical and Applied Mechanics (2021).

[58] H.M. Yin, Yingtao Zhao, Introduction to the Micromechanics of Composite Materials, CRC PR INC, 2016.
[59] T. Kanit, S. Forest, I. Galliet, V. Mounoury, D. Jeulin, Determination of the size of the representative volume element for random composites: statistical and numerical approach, International Journal of Solids and Structures 40 (13–14) (jun 2003) 3647–3679.
[60] H.M. Yin, W.G. Buttlar, G.H. Paulino, H. Di Benedetto, Assessment of existing micromechanical models for asphalt mastics considering viscoelastic effects, Road Materials and Pavement Design 9 (1) (jan 2008) 31–57.
[61] Ian M. Smith, Gernot Beer, Christian Duenser, The Boundary Element Method with Programming, Springer, Vienna, 2008.
[62] Johan Helsing, Fast and accurate numerical solution to an elastostatic problem involving ten thousand randomly oriented cracks, International Journal of Fracture 100 (4) (2000) 321–327.
[63] Yuhong Fu, Kenneth J. Klimkowski, Gregory J. Rodin, Emery Berger, James C. Browne, Jürgen K. Singer, Robert A. Van De Geijn, Kumar S. Vemaganti, A fast solution method for three-dimensional many-particle problems of linear elasticity, International Journal for Numerical Methods in Engineering 42 (7) (August 1998) 1215–1229.
[64] Yuhong Fu, Gregory J. Rodin, Fast solution method for three-dimensional Stokesian many-particle problems, Communications in Numerical Methods in Engineering 16 (2) (February 2000) 145–149.
[65] Yi-Shao Lai, Gregory J. Rodin, Fast boundary element method for three-dimensional solids containing many cracks, Engineering Analysis with Boundary Elements 27 (8) (sep 2003) 845–852.
[66] Y.J. Liu, A fast multipole boundary element method for 2D multi-domain elastostatic problems based on a dual BIE formulation, Computational Mechanics 42 (5) (March 2008) 761–773.
[67] Y.J. Liu, L. Shen, A dual BIE approach for large-scale modelling of 3-d electrostatic problems with the fast multipole boundary element method, International Journal for Numerical Methods in Engineering 71 (7) (2007) 837–855.
[68] T.J.R. Hughes, J.A. Cottrell, Y. Bazilevs, Isogeometric analysis: CAD, finite elements, NURBS, exact geometry and mesh refinement, Computer Methods in Applied Mechanics and Engineering 194 (39-41) (oct 2005) 4135–4195.
[69] Seyed Shahram Ghorashi, Navid Valizadeh, Soheil Mohammadi, Extended isogeometric analysis for simulation of stationary and propagating cracks, International Journal for Numerical Methods in Engineering 89 (9) (sep 2011) 1069–1101.
[70] Ruben Sevilla, Sonia Fernández-Méndez, Antonio Huerta, NURBS-enhanced finite element method (NEFEM), Archives of Computational Methods in Engineering 18 (4) (oct 2011) 441–484.
[71] Grégory Legrain, A NURBS enhanced extended finite element approach for unfitted CAD analysis, Computational Mechanics 52 (4) (apr 2013) 913–929.
[72] F.L. Sun, Y.P. Gong, C.Y. Dong, A novel fast direct solver for 3D elastic inclusion problems with the isogeometric boundary element method, Journal of Computational and Applied Mathematics 377 (oct 2020) 112904.
[73] Bo Yu, Geyong Cao, Wendong Huo, Huanlin Zhou, Elena Atroshchenko, Isogeometric dual reciprocity boundary element method for solving transient heat conduction problems with heat sources, Journal of Computational and Applied Mathematics 385 (mar 2021) 113197.
[74] Y.H. Wu, C.Y. Dong, H.S. Yang, Isogeometric indirect boundary element method for solving the 3D acoustic problems, Journal of Computational and Applied Mathematics 363 (jan 2020) 273–299.
[75] Matthias Taus, Gregory J. Rodin, Thomas J.R. Hughes, Michael A. Scott, Isogeometric boundary element methods and patch tests for linear elastic problems: formulation, numerical integration, and applications, Computer Methods in Applied Mechanics and Engineering 357 (dec 2019) 112591.

[76] Gernot Beer, Vincenzo Mallardo, Eugenio Ruocco, Benjamin Marussig, Jürgen Zechner, Christian Dünser, Thomas-Peter Fries, Isogeometric boundary element analysis with elastoplastic inclusions. Part 2: 3-D problems, Computer Methods in Applied Mechanics and Engineering 315 (mar 2017) 418–433.
[77] N. Sukumar, D.L. Chopp, N. Moës, T. Belytschko, Modeling holes and inclusions by level sets in the extended finite-element method, Computer Methods in Applied Mechanics and Engineering 190 (46-47) (sep 2001) 6183–6200.
[78] Chengbin Du, Zongquan Ying, Shouyan Jiang, Extended finite element method and its application in heterogeneous materials with inclusions, IOP Conference Series: Materials Science and Engineering 10 (jun 2010) 012083.
[79] Brett A. Benowitz, Haim Waisman, A spline-based enrichment function for arbitrary inclusions in extended finite element method with applications to finite deformations, International Journal for Numerical Methods in Engineering 95 (5) (jun 2013) 361–386.
[80] S. Nemat-Nasser, M. Hori, Micromechanics: Overall Properties of Heterogeneous Materials, Elsevier, 1998.
[81] Xinchuan Qin, The Exploration of Boundary Element Method in Electromagnectics, Science Press, 2017.
[82] Paul Dirac, The Principles of Quantum Mechanics, Oxford University Press, 1930.
[83] Francis George Denton Duff, Derek Naylor, Differential equations of applied mathematics, American Journal of Physics 34 (12) (1966) 1205–1206.
[84] William Thomson (Lord Kelvin), Note on the integration of the equations of equilibrium of an elastic solid, Cambridge and Dublin Mathematical Journal 3 (1848) 87–89.
[85] L.J. Walpole, An inclusion in one of two joined isotropic elastic half-spaces, IMA Journal of Applied Mathematics 59 (2) (1997) 193–209.
[86] Raymond D. Mindlin, Force at a point in the interior of a semi-infinite solid, Journal of Applied Physics 7 (5) (1936) 195–202.
[87] L. Rongved, Force interior to one of two joined semi-infinite solids, in: Proceedings of the Second Midwestern Conference on Solid Mechanics, in: Purdue University–Indiana Research Series, vol. 1, 1955, p. 13.
[88] H.Y. Yu, S.C. Sanday, Elastic field in joined semi-infinite solids with an inclusion, Proceedings of the Royal Society of London A: Mathematical, Physical and Engineering Sciences 434 (1991) 521–530, The Royal Society.
[89] L.J. Walpole, An elastic singularity in joined half-spaces, International Journal of Engineering Science 34 (6) (1996) 629–638.
[90] M.Z. Wang, Application of the finite part of a divergent integral in the theory of elasticity, Applied Mathematics and Mechanics (English Edition) 12 (1985).
[91] J.W. Ju, L.Z. Sun, A novel formulation for the exterior-point Eshelby's tensor of an ellipsoidal inclusion, Journal of Applied Mechanics 66 (2) (1999) 570–574.
[92] S. Trotta, G. Zuccaro, S. Sessa, F. Marmo, L. Rosati, On the evaluation of the Eshelby tensor for polyhedral inclusions of arbitrary shape, Composites. Part B, Engineering 144 (jul 2018) 267–281.
[93] Igor Sevostianov, Mark Kachanov, Relations between compliances of inhomogeneities having the same shape but different elastic constants, International Journal of Engineering Science 45 (10) (2007) 797–806.
[94] Wen-Nan Zou, Quan-Shui Zheng, The second Eshelby problem and its solvability, Acta Mechanica Sinica 28 (5) (2012) 1331.
[95] H.M. Yin, L.Z. Sun, J.S. Chen, Magneto-elastic modeling of composites containing chain-structured magnetostrictive particles, Journal of the Mechanics and Physics of Solids 54 (5) (2006) 975–1003.
[96] Hong-Bao Li, Guo-Ming Han, Herbert A. Mang, A new method for evaluating singular integrals in stress analysis of solids by the direct boundary element method, International Journal for Numerical Methods in Engineering 21 (11) (nov 1985) 2071–2098.

[97] Xiaowei Gao, 3D non-linear and multi-region boundary element stress analysis, PhD thesis, University of Glasgow, 1999.

[98] George Voyiadjis, Mohammadreza Yaghoobi, Size Effects in Plasticity: From Macro to Nano, Academic Press, 2019.

[99] H.M. Yin, D.J. Yang, G. Kelly, J. Garant, Design and performance of a novel building integrated PV/thermal system for energy efficiency of buildings, Solar Energy 87 (2013) 184–195.

[100] Po-Hua Lee, Huiming Yin, Size effect on functionally graded material fabrication by sedimentation, Journal of Nanomechanics and Micromechanics 5 (2) (2014) A4014008.

[101] Andrei A. Gusev, Representative volume element size for elastic composites: a numerical study, Journal of the Mechanics and Physics of Solids 45 (9) (1997) 1449–1459.

[102] G.R. Liu, S.S. Quek, Chapter 7 - FEM for two-dimensional solids, in: G.R. Liu, S.S. Quek (Eds.), The Finite Element Method, second edition, Butterworth-Heinemann, Oxford, 2014, pp. 161–217.

[103] G.R. Liu, S.S. Quek, Chapter 9 - FEM for 3D solid elements, in: G.R. Liu, S.S. Quek (Eds.), The Finite Element Method, second edition, Butterworth-Heinemann, Oxford, 2014, pp. 249–287.

[104] E. Kröner, Modified Green functions in the theory of heterogeneous and/or anisotropic linearly elastic media, in: Micromechanics and inhomogeneity, Springer, 1990, pp. 197–211.

[105] H.M. Yin, L.Z. Sun, Magnetoelastic modelling of composites containing randomly dispersed ferromagnetic particles, Philosophical Magazine 86 (28) (2006) 4367–4395.

[106] H.M. Yin, G.H. Paulino, W.G. Buttlar, L.Z. Sun, Effective thermal conductivity of functionally graded particulate nanocomposites with interfacial thermal resistance, Journal of Applied Mechanics 75 (5) (2008) 051113.

[107] Thomas H. Cormen, Introduction to Algorithms, MIT Press, 2009.

[108] G.K. Sfantos, M.H. Aliabadi, Multi-scale boundary element modelling of material degradation and fracture, Computer Methods in Applied Mechanics and Engineering 196 (7) (jan 2007) 1310–1329.

[109] Noriyoshi Kumazawa, Hiroshi Okada, Yasuyoshi Fukui, Homogenization analysis for particulate composite materials using the boundary element method, Computer Modeling in Engineering & Sciences 5 (2) (2004) 135–150.

[110] G. Song, L.L. Zhang, H.M. Yin, An inclusion based boundary element method (iBEM) for virtual experiments and cross-scale modeling of particulate composites, International Journal for Heat and Mass Transfer (2022), (in process).

[111] G.K. Batchelor, An Introduction to Fluid Dynamics, Cambridge University Press, 1967.

[112] Leif Rongved, Force at point in the interior of a semi-infinite solid with fixed boundary, Journal of Applied Mechanics 22 (1955) 545–546.

[113] J.W. Ju, T.M. Chen, Micromechanics and effective moduli of elastic composites containing randomly dispersed ellipsoidal inhomogeneities, Acta Mechanica 103 (1) (1994) 103–121.

[114] J.W. Ju, T.M. Chen, Effective elastic moduli of two-phase composites containing randomly dispersed spherical inhomogeneities, Acta Mechanica 103 (1) (1994) 123–144.

[115] Thomas M. Michelitsch, Huajian Gao, Valery M. Levin, Dynamic Eshelby tensor and potentials for ellipsoidal inclusions, Proceedings of the Royal Society of London A: Mathematical, Physical and Engineering Sciences 459 (2003) 863–890, The Royal Society.

[116] Y. Mikata, S. Nemat-Nasser, Elastic field due to a dynamically transforming spherical inclusion, Journal of Applied Mechanics 57 (4) (1990) 845–849.

[117] Anil Gupta, John M. Sullivan, Hugo E. Delgado, An efficient BEM solution for three-dimensional transient heat conduction, International Journal of Numerical Methods for Heat & Fluid Flow 5 (4) (April 1995) 327–340.

[118] H.M. Yin, G.H. Paulino, W.G. Buttlar, L.Z. Sun, Heat flux field for one spherical inhomogeneity embedded in a functionally graded material matrix, International Journal of Heat and Mass Transfer 51 (11–12) (June 2008) 3018–3024.

[119] J.S. Hsiao, An efficient algorithm for finite-difference analyses of heat transfer with melting and solidification, Numerical Heat Transfer 8 (6) (January 1985) 653–666.
[120] Alexander M. Thiele, Zhenhua Wei, Gabriel Falzone, Benjamin A. Young, Narayanan Neithalath, Gaurav Sant, Laurent Pilon, Figure of merit for the thermal performance of cementitious composites containing phase change materials, Cement & Concrete Composites 65 (January 2016) 214–226.
[121] Masataka Tanaka, Toshiro Matsumoto, Qing Feng Yang, Time-stepping boundary element method applied to 2-d transient heat conduction problems, Applied Mathematical Modelling 18 (10) (October 1994) 569–576.
[122] Richard A. Schapery, Correspondence principles and a generalized J integral for large deformation and fracture analysis of viscoelastic media, International Journal of Fracture 25 (3) (1984) 195–223.
[123] Glaucio H. Paulino, Z.-H. Jin, Correspondence principle in viscoelastic functionally graded materials, Journal of Applied Mechanics 68 (1) (2001) 129–132.
[124] Harald Stehfest, Algorithm 368: Numerical inversion of Laplace transforms [d5], Communications of the ACM 13 (1) (1970) 47–49.
[125] Pablo A. Prieto-Muñoz, Huiming M. Yin, Rene B. Testa, Mechanics of an adhesive anchor system subjected to a pullout load. II: viscoelastic analysis, Journal of Structural Engineering 140 (2) (2014) 04013053.
[126] A.H.-D. Cheng, H. Antes, On free space Green's function for high order Helmholtz equations, in: Boundary Element Methods, Springer, Berlin, Heidelberg, 1992, pp. 67–71.
[127] L.S. Fu, T. Mura, Volume integrals of ellipsoids associated with the inhomogeneous Helmholtz equation, Wave Motion 4 (2) (1982) 141–149.
[128] M.S. Howe, Acoustics of Fluid-Structure Interactions, Cambridge University Press, 2004.
[129] A. Milton, S. Irene, Handbook of Mathematical Functions: With Formulas, Graphs, and Mathematical Tables, DOVER PUBN INC, 1965.
[130] Arnold Sommerfeld, Partial Differential Equations in Physics: Lectures on Theoretical Physics, Academic Press, New York, 1967.
[131] Bahaa Saleh, Fundamentals of Photonics, Wiley, New York, 1991.
[132] Robert L. Nowack, Introduction to seismology, https://web.ics.purdue.edu/~nowack/geos557/lecture11-dir/lecture11.htm, 2003.
[133] G.F. Dargush, P.K. Banerjee, Boundary element methods in three-dimensional thermoelasticity, International Journal of Solids and Structures 26 (2) (1990) 199–216.
[134] H.M. Yin, L.Z. Sun, Glaucio H. Paulino, Micromechanics-based elastic model for functionally graded materials with particle interactions, Acta Materialia 52 (12) (2004) 3535–3543.
[135] John R. Reitz, Frederick J. Milford, Robert W. Christy, Foundations of Electromagnetic Theory, Addison-Wesley Publishing Company, 2008.
[136] David J. Griffiths, Introduction to Electrodynamics, Cambridge University Pr., 2017.
[137] Y.-H. Pao, Electromagnetic forces in deformable continua, in: Mechanics Today. Volume 4 (A78-35706 14-70), Pergamon Press, Inc., New York, 1978, pp. 209–305, NSF-supported research.
[138] G. Song, L.L. Zhang, H.M. Yin, Equivalent inclusion method-based virtual experiments of magnetic particle alignment in a viscous fluid, Acta Mechanica (2022), (in process).
[139] Tae Gon Kang, Martien A. Hulsen, Jaap M.J. den Toonder, Patrick D. Anderson, Han E.H. Meijer, A direct simulation method for flows with suspended paramagnetic particles, Journal of Computational Physics 227 (9) (2008) 4441–4458.
[140] Y. Gao, M.A. Hulsen, T.G. Kang, J.M.J. den Toonder, Numerical and experimental study of a rotating magnetic particle chain in a viscous fluid, Physical Review E 86 (4) (2012) 041503.
[141] Abderrahim Boudenne, Yevgen Mamunya, Volodymyr Levchenko, Bertrand Garnier, Eugene Lebedev, Improvement of thermal and electrical properties of silicone–Ni composites using magnetic field, European Polymer Journal 63 (C) (February 2015) 11–19.
[142] Sha Xiao, Zhongqi Quentin Yue, Hongtian Xiao, Boundary element analysis of transversely isotropic bi-material halfspaces with inclined planes of isotropy and interfaces, International Journal for Numerical and Analytical Methods in Geomechanics 43 (17) (2019) 2599–2627.

[143] Huiming Yin, Micromechanics-based magneto-elastic constitutive modeling of particulate composites, 2004.
[144] S. Xiao, Z.Q. Yue, H.T. Xiao, Boundary element analysis of elastic fields in non-horizontally layered halfspace whose horizontal boundary subject to tractions, Engineering Analysis with Boundary Elements 95 (2018) 105–123.
[145] Zhong Qi Yue, On generalized kelvin solutions in a multilayered elastic medium, Journal of Elasticity 40 (1) (1995) 1–43.
[146] Andrzej Katunin, Critical self-heating temperature during fatigue of polymeric composites under cyclic loading, Composites Theory and Practice 12 (1) (2012) 72–76.
[147] A. Katunin, M. Fidali, Self-heating of polymeric laminated composite plates under the resonant vibrations: theoretical and experimental study, Polymer Composites 33 (1) (2012) 138–146.
[148] S.B. Ratner, V.I. Korobov, Self-heating of plastics during cyclic defformation, Polymer Mechanics 1 (3) (1965) 63–68.
[149] Siyu Zhu, Chunlin Wu, Huiming Yin, Virtual experiments of particle mixing process with the SPH-DEM model, Materials 14 (9) (2021) 2199.
[150] Lee R. Aarons, S. Balachandar, Yasuyuki Horie, The mixing of cohesive granular materials featuring a large size range in the absence of gravity, Powder Technology 235 (2013) 18–26.
[151] Zhenyu Shou, Fangliang Chen, Huiming Yin, Self-heating of a polymeric particulate composite under mechanical excitations, Mechanics of Materials 117 (2018) 116–125.
[152] David A. Dillard, Jeffrey A. Hinkley, W. Steven Johnson, Terry L.St. Clair, Spiral tunneling cracks induced by environmental stress cracking in LaRC-TPI adhesives, The Journal of Adhesion 44 (1–2) (1994) 51–67.
[153] Z. Cedric Xia, John W. Hutchinson, Crack patterns in thin films, Journal of the Mechanics and Physics of Solids 48 (6–7) (2000) 1107–1131.
[154] Juan C. Simo, Thomas J.R. Hughes, Computational Inelasticity, vol. 7, Springer Science & Business Media, 2006.
[155] T.S.A. Ribeiro, Gernot Beer, Christian Dünser, Efficient elastoplastic analysis with the boundary element method, Computational Mechanics 41 (5) (2008) 715–732.
[156] G.J. Weng, A micromechanical theory of grain-size dependence in metal plasticity, Journal of the Mechanics and Physics of Solids 31 (3) (1983) 193–203.
[157] J.W. Ju, L.Z. Sun, Effective elastoplastic behavior of metal matrix composites containing randomly located aligned spheroidal inhomogeneities. Part I: micromechanics-based formulation, International Journal of Solids and Structures 38 (2) (2001) 183–201.
[158] Alok Sutradhar, Glaucio Paulino, Leonard J. Gray, Symmetric Galerkin Boundary Element Method, Springer Science & Business Media, 2008.

Index

J

K

L

M

Q

R

S

T

V

W

FSC
www.fsc.org
MIX
Papier | Fördert gute Waldnutzung
FSC® C083411

Kolibri 360 GmbH
Ferdinand-Jühlke-Straße 7
99095 Erfurt, Deutschland
produktsicherheit@kolibri360.de